The Exergy Method of Thermal Plant Analysis

To Jola *and* Joanna
Eva
Mark

The Exergy Method of Thermal Plant Analysis

T. J. Kotas
Department of Mechanical Engineering, Queen Mary College, University of London

Butterworths
London Boston Durban Singapore Sydney Toronto Wellington

First published 1985

Printed and bound in Great Britain by
Anchor Brendon Ltd, Tiptree, Essex

British Library Cataloguing in Publication Data

Kotas, T.J.
The Exergy Method of thermal plant analysis.
1. Power (Mechanics) 2. Thermodynamics
I. Title
621.4 TJ163.9

ISBN 0-408-01350-8

Library of Congress Cataloging-in-Publication Data

Kotas, T. J. (Tadeusz Jozef)
The exergy method of thermal plant analysis.

Bibliography: p.
Includes index.
1. Thermodynamics. 2. System analysis. I. Title.
II. Title: Exergy method.
TJ265.K63 1985 621.402′1 85-17494
ISBN 0-408-01350-8

Foreword

R. S. Silver, CBE, MA, DSc, FInstP, FIMechE, FRSE
Professor Emeritus (James Watt Chair of Mechanical Engineering), University of Glasgow

In prehistoric times, man was aware of the phenomena of temperature and chemical change and used them to warm himself and his immediate surroundings. He was also a tool-user, and in early historic times he applied these phenomena to supersede stone tools by metal tools, becoming increasingly skilful in the metallurgical processes which could thus be generated. These limitations of thermal utility to heating and chemical change remained through all the thousands of years of human history until only about 600 years ago, when man discovered it was possible to throw his most primitive tool, a missile, at an enemy, through the use of rapid combustion in a cannon. It took 300 years more before he found the possibility of thermal driving of other tools.

Now although the utility of warmth and of metallurgical conversions are undoubted it is no exaggeration to say that the whole material growth of civilisation depended far more critically on the use of tools, spade, plough, hammer, chisel, knife, shuttle, lever, lifting gear, wheelbarrow, etc. The basic experience of the user of all such tools is that he requires to *exert* a force to move the tool to do the desired job. Thus the rate of material growth of civilisation is critically dependent on the ability of man to exert, or contrive to exert, forces to drive tools. Until less than 300 years ago, all use of tools was obtained either from muscular exertion, by man himself or by animals, or from contrivances using forces exerted by wind or by water. The discovery that thermal phenomena could be applied to drive tools proved vitally important because the amount of driving power obtained from a small supply of fuel was very far in excess of what could be obtained by the muscular exertion of many men or animals or from wind or water. With that great excess of power so readily available the rate of material growth of civilisation suddenly changed, starting what we call the Industrial Revolution. Now, almost all the exertion in the civilised world is thermally driven, even to the electrical toothbrush.

The science of thermodynamics developed with the primary objective of understanding the inter-relations between chemical, thermal, and mechanical phenomena. All modern industry has to apply that understanding in the engineering of chemical and power plant. At one time it was quite satisfactory to work out an energy balance sheet for a process on First Law (conservation of energy) alone and check subsequently at critical points to ensure that no Second Law infringements were included. Usually, at a first trial of a complex process some infringement would be found and a reiterative programme was required. But increasing complexity of all process and power plant calls for precise thermodynamic analysis to ensure the optimum use of energy resources. For many years now, therefore, several thermodynamicists have been evolving a mode of analysis which combines both First and Second Law constraints simultaneously, and this is what Dr. Kotas describes in the present book.

While the basic concept of the method has had various names offered for it, it has gradually come about that the name exergy is most widely accepted and I am glad that Dr. Kotas has used it. I have myself always favoured the name, since I first encountered it in the work of Dr Myron Tribus in 1962. I confess that as soon as I heard it I linked it in my mind with *exertion*, the basic experience in using a tool. However, I have since learned that the etymology of this term as conceived by its originator, Professor Z. Rant, is quite different. It is made up of two parts: ex-Latin prefix for *from* or *out*, akin to Greek $\varepsilon\xi$(ex) or $\varepsilon\xi o$(exo), and $\varepsilon\rho\gamma o\nu$(ergon) – Greek for *work*. Thus the implication is that exergy is the work which can be obtained from a given system.

The scope which Dr. Kotas has covered in his book is sufficient to prepare an engineering student to apply exergy analysis to several of the power and process plants he is most likely to encounter. More important, the fundamentals are so well covered that the reader should be able to develop his own schemes for application to new plants and processes which will undoubtedly occur. This book should strongly influence the wise choice of process design in many industries.

Preface

The Exergy Method is a relatively new analysis technique in which the basis of evaluation of thermodynamic losses follows from the Second Law rather than the First Law of Thermodynamics. Thus it belongs to that category of analyses known as Second Law analyses. Another name which has been used, mostly in the past, is Availability Analysis.

This book is intended for undergraduate and postgraduate Mechanical and Chemical Engineering students and for practising engineers. The topics covered include various aspects of power generation, refrigeration and cryogenic processes, distillation, and chemical processes, including combustion. Thermoeconomics, to which the whole of Chapter 6 is devoted, is a relatively new area of application of the Exergy Method and may be said to be still in a state of flux—more so than the other aspects of the Exergy Method. However, the technique has developed sufficiently to be useful in analysis and optimisation of thermal and chemical plants.

The book lends itself directly to use by the more advanced undergraduate students, postgraduate students and anyone who is already familiar with the fundamentals of Applied Thermodynamics and the traditional techniques of application of this discipline to thermal or chemical systems or both. In teaching undergraduate students, I feel that the concepts and techniques of the Exergy Method should be integrated into a course of Engineering Thermodynamics to complement and reinforce the traditional work. Those who adopt this approach will find this book easily adaptable for this purpose. Chapter 1 gives some ideas for the introduction of the fundamentals in the first year of an undergraduate course. The introduction of the Second Law through the Entropy Postulate leads naturally to the formulation of the concept of entropy production. Calculating entropy production for a system, or entropy production rate in steady flow processes, in some simple numerical examples gives valuable experience of a quantitative assessment of the degree of thermodynamic perfection of a process and an opportunity to handle such basic concepts as *isolated system*, *environment*, *thermal energy reservoir*, *control surface*, *etc*. The concept of entropy production is also applied, as a subsidiary quantity, in the derivation of some fundamental exergy relations.

The bulk of the material in Chapters 2 to 6 can be dealt with in the second and third year of an undergraduate course of Engineering Thermodynamics. In the second year course the simpler concepts of the Exergy Method can be introduced and the methods of analysis applied to the less complex physical processes such as compression and expansion, heat transfer, mixing, and simple separation processes. Multi-component plants such as the more elementary types of power plants and refrigeration plants can then be analysed. In preparation for the introduction in the third year of the concept of chemical exergy, the maximum work of a chemical reaction and the concept of chemical potential should be covered. A simple treatment of these topics is given in Chapter 1.

The third year course should start with the extension of the exergy analysis to include chemically reacting systems. The required depth of treatment might be greater for Chemical Engineering students than for those in Mechanical Engineering; the more advanced treatment of chemical exergy and related concepts is given in Appendix A. The third year course should also include in-depth treatment of the application of the Exergy Method to areas, according to particular specialisation, such as: expanders and compressors, heat exchangers, refrigeration and cryogenic processes, distillation processes, chemical processes, combustion, power generation, and thermoeconomics. Quite clearly, because of the different ways in which Engineering Thermodynamics is taught, the scheme outlined above is intended only for general guidance.

In the course of writing this book I have received much help and encouragement from my colleagues at Queen Mary College and from fellow thermodynamicists from outside the college. In particular I should like to thank Professor W. A. Woods of the Department of Mechanical Engineering, Queen Mary College for his support and encouragement and to Mr Jeff Cooper of the same Department for his comments on Chapter 1. I am also very much obliged to Professor Emeritus R. S. Silver of Glasgow University and Professor J. Szargut of Silesian Technological University, Poland, for their helpful criticisms and suggestions on Chapter 6.

Dr J. R. Flower of the Department of Chemical Engineering of Leeds University has read some sections of the book and kindly offered his comments on sections on distillation, for which I am very grateful.

My special thanks go to my friend and colleague Dr Raj Raichura of the Department of Nuclear Engineering, Queen Mary College, who read through the draft manuscript and contributed many valuable suggestions for improvement. I am also grateful to him for the many discussions, which helped to clarify many an obscure point and engendered some new ideas.

My thanks are also due to Dr G. P. Moss of the Department of Chemistry, Queen Mary College, for checking names of organic compounds in Table A.4, to my postgraduate student, Mr Deogratias Kibiikyo for the computation of Tables D.1, D.2 and D.3, to Mrs Marian Parsons for drawing with skill and care the illustrations and to Mrs Audrey Hinton for her patience and perseverance while typing a very difficult manuscript.

Finally, I wish to acknowledge my indebtedness to all authors on whose work I have drawn. In particular, I have benefited from the books of Professors J. Szargut, R. Petela and V. M. Brodyanskii. While writing Chapter 6 I found papers by Dr-Ing. J. Beyer and Professors R. B. Evans and Y. M. El-Sayed especially valuable. The benefits derived from these and other authors are gratefully acknowledged in references to their works, and any omissions in this respect are indeed unintentional.

Tadeusz J. Kotas

Department of Mechanical Engineering
Queen Mary College
University of London

May 1984

Contents

Notation

A	Area
a^c	Capital recovery factor (ee)†, Section 6.6)
C	Velocity
C^c	Capital cost
c_P	Specific isobaric heat capacity
c_v	Specific isochoric heat capacity
$\tilde{c}_P^h$	Mean molar isobaric heat capacity for enthalpy calculation
$\tilde{c}_P^s$	Mean molar isobaric heat capacity for entropy calculation
$\tilde{c}_P^\varepsilon$	Mean molar isobaric exergy capacity
c^ε	Unit cost of exergy
c^{en}	Unit cost of energy
c^{sp}	Specific cost of energy
c^I	Unit cost of irreversibility
C_T	Total annual cost of plant operation
$(CP)_{ref}$	Coefficient of performance of a refrigerator
$\mathbf{E}$, $\mathbf{e}$	Energy, specific energy
$\mathbf{E}_k$, $\mathbf{e}_k$	Kinetic energy, specific kinetic energy
$\mathbf{E}_p$, $\mathbf{e}_p$	Potential energy and specific potential energy
f	Function
G, g	Gibbs function, specific Gibbs function
ΔG_0	Gibbs function of reaction
$\Delta\tilde{g}_f$	Molar Gibbs function of formation
g_E	Acceleration due to gravity
H, h	Enthalpy, specific enthalpy
H_d, h_d	Enthalpy of devaluation, specific enthalpy of devaluation
h_H	Hydraulic total head
H_{ph}, h_{ph}	Physical enthalpy, specific physical enthalpy
h_T	Total specific enthalpy $(h + C_0^2/4 + g_E Z_0)$
Δh_{HL}	Hydraulic head loss
ΔH_0	Enthalpy of reaction
I, i	Irreversibility, specific irreversibility

† Throughout this book, equations and expressions will be referred to solely by their number, eg (4.6), or letter (a) in brackets. References are number superscripts in square brackets.

$I^{\Delta P}$	Irreversibility due to pressure losses
$I^{\Delta T}$	Irreversibility due to heat transfer over a finite temperature difference
I^{Q}	Irreversibility due to heat exchange with the environment
i_R	Interest rate
K_P	Equilibrium constant
L_T	Lagrangian
m	Mass
$\tilde{m}$	Molar mass
n	Polytropic index; number of moles
N or n	Number of items
P	Thermodynamic pressure
Q, q	Heat transfer, heat transfer per mass
$\dot{Q}_A$	Heat flux, $\dot{Q}/A$
r	Frictional reheat
R_F	Reheat factor
R	Specific ideal gas constant
$\tilde{R}$	Molar (universal) gas constant
R_Q	Ratio defined by (6.36)
S, s	Entropy, specific entropy
t	Time
t_{op}	Time of plant operation per year
T	Thermodynamic (absolute) temperature
U, u	Internal energy, specific internal energy
U_{ov}	Overall heat transfer coefficient
V, v	Volume, specific volume
W, w	Work, specific work
W_x, w_x	Shaft work, specific shaft work
$\dot{W}_x$	Shaft power
$\dot{W}_{el}$	Electrical power
x	Cartesian co-ordinate, mole fraction
x_i	Variable parameter
y	Cartesian co-ordinate, dryness fraction
Z_0	Height above datum level
$\dot{Z}$	Cost rate of capital investment

Greek symbols

α	Angle; specific non-flow exergy function $(u+P_0v-T_0s)$
α_P	Coefficient of thermal expansion (cubic)
β	Specific exergy function $(h-T_0s)$
γ	Ratio of heat capacities, c_P/c_v; activity coefficient
δ	Efficiency defect, $1-\psi$
δ_i	Efficiency defect of the i-th sub-region (component)
E, ε	Exergy, specific exergy
E^r, ε^r	Relative exergy, specific relative exergy (Appendix A)
E_{ph}, ε_{ph}	Physical exergy, specific physical exergy
E_0, ε_0	Chemical exergy, specific chemical exergy
E_k, ε_k	Kinetic exergy, specific kinetic exergy
E_p, ε_p	Potential exergy, specific potential exergy
$E^{\Delta P}$, $\varepsilon^{\Delta P}$	Pressure component of exergy and of specific exergy

$E^{\Delta T}$, $\varepsilon^{\Delta T}$	Thermal component of exergy and of specific exergy
E^{Q}, ε^{Q}	Thermal exergy, specific thermal exergy, in open systems
$\zeta_{k,i}$	Capital cost coefficient, (6.56)
η_{Carnot}	Carnot efficiency
η_{comb}	Combustion efficiency
η_{el}	Electrical efficiency
η_{m}	Mechanical efficiency
η_{s}	Isentropic efficiency
η_{iso}	Isothermal efficiency
θ	Temperature on an arbitrary scale
κ_T	Isothermal compressibility
λ	Lagrange multiplier; angle
μ	Chemical potential
ν	Stoichiometric coefficient
Ξ, ξ	Non-flow exergy, specific non-flow exergy
Ξ^{Q}, ξ^{Q}	Thermal exergy, specific thermal exergy, in closed systems
Π, π	Entropy production, specific entropy production; 3.14 . . .
ρ	Density
$\sigma_{k,i}$	Coefficient of structural bonds, (6.1)
τ	Dimensionless exergetic temperature, $(1-T_0/T)$
φ	Ratio $\varepsilon_F^0/(\text{NCV})^0$ for industrial fuels
φ_0	Relative humidity
$\chi_{j,i}$	Coefficient of external bonds, (6.9)
ψ	Rational efficiency
ψ'	Partial incremental rational efficiency, (6.48)
ω_i	Exergy ratio, E''/E'

Subscripts

CR	Control region
CS	Control surface
e	Exit from the control region
el	Chemical element; electric
f	Formation; saturated liquid state
F	Fuel
g	Saturated vapour state
HT	Heat transfer
i	Property of the *i*-th component of a mixture; *i*-th item
i	Inlet to the control region
IN	Entering the control region; input
IRREV	Irreversible
ISOL	Isolated
j	*j*-th co-reactant
k	*k*-th product, *k*-th plant element
M	Mixture
MAX	Maximum
MIN	Minimum
NET	Net
0	Environmental state
00	Dead state

OUT	Leaving the control region; output
P	Product, constant pressure
R	Reactant
r	r-th thermal energy reservoir
REV	Reversible
RS	Reference substance
s	s-th sub-region; solid
sh	Shaft
T	Total quantity, constant temperature
ΔP	Pressure component
ΔT	Thermal component

Superscripts

$\sim$	Molar quantity
$-$	Dimensionless
0	Standard environmental state
00	Standard dead state
$'$	Entering the control region
$''$	Leaving the control region
ΔP	Pressure component
ΔT	Thermal component
v	Per volume

Abbreviations

CEB	Coefficient of external bonds
CSB	Coefficient of structural bonds
CS	Control surface
CR	Control region
EMR	Exit matter reservoir
IMR	Inlet matter reservoir
MER	Mechanical energy reservoir
NCV	Net calorific value
NRNR	Non-renewable natural resource
PMM2	Perpetual motion machine of the second kind
RHE	Reversible heat engine
RHP	Reversible heat pump
SFEE	Steady flow energy equation
S-R	Sub-region
TER	Thermal energy reservoir

Special notations

d	An infinitesimal change in a property
đ	An infinitesimal amount of transfer by some mechanism
$\rightarrow$min	Tends to a minimum
$\left(\dfrac{\partial Y}{\partial X}\right)_{x_i=\text{var}}$	$\dfrac{(\partial Y/\partial x_i)}{(\partial X/\partial x_i)}$
$\dot{Q}, \dot{m}, \dot{\Pi}$	Rates of transfer, flow or production (*not* time derivatives)

$\sum_i x_i$	Sum of $x_1 + x_2 + x_3 + \cdots + x_N$
$\prod_{i=a}^{b} x_i$	Product of $x_a x_{a+1} \cdots x_{b-1} x_b$
$[X_{REV}]_1^2$	The magnitude of an interaction X in a reversible process between states 1 and 2

Introduction

Traditional techniques of process evaluation

One of the main applications of Engineering Thermodynamics is the study of process effectiveness. Traditional techniques are of two types:

(a) Use of an energy balance on the system, usually to determine 'unaccounted for' heat transfer between system and environment;
(b) Calculation of a criterion of performance relevant to the system under consideration.

Energy balances treat all forms of energy as equivalent, without differentiating between the different grades of energy crossing the system boundary. Thus heat transfer to the environment from a pipe carrying high temperature steam will be treated in the same way as low grade thermal energy rejected in the condenser of a steam plant. Results from energy balances on cryogenic systems can be baffling since here the *loss* of thermal energy ('production of cold') is desirable while gaining thermal energy as a result of, say, heat transfer from the environment is to be avoided.

In generaly energy balances provide no information about internal losses. An energy balance for an adiabatic system such as a throttling valve, a heat exchanger or a combustion chamber, could lead one to believe that these processes are free of losses of any kind.

Calculating criteria of performance appropriate to the process under consideration can be helpful in assessing degree of thermodynamic perfection. However, this is not always the case. For example, there are no traditional criteria for assessing the performance or thermodynamic losses of processes occurring in a throttling valve, a steam ejector, an open type (mixing) feed heater or an adiabatic combustion chamber. For other processes, criteria of performance may be available but the information which they provide on the performance of the system may be incomplete or inadequate. For example, heat exchanger effectiveness does not indicate the effect of pressure losses on performance. The overall thermal efficiency of a CHP plant, in which the heating and power outputs are simply added to yield the numerator, is primarily an assessment of the boiler flue losses rather than of plant efficiency.

These traditional methods of process analysis are based in the main on the First Law of Thermodynamics. The Second Law is only brought into the formulation of such criteria of performance as isentropic efficiencies, even though it governs the limits of convertibility between different forms of energy and determines the relative grades or quality of these energy forms. Hence it is the failure to consider changes in energy quality during a process which makes traditional thermodynamic analysis methods so unsatisfactory.

The Exergy Method of thermodynamic analysis

The Exergy Method is an alternative, relatively new technique based on the concept of exergy, loosely defined as a universal measure of the work potential or quality of different forms of energy in relation to a given environment. An exergy balance applied to a process or a whole plant tells us how much of the usable work potential, or exergy, supplied as the input to the system under consideration has been consumed (irretrievably lost) by the process. The loss of exergy, or *irreversibility*, provides a generally applicable quantitative measure of process inefficiency. Analysing a multi-component plant indicates the total plant irreversibility distribution among the plant components, pinpointing those contributing most to overall plant inefficiency.

Unlike the traditional criteria of performance, the concept of irreversibility is firmly based on the two main laws of thermodynamics. The exergy balance for a control region, from which the irreversibility rate of a steady flow process can be calculated, can be derived by combining the steady flow energy equation (First Law) with the expression for the entropy production rate (Second Law). Although the Second Law is not used explicitly in the Exergy Method, its application to process analysis demonstrates the practical implications of the Second Law. Thus studying different forms of irreversibilities and their effect on plant performance, gives a better and more useful understanding of the Second Law than studying its statements and corollaries.

A brief historical outline

Over the years two basic concepts, exergy and irreversibility, gave rise to a variety of derived concepts, techniques and criteria of performance. Although the Exergy Method is usually regarded as a new technique, the first attempts at an assessment of various energy forms according to their convertibility are directly connected with the beginning of the formulation of the Second Law of Thermodynamics. The earliest contributions, from 1868, to the concept of 'availability' of energy for conversion to work are due to Clausius, Tait, Thomson, Maxwell and Gibbs, although Gouy's paper[0.1] 'On usable energy' published in 1889 has long been regarded by continental European thermodynamicists as a 'document of birth' of the concept of available energy. In 1898 Stodola[0.2], apparently quite independently, developed the concept of 'free technical energy', applicable to steady flow processes. Further development was slow until the 1930s when interest in the practical application of the concepts was stimulated by industrial growth and new technological developments. Papers by Bošnjaković[0.3,0.4] published in 1938 and 1939 mark a new era in the development of the Second Law Analysis. These papers made an important contribution to the formulation of new criteria of performance and techniques of assessment of thermodynamic perfection of processes. Progress was interrupted by World War II but revived in the 1950s with renewed vigour. The number of papers published since then on various aspects of the Second Law Analysis is too great to mention here*. Rant's 1956 paper[0.5], however,

* Reference [0.6] gives a more extensive history while [0.7] deals in particular with the developments in 1970 to 1980.

had a major impact on the terminology of this subject; he coined the term 'exergy' which has now gained general acceptance, replacing terms used in different languages such as availability, available energy, usable energy, and work capability.

Exergy Analysis (under its various names) has been given some space, usually a section or at most a chapter, in textbooks on Engineering Thermodynamics in the past but there are few textbooks which are *wholly* devoted to it; two are particularly noteworthy. 'Exergy' by Professors J. Szargut and R. Petela[3.3], published in 1965 in Polish, is based mainly on the work of the authors and their co-workers and deals with a variety of topics relating to power plant and metallurgical processes. The theory of reference substance, on which the tables of values of standard chemical exergy of substances and enthalpy of devaluation are based, is also covered.

'Exergy Method of Thermodynamic Analysis' by Professor V. M. Brodyanskii[2.2], published in 1973 in Russian, covers a broad range of topics with particular reference to refrigeration, cryogenic engineering and distillation processes.

Chapter 1 Review of the fundamentals

To avoid problems arising from different terminologies which have been used, this chapter first defines and briefly discusses the principal concepts of classical thermodynamics. A revision of the basic laws of thermodynamics puts particular emphasis on the Second Law which is introduced through the Entropy Postulate leading directly to the concept of entropy production which is required as a subsidiary concept in the subsequent development of some basic exergy relations. Further, in this chapter, different forms of irreversibilities are discussed and examples given. Finally the concepts of the maximum work of a chemical reaction and chemical potential are reviewed in preparation for the introduction of the concept of chemical exergy.

1.1 Basic concepts

System: A system is an identifiable collection of matter whose behaviour is the subject of study. For identification, the system is enclosed by a *system boundary*, which may be purely imaginary or may coincide with a real boundary. The term *closed system* is sometimes used to emphasise that there is no flow of matter across the system boundary. The type of thermodynamic analysis used is known as a system analysis or control mass analysis.

When motions are involved, the system definition must include a *reference frame* to which velocities and displacements are related. The most commonly used reference frame is the *inertial reference frame* in which a free particle moves at constant velocity.

Surroundings: Everything outside the boundary of the system is called the surroundings.

Isolated system: If changes in the surroundings produce no changes in the system the system is known as an isolated system. A system combining a system and its surroundings is an isolated system, often called the *universe* which is not a universe in the cosmological sense but only in the thermodynamic sense.

Property: A thermodynamic property is any measurable characteristic of a system whose value depends on the condition of the system.

Thermodynamic state: The state of a system is that condition of the system which is described fully by its observable properties. In identical states, the properties have the same values. Thus properties are functions of the state of the system and not of a process which the system might undergo.

State of equilibrium: An isolated system which has no tendency to undergo a change of state even after a long time is in a state of equilibrium.

Process: When the state of the system changes it is said to undergo a process.

Extensive properties: For a system divided into N sub-systems by real or imaginary boundaries, the value of an extensive property, X, for the whole system is the sum of the values of that property for all sub-systems. Denoting the extensive property of the i-th sub-system as X_i, then:

$$X = \sum_i X_i \tag{1.1}$$

Mass and volume are extensive properties. Extensive properties have values regardless of whether the system is in equilibrium or not.

Intensive properties: These properties are independent of the size of the system and only have meaning for systems in equilibrium states; pressure and temperature are intensive properties. A special type of intensive property is a *specific property* which is the ratio of an extensive property to the corresponding mass.

Homogeneous system: If the value of any intensive property anywhere in the system is the same, the system is homogeneous.

Cyclic process: When a system goes through some changes of state (or processes) and finally returns to its initial state, it has gone through a *cycle* or *cyclic process.*

Reversible process: A process is reversible if, after it has taken place, means can be found to restore the system and its surroundings to their initial states with no residual effects in either of them. Full reversibility is not possible in a real process.

Quasi-static process: In such processes the system is *infinitesimally close* to equilibrium at all times and all states through which the system passes can be described by thermodynamic coordinates referring to the system as a whole. A quasi-static process is an idealisation which can be approached in practice with any degree of accuracy which may be desired.

Control region: A control region, also known as control volume or open system, is any defined region in space under analysis. The extent of the control region is defined by the *control surface.*

Work and heat: These phenomena are describable at the system boundary and exist only while the system (or control region) and the surroundings interact and thus both work and heat are called *interactions.* Since these interactions result in energy transfers across the boundary, work and heat may be regarded as *energy in transit.* Although they have this in common, there are also important distinctions between them. Work is an interaction between two systems such that the sole effect of the action of one system on the other can be shown to be equivalent to the raising or lowering of a weight. Heat results in an energy transfer between two systems in thermal contact by virtue of their temperature difference. These differences, and their important consequences, will be discussed in Chapter 2.

Thermal energy reservoir: A thermal energy reservoir (TER) is a body of constant

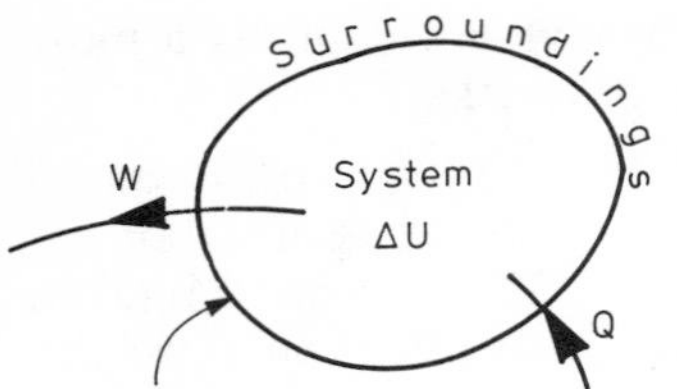

Fig 1.1 Energy transfers between the system and its surroundings. The arrows define positive directions of heat and work transfers.

volume in stable equilibrium with a very large heat capacity so that it may act as a heat source or heat sink without suffering a finite change in temperature. In practical thermodynamic analysis, any constant temperature body, such as a heated room or a refrigerated cold chamber, whose sole function is either to receive or to deliver thermal energy through heat interaction may be treated as a TER within which all processes are assumed to be quasi-static.

Mechanical energy reservoir: This is a system capable of storing fully ordered mechanical energy as potential energy (raised weight in a gravitational field) or kinetic energy (spinning flywheel). A mechanical energy reservoir (MER) is an idealised system in which the energy is stored, received, and delivered in a reversible manner.

Heat engine: A heat engine operates in a cyclic manner and exchanges thermal and mechanical energy with other systems.

Adiabatic boundary: A system boundary or control surface which does not permit a heat interaction to take place is adiabatic.

Diathermal boundary: A system boundary or control surface which permits a heat interaction to take place is diathermal.

Sign convention (*Fig 1.1*)*:* In this book heat transfer to the system is positive and heat transfer from the system is negative. Work done by a system is positive and work done on the system is negative.

State postulate: The number of independently variable thermodynamic properties for a particular system is the number of relevant reversible work modes plus one. The reversible work modes, ie ways by which a system can transfer energy as work, depend on the properties of the substance comprising the system. For example if the substance is compressible and magnetic its work modes are associated with volume changes due to applied pressure and changes in magnetisation in a changing magnetic field. Three independently variable properties are required to define its state. This book will not deal with magnetic effects, electric properties of substances, or with surface tension. Only the work mode due to compressibility will be considered; the state of such a *simple compressible substance* can be defined for a non-reacting system in terms of only two independently variable properties.

Equation of state: There are a number of equations of state relating the properties of simple compressible substances; the simplest is the *ideal gas* equation:

$$Pv = RT \tag{1.2}$$

where R is the specific ideal-gas constant. When, for the range of properties under consideration, an ideal gas can be assumed to have constant specific heat capacities it is called a *perfect gas*. A real gas departs from ideal gas behaviour, requiring a more complex equation of state or tabular presentation of its properties.

Molar properties: A *mole* of any give molecular species is the amount of substance which contains as many elementary entities as there are atoms in 0.012 kg of carbon-12. This definition is cumbersome so, for practical purposes, the mass of one mole is taken to be given numerically in grammes by the relative molecular mass (molecular weight) of the substance. For example, the mole of diatomic oxygen is approximately 32 g or, using the kilomole (kmol), the molar mass is approximately 32 kg/kmol.

For a quantity of substance of mass m and molar mass $\tilde{m}$ the number of moles is:

$$n = \frac{m}{\tilde{m}} \tag{1.3}$$

If in a mixture of gases of n moles there are n_i moles of the i-th component, the mole fraction of that component is:

$$x_i = \frac{n_i}{n} \tag{1.4}$$

For each specific quantity such as, specific enthalpy*, specific entropy* or specific gas constant there is a corresponding molar quantity, given per mole of the substance. The molar form of the ideal gas equation, (1.2), is:

$$P\tilde{v} = \tilde{R}T \tag{1.5}$$

where $\tilde{v}$ is the molar volume and $\tilde{R}$ is the molar ideal-gas constant (universal gas constant).

Gibbs–Dalton law

The pressure (P) *and internal energy* (U) *of a mixture of gases are respectively equal to the sums of the pressures* (P_i) *and internal energies* (U_i) *of the individual components when each occupies a volume equal to that of the mixture at the temperature of the mixture.* Thus:

$$P = \sum_i P_i \Big]_{V,T} \tag{1.6}$$

$$U = \sum_i U_i \Big]_{V,T} \tag{1.7}$$

From this it can be shown that analogous relationships apply to enthalpy and entropy, ie:

$$H = \sum_i H_i \Big]_{V,T} \tag{1.8}$$

$$S = \sum_i S_i \Big]_{V,T} \tag{1.9}$$

For an ideal gas it can be shown from (1.7) and (1.8) that the specific heat capacities for a

* Enthalpy is defined by (1.27) whilst entropy is defined in Section 1.5.

mixture can be obtained from:

$$c_v=\sum_i \frac{m_i}{m} c_{vi} \tag{1.10}$$

$$c_P=\sum_i \frac{m_i}{m} c_{Pi} \tag{1.11}$$

where m_i/m is the mass fraction of the i-th constituent of the mixture. In molar terms, (1.10) and (1.11) take the form

$$\tilde{c}_v=\sum_i x_i\tilde{c}_{vi} \tag{1.12}$$

$$\tilde{c}_P=\sum_i x_i\tilde{c}_{Pi} \tag{1.13}$$

Law of partial volumes

Another empirical law applicable to mixtures of gases is Amagat's law or the law of partial volumes:

The volume (V) of a mixture of gases is the sum of the partial volumes (V_i) of its components.

The partial volume of a component of a mixture is the volume it would occupy if it existed on its own at the pressure and temperature of the mixture, ie:

$$V=\sum_i V_i\Big]_{P,T} \tag{1.14}$$

Both the Gibbs–Dalton and Amagat's law are only approximately true for real gases but are obeyed exactly by mixtures of ideal gases for which the two laws are not independent.

For an isothermal process of an ideal gas (PV=const), in which one state corresponds to (P, V_i) and another to (P_i, V):

$$\frac{V_i}{V}=\frac{P_i}{P} \tag{1.15}$$

Since equal volumes of ideal gases at the same pressure and temperature contain equal numbers of moles, the mole fraction of a component of a mixture can be expressed as:

$$\frac{V_i}{V}=x_i \tag{1.16}$$

1.2 Zeroth law

If two systems A and B are in equilibrium with system C when in thermal contact with C, then they are in equilibrium with each other.

This law, also known as principle of thermal equilibrium, is the basis of temperature measurement.

1.3 First law

There is an extensive property, internal energy (U), such that a change in its value is given for a system not in motion by the difference between the heating (Q) done to the system and the work (W) done by the system during any change of state.

$$Q - W = \Delta U \tag{1.17}$$

where:

$$\Delta U = U_{\text{final}} - U_{\text{initial}}$$

Eq (1.17) is known as the energy balance for a non-flow process or the non-flow energy equation.

System in motion

For a more general case of a system in motion, changes in kinetic energy $\Delta \mathbf{E}_k$ and potential energy $\Delta \mathbf{E}_p$ (evaluated with respect to the reference frame of the system) must be added to the change in internal energy ΔU in Eq (1.17). These three terms together constitute then the change in the energy of the system, $\Delta \mathbf{E}$. Thus the more general form of the energy balance is:

$$Q - W = \Delta \mathbf{E} \tag{1.18}$$

where:

$$\Delta \mathbf{E} = \Delta U + \Delta \mathbf{E}_k + \Delta \mathbf{E}_p \tag{1.19}$$

Differential form

For an infinitesimal part of the process, the differential form of the energy equation (1.18) takes the following form:

$$đQ - đW = \mathrm{d}\mathbf{E} \tag{1.20}$$

Note that đ is used here in place of the differential operator d to indicate that Q and W cannot be evaluated without information on the process path, ie they are not properties of the system.

Conservation of energy

The LHS of (1.18) gives the net energy input to the system, ie the increase in the energy of the system. Since the net energy input to the system comes from the surroundings:

$$\Delta \mathbf{E}_{\text{system}} + \Delta \mathbf{E}_{\text{surroundings}} = 0 \tag{1.21}$$

This is a statement of the well-known principle of conservation of energy, which can be expressed as:

The energy of a system and its surroundings, considered together, is constant.
Or, since we can regard the system with its surroundings as an isolated system:
The energy of an isolated system remains constant.

First Law for a cyclic process

A statement of the First Law for a cyclic process can be obtained by summing the quantities in (1.17) around a cycle. The change in internal energy of a process with the same initial and final state is zero, as it must be for any property of the system. For a cycle with a number of work and heat transfers:

$$\oint Q = \oint W \tag{1.22}$$

Or, integrating the differential form of the energy balance (1.20):

$$\oint đQ = \oint đW \tag{1.23}$$

Hence, the First Law for a cyclic process can be stated as:

When a system undergoes a cyclic process, the net work done on the surroundings is equal to the net heat transfer to the system.

Limitations of the First Law

The First Law deals with the amounts of energy of various forms transferred between the system and its surroundings and with the changes in the energy stored in the system. It treats work and heat interactions as equivalent forms of energy in transit and offers no indication about the possibility of a spontaneous process proceeding in a certain direction. The Second Law of Thermodynamics is required to establish the difference in quality between mechanical and thermal energy and to indicate the directions of spontaneous processes.

1.4 Equations of mass balance and energy balance for a control region

Transition from a system to a control region

Most important engineering processes are associated with the flow of matter from one plant component to another. Thus it is not convenient to study the processes using closed system analysis, since a system boundary does not permit the flow of matter across it, and it is necessary therefore to reformulate relations to make them directly applicable to analysis of a control region. Consider a system, shown in Fig 1.2, in two positions as it moves through the control region.

Mass balance for a control region

Since the mass of the system in the two positions must be the same:

$$\mathrm{d}m_{\mathrm{CR}} + \mathrm{d}m_{\mathrm{e}} = \mathrm{d}m_{\mathrm{i}}$$

Rearranging, dividing by the time interval $\mathrm{d}t$, and expressing on a rate basis:

$$\frac{\mathrm{d}m_{\mathrm{CR}}}{\mathrm{d}t} + \dot{m}_{\mathrm{e}} - \dot{m}_{\mathrm{i}} = 0 \tag{1.24}$$

Under steady flow operating conditions $\mathrm{d}m_{\mathrm{CR}}/\mathrm{d}t = 0$, and hence:

$$\dot{m}_{\mathrm{e}} = \dot{m}_{\mathrm{i}} \tag{1.25}$$

Energy equation for a control region

In applying the energy equation (1.20) to a system undergoing an infinitesimal process as it moves through the control region, it is assumed, for the sake of generality, that the volume of the control region changes, giving rise to the work term $đW_{\mathrm{CS}}$ which represents the work done by the system on the piston. Therefore, the work done by the

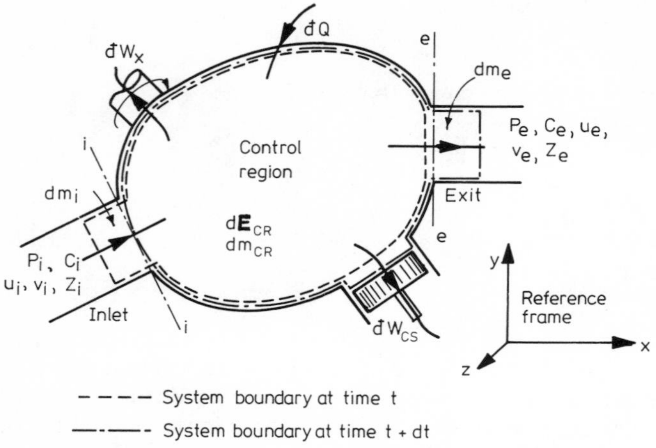

Fig 1.2 Movement of a system through a control region.

system comprises:

(i) Shaft work, $đW_x$.
(ii) Net displacement work, $P_e v_e dm_e - P_i v_i dm_i$.
(iii) Work due to change in the volume of the control region, $đW_{CS}$.

Heat transfer to the system is $đQ$ and the net increase in the total energy of the system is:

$$\mathbf{dE}_{CR} + (u_e + \tfrac{1}{2}C_e^2 + g_E Z_e)dm_e - (u_i + \tfrac{1}{2}C_i^2 + g_E Z_i)dm_i$$

Combining these terms into an energy balance and rearranging:

$$đQ - đW_x - đW_{CS} = \mathbf{dE}_{CR} + [(u_e + P_e v_e) + \tfrac{1}{2}C_e^2 + g_E Z_e]dm_e - [(u_i + P_i v_i) + \tfrac{1}{2}C_i^2 + g_E Z_i]dm_i \tag{1.26}$$

Defining specific enthalpy:

$$h = u + Pv \tag{1.27}$$

Dividing (1.26) by dt and expressing on a rate basis:

$$\dot{Q} - \dot{W}_x - \dot{W}_{CS} = [h_e + \tfrac{1}{2}C_e^2 + g_E Z_e]\dot{m}_e - [h_i + \tfrac{1}{2}C_i^2 + g_E Z_i]\dot{m}_i + \frac{\mathbf{dE}_{CR}}{dt} \tag{1.28a}$$

(1.28a) is the energy balance for a control region undergoing a non-steady process.

In general, the control region may interact thermally with a number of TERS and there may be several streams of matter entering and leaving the control region. The more general version of (1.28a) is:

$$\sum_r \dot{Q}_r - \dot{W}_x - \dot{W}_{CS} = \sum_{OUT} \dot{m}_e[h_e + \tfrac{1}{2}C_e^2 + g_E Z_e] - \sum_{IN} \dot{m}_i[h_i + \tfrac{1}{2}C_i^2 + g_E Z_i] + \frac{\mathbf{dE}_{CR}}{dt} \tag{1.28b}$$

Steady state, steady flow case

Under these conditions:

$$\dot{W}_{CS}=0 \qquad \frac{d\mathbf{E}_{CR}}{dt}=0 \qquad \dot{m}_e=\dot{m}_i$$

which turns (1.28a) into the steady flow energy equation

$$\dot{Q}-\dot{W}_x=\dot{m}[(h_e-h_i)+\tfrac{1}{2}(C_e^2-C_i^2)+g_E(Z_e-Z_i)] \tag{1.29}$$

1.5 Second law

Entropy postulate

There is an extensive property of a system called entropy, S. The entropy of an isolated system can never decrease.

$$(\Delta S)_{ISOL}\geqslant 0 \tag{1.30}$$

where the equality sign corresponds to the ideal case of a reversible process.

Combined system

If the system under study is not an isolated system, it is necessary to consider all systems participating in the process so that the new, extended system can be regarded as an isolated system. By considering a combined system consisting of the system and the surroundings, (1.30) can be written:

$$(\Delta S)_{SYSTEM}+(\Delta S)_{SURROUNDINGS}\geqslant 0 \tag{1.31}$$

Implications of the Second Law

The implications of the Second Law are manifold. The condition of the increase of entropy can be used to predict what processes, chemical reactions, transformations between various energy forms, or directions of heat transfer can and cannot occur. From the condition that a state of equilibrium of an isolated, two part system corresponds to a maximum of entropy of the system, it can be shown[1.1] that the conditions of thermal, mechanical and chemical equilibrium correspond respectively to equality of temperature, pressure and chemical potential (defined in Section 1.13). In addition, the Second Law governs the limits to energy conversion between different energy forms, leading to concept of energy quality. Some of these points will be discussed in depth in Chapter 2.

Microscopic view

From a consideration of the microscopic nature of matter through the statistical approach, entropy can be shown to be a measure of microscopic randomness and the resulting uncertainty about the microscopic state. Real processes tend to make the distribution of kinetic energy of molecules of the system more random, making a smaller proportion of this energy available for conversion to useful, organised work. Thus, it can be said that entropy is a measure of the 'unavailability' of internal energy[1.2].

Gibbs equation

Applying the entropy postulate in a thermodynamic analysis of macroscopic systems requires quantitative evaluation of the entropy changes of the system under consideration. When the system consists of a simple compressible substance in a state of equilibrium, any two suitable independent properties can be used to determine a third property. Thus:

$$U = U(S, V) \tag{1.32}$$

Expressing U in a total differential form:

$$\mathrm{d}U = \left(\frac{\partial U}{\partial S}\right)_V \mathrm{d}S + \left(\frac{\partial U}{\partial V}\right)_S \mathrm{d}V \tag{1.33}$$

and defining thermodynamic temperature and pressure:

$$T \equiv \left(\frac{\partial U}{\partial S}\right)_V \qquad P \equiv -\left(\frac{\partial U}{\partial V}\right)_S \tag{1.34}$$

gives:

$$\mathrm{d}U = T\mathrm{d}S - P\mathrm{d}V \tag{1.35}$$

and hence:

$$\mathrm{d}S = \frac{1}{T}\mathrm{d}U + \frac{P}{T}\mathrm{d}V \tag{1.36}$$

is known as the Gibbs equation. This can be used to calculate changes of entropy of a simple compressible substance of invariable composition.

Although the definitions of thermodynamic temperature and pressure given by (1.34) may appear arbitrary, it can be shown[1.2] that for two parts of a system initially not in equilibrium, the equality of a property given by $(\partial U/\partial S)_V$ is essential for the thermal equilibrium of the system. Similarly $(\partial U/\partial V)_S$ governs mechanical equilibrium. These considerations alone are not sufficient to give the definitions of the thermodynamic temperature and pressure in the form given above. The precise form is based on the need for the thermodynamic pressure and temperature to be in agreement with the corresponding empirical concepts.

Referring back to the differential form of the energy equation (1.20) it can be shown that for an infinitesimal quasi-static process the work is given by:

$$đW = P\mathrm{d}V \tag{1.37}$$

From (1.20) and (1.36):

$$đQ_{\mathrm{REV}} - P\mathrm{d}V = \mathrm{d}U \tag{1.38}$$

Comparing (1.35) and (1.38):

$$\mathrm{d}S = \frac{đQ_{\mathrm{REV}}}{T} \tag{1.39}$$

where the subscript 'REV' indicates that this relation is applicable only to an internally reversible heat transfer process. (1.39) is an alternative definition of entropy.

Entropy change for a perfect gas

For a perfect gas, integrating (1.36) with the aid of (1.2), leads to an expression for a change in specific entropy:

$$s_2 - s_1 = c_P \ln \frac{T_2}{T_1} - R \ln \frac{P_2}{P_1} \tag{1.40}$$

Alternative, but similar expressions can be obtained in terms of other pairs of properties from P, v and T.

1.6 Maximum efficiency of a heat engine

The energy conversion efficiency for a heat engine operating cyclically between two thermal energy reservoirs is:

$$\eta \equiv \frac{W_{NET}}{Q_{IN}} \tag{1.41}$$

where W_{NET} is the network delivered and Q_{IN} is the corresponding heat transfer to the engine per cycle. To determine the maximum limiting efficiency of a heat engine, consider the system in Fig 1.3 in which a heat engine operates between a high

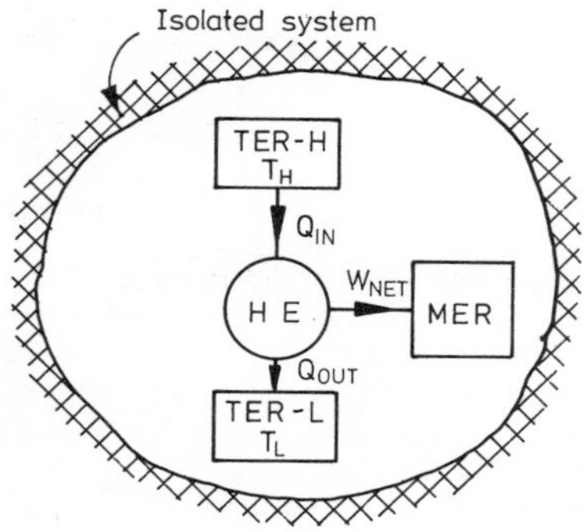

Fig 1.3 A heat engine (HE) operating between two thermal energy reservoirs (TER-H and TER-L) and delivering work to a mechanical energy reservoir (MER).

temperature thermal energy reservoir TER-H and a low temperature reservoir TER-L and delivers work to a mechanical energy reservoir MER[1.2]. The heat engine and the reservoirs form an isolated system for which, according to the Second Law, the entropy cannot decrease. The heat transfers from TER-H and to TER-L occur reversibly and at constant temperatures T_H and T_L respectively. Hence, from (1.39), the expressions for entropy changes over a cycle of the two thermal energy reservoirs are:

$$\Delta S_H = -\frac{Q_{IN}}{T_H} \qquad \Delta S_L = \frac{Q_{OUT}}{T_L}$$

Note that the entropy change for the heat engine itself must be zero since we are considering a cycle. Also MER, by definition, does not suffer any change in entropy. Hence, applying (1.30):

$$\left(\frac{Q_{OUT}}{T_L} - \frac{Q_{IN}}{T_H}\right) \geqslant 0 \tag{1.42}$$

For the reversible mode of operation of the engine the sign of equality applies in (1.42), giving:

$$\frac{Q_{IN}}{Q_{OUT}}=\frac{T_H}{T_L} \tag{1.43}$$

From the First Law for a cyclic process (1.22):

$$W_{NET}=Q_{IN}-Q_{OUT} \tag{1.44}$$

Thus, for any heat engine:

$$\eta=\frac{Q_{IN}-Q_{OUT}}{Q_{IN}} \tag{1.45}$$

and for a reversible heat engine:

$$\eta_{Carnot}=\frac{T_H-T_L}{T_H} \tag{1.46}$$

which is commonly known as the *Carnot efficiency* although it is applicable to any reversible heat engine. Using the sign of inequality in (1.42), which corresponds to the irreversible mode of operation of the engine, for $Q_{IN}>0$:

$$\frac{Q_{IN}}{Q_{OUT}}<\frac{T_H}{T_L} \tag{1.47}$$

From (1.45), (1.46) and (1.47) it follows that:

$$\eta_{IRREV}<\eta_{REV} \tag{1.48}$$

which shows that the maximum efficiency of any heat engine operating between two TERS of given temperatures is the Carnot efficiency.

1.7 Reversibility and irreversibility

The concepts of reversibility and irreversibility are important in thermodynamics and crucial to the Exergy Method. Understanding the nature of irreversibilities and how to minimise them in practice is essential for an engineering thermodynamicist.

A reversible process is an idealisation which can never be realised fully but which is useful conceptually because it is easier to describe in mathematical terms than an irreversible process. The reversible process can be used conveniently as a standard of perfection by which real processes can be judged.

An irreversible process is unavoidably accompanied by an increase in entropy in the combined system (or universe). From the microscopic and statistical point of view, this indicates that there has been a change from a more organised form of energy to one characterised by a greater degree of randomness.

Two groups of phenomena are found in irreversible processes. One involves direct dissipation of work into internal energy of the system, ie fully organised macroscopic work is converted to the microscopic energy associated with the random motion of the molecules. This dissipation is caused by solid or fluid friction, mechanical or electrical hysteresis, ohmic resistance *etc*.

The other group of phenomena are associated with *spontaneous non-equilibrium processes*, in which a system tends to move in an unrestrained manner from a state of non-equilibrium to one of equilibrium. Typical phenomena in this group are spontaneous chemical reactions, free diffusion, unrestrained expansion and equalisation of temperature.

Irreversible processes have a mixture of the phenomena belonging to the two groups. For example in combustion there will be irreversibilities due to the mixing of the reactants, spontaneous chemical reaction, fluid friction and heat conduction over a finite temperature difference. Further, one form of irreversibility may give rise to another; in a brake system, for example, when work is dissipated through friction between the brake shoe and the drum, the adjacent layers of material become hot. This leads to the equalisation of temperature distribution which is a non-equilibrium phenomena.

Thus the condition for reversibility of a process is that:

(i) the system passes through a series of equilibrium states, ie the process is performed quasi-statically, and
(ii) dissipative phenomena are absent from the system.

1.8 Three examples of irreversible processes

The concept of irreversibility is based on the Second Law and, therefore, any means devised to test a process for irreversibility must involve its application.

One possible test involves the direct application of the entropy postulate; the entropy change for the system and its surroundings is positive, then the process is irreversible. Another test uses the definition of reversibility given in Section 1.1. The systems participating in the process are restored to their original states by some convenient reversible processes using ideal, cyclically operating devices compatible with the Second Law. If, on completing this restoration, the net effect is that the surroundings have to deliver some work, then the process is irreversible.

The three examples of systems undergoing simple irreversible processes which follow illustrate the application of the two tests and throw some light on the nature of some of the irreversible processes and their relation to the Second Law.

Heat transfer through a finite temperature difference

Consider two thermal energy reservoirs TER-H and TER-L, at temperatures T_H and T_L with $T_H > T_L$ which are brought into thermal contact by a thermal conductor of finite resistance to heat transfer and of negligible heat capacity (Fig 1.4(a)). After heat transfer Q has taken place between the reservoirs, the conductor is removed. The total entropy increase for the two reservoirs together considered as an isolated system is:

$$(\Delta S)_{ISOL} = \left(\frac{Q}{T_L} - \frac{Q}{T_H}\right) > 0 \qquad \text{(a)}$$

As long as $T_H > T_L$ the entropy change of the combined system will be positive; heat transfer over a finite temperature difference is irreversible.

Applying the second irreversibility test, the two reservoirs are restored to their initial states by returning the quantity Q of thermal energy to TER-H and removing Q from TER-L using two idealised devices, a reversible heat engine, RHE, and a reversible heat pump, RHP (Fig 1.4(b)). RHE and RHP operate between the two reservoirs and the surroundings, comprising the environment at temperature T_0, and an MER. From Eqs (1.41) and (1.46)

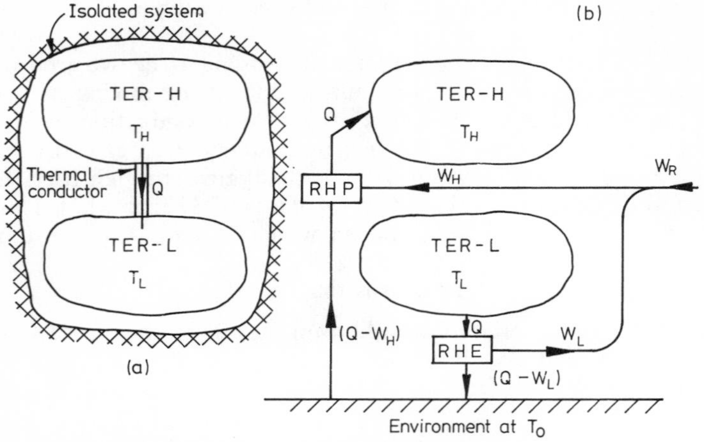

Fig 1.4 Irreversible heat transfer between two thermal energy reservoirs.

the work terms (absolute values) W_H and W_L involved in the restoration processes are:

$$W_L = Q\,\frac{T_L - T_0}{T_L} \tag{b}$$

$$W_H = Q\,\frac{T_H - T_0}{T_H} \tag{c}$$

The net restoring work which has to be provided by the surroundings to restore TER-H and TER-L to their respective initial states is:

$$W_R = -(W_L - W_H) \tag{d}$$

Substituting from (b) and (c):

$$W_R = T_0\left(\frac{Q}{T_L} - \frac{Q}{T_H}\right) \tag{e}$$

Since T_H must always be greater than T_L, W_R is positive, proving that the process is irreversible.

It will also be found (Fig 1.4(b)) that the net heat transfer to the environment during the restoration process is:

$$(Q - W_L) - (Q - W_H) = -(W_L - W_H) = W_R \tag{f}$$

Unrestrained expansion

Another typically irreversible process is unrestrained expansion. The energy equation for a gas expanding into an evacuated space (Fig. 1.5) is:

$$Q - W = U_2 - U_1 \tag{g}$$

Taking the wall of the vessel to be rigid and adiabatic (ie an isolated system):

$$Q = 0 \qquad W = 0$$

Hence:

$$U_2 = U_1$$

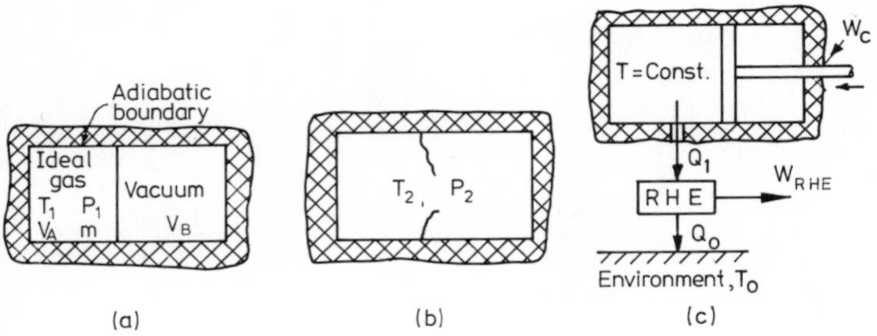

Fig 1.5 Unrestrained expansion of an ideal gas.

and for an ideal gas:

$$T_2 = T_1 \tag{h}$$

The change in entropy due to this process, from an expression similar to (1.40), is:

$$(\Delta S)_{ISOL} = \frac{P_1 V_A}{T_1} \ln \frac{V_A + V_B}{V_A} > 0 \tag{i}$$

The increase in entropy proves that an unrestrained expansion is an irreversible process.

For the second test, recompress the gas reversibly and isothermally to its original volume, and simultaneously use the heat rejected to operate a reversible heat engine between the system at temperature T_1 and the environment at temperature T_0. The work of compression is:

$$W_C = P_1 V_A \ln \frac{V_A + V_B}{V_A} \tag{j}$$

and the work recovered by the heat engine:

$$W_{RHE} = Q_1 \frac{T_1 - T_0}{T_1} \tag{k}$$

But, for an isothermal compression process of an ideal gas:

$$Q_1 = W_C \tag{l}$$

and so:

$$W_{RHE} = \frac{T_1 - T_0}{T_1} P_1 V_A \ln\left(\frac{V_A + V_B}{V_A}\right) \tag{m}$$

Hence the net restoring work to be delivered by the surroundings is:

$$W_R = -[W_{RHE} - W_C] = \frac{P_1 V_A}{T_1} T_0 \ln\left(\frac{V_A + V_B}{V_A}\right) \tag{n}$$

W_R is equal to the heat Q_0 rejected by the RHE to the environment. Once again, an attempt to restore a system to its initial state results in the surroundings loosing a quantity of organised-type of energy and receiving an equal amount of thermal energy at temperature T_0. The process is irreversible.

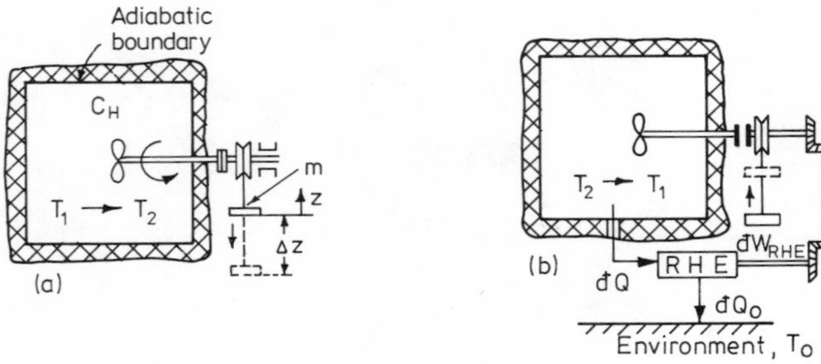

Fig 1.6 Viscous dissipation of mechanical energy.

A dissipative process

As an example of dissipative irreversibility, consider a stirring process. A rigid, adiabatic vessel contains fluid of constant heat capacity C_H. The stirrer is driven, through a coupling, by a falling weight on a string wound around the pulley (Fig 1.6(a)). During the process, loss of potential energy of the pulley, $mg_E\Delta Z$, is used as stirring work W_F:

$$W_F = -C_H(T_2 - T_1) \tag{o}$$

Assuming that components of the system outside the adiabatic vessel suffer no frictional losses, and that the pressure of the fluid inside the vessel remains constant, the entropy change of the fluid is:

$$(\Delta S)_{ISOL} = C_H \ln \frac{T_2}{T_1} > 0 \tag{p}$$

The positive value of the entropy change is an indication of the irreversibility of the process.

The fluid and weight are restored to their initial states using the arrangement shown in Fig 1.6(b). The fluid is cooled reversibly from T_2 to T_1 using the thermal energy withdrawn from it in a reversible heat engine. The work output from the latter is used towards restoring some of the potential energy of the weight. The elemental work delivered by the RHE is:

$$đW_{RHE} = \frac{T - T_0}{T}\, đQ \tag{r}$$

Introducing $đQ = C_H \mathrm{d}T$ into (r) and integrating between the limits $T = T_2$ and $T = T_1$:

$$W_{RHE} = C_H\left[(T_2 - T_1) - T_0 \ln \frac{T_2}{T_1}\right] \tag{s}$$

The net restoring work which has to be provided by the surroundings will be given by the difference between the loss of potential energy ($= -W_F$) in the original process and the work recovered through the RHE in the reversible cooling process. Thus:

$$W_R = (-W_F) - W_{RHE} \tag{t}$$

Substituting in (t) from (o) and (s) gives the restoring work:

$$W_R = C_H T_0 \ln \frac{T_2}{T_1} \qquad \text{(u)}$$

As this is always positive, the process must be irreversible.

Discussion of the illustrative examples

The first two examples are typical non-equilibrium processes and the third a dissipative process; the nature of the irreversibilities suffered by the systems in the non-equilibrium processes is, perhaps, the more difficult to appreciate since there is no work lost. In both cases energy is conserved and, therefore, the First Law can be of little real help here in the analysis. In the third example, all the potential energy of the weight has been dissipated in the stirring process. Note, however, that part of the resulting increase in the internal energy of the fluid has been recovered as mechanical energy by the RHE, reducing thus the net amount work to be provided by the surroundings.

All real processes are irreversible, so the question is not (usually) whether the process is irreversible but what, quantitatively, is the degree of irreversibility. This information is obtained in the above examples in the form of the restoring work, W_R. The bigger the value of W_R the greater is the degree of irreversibility.

Applying the technique involving reversible heat engines and other idealised devices to the analysis of real processes is cumbersome and, in general, quite impractical. A clue to an alternative and more direct method of quantitative assessment of process irreversibility is offered by the relationship between the expressions for W_R and $(\Delta S)_{ISOL}$ calculated for all three cases. Comparing pairs of expressions—(a) with (e), (i) with (n) and (p) with (u)—shows that the restoring work can be related to the entropy change of the isolated system:

$$W_R = T_0 (\Delta S)_{ISOL} \qquad \text{(v)}$$

These three special cases provide some insight into the nature of irreversibilities and a pointer as how to assess them quantitatively. A general and practical technique for treating all such cases is given in Chapter 3.

1.9 Entropy production in a system

As can be judged from the statement of the entropy postulate, the change in entropy of an isolated system is of much greater thermodynamic significance than in a non-isolated system.

This entropy change is always positive and reflects the irreversibility of the process taking place within the isolated system. Thus the entropy increase within an isolated system will be called *the entropy production* and will be denoted by Π, rather than the more cumbersome form $(\Delta S)_{ISOL}$ used up to now.

To calculate entropy production, it will be useful to have a relation in which the entropy change associated with heat transfer across the system boundary is expressed as a separate term. Consider a system shown in Fig 1.7 undergoing an infinitesimal irreversible process. The system suffers an entropy change dS, while at the same time receiving $đQ$ from a TER and delivering an amount of work $đW$ to the MER; TER and the MER form hypothetical surroundings of the system and all three together, for the purpose of the analysis, form an isolated system.

The entropy change of the isolated system is equal to the entropy changes of its

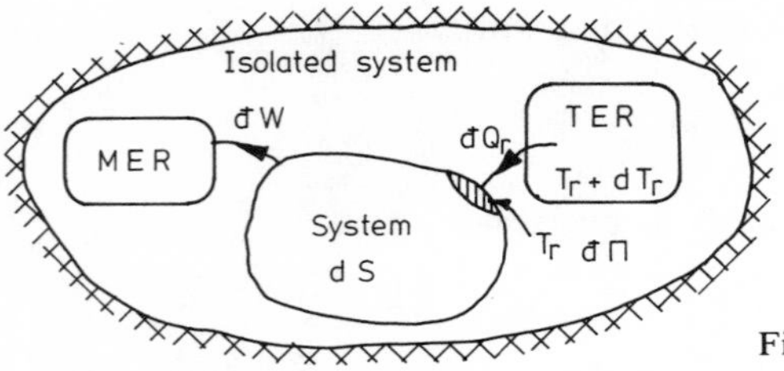

Fig 1.7 Entropy production in a system

constituent parts:

$$đ\Pi = dS + (dS)_{TER} + (dS)_{MER} \tag{1.49}$$

đ rather than d is used here because entropy production is a function of the process and not of the state of the system.

To limit all irreversible effects to the system, the heat transfer is arranged to take place *across an infinitesimal temperature difference*, ie reversibly. Hence the entropy change of the TER is:

$$(dS)_{TER} = -\frac{đQ_r}{T_r} \tag{1.50}$$

$đQ$ being positive with respect to the system. The work transfer to the MER is by definition reversible, hence:

$$(dS)_{MER} = 0 \tag{1.51}$$

Thus from (1.49)–(1.51) and in accordance with the entropy postulate (1.30):

$$đ\Pi = dS - \frac{đQ_r}{T_r} \geqslant 0 \tag{1.52}$$

For a finite process between states 1 and 2, integration with T=const yields:

$$\Pi = (S_2 - S_1) - \frac{Q_r}{T_r} \geqslant 0 \tag{1.53}$$

The first term in the middle of (1.53) is the actual change in entropy of the system during the process. The second term is the reversible entropy change due to the heat transfer. Because this part of the increase in entropy of the system is, under reversible heat transfer conditions, equal to the decrease in entropy of the TER it is convenient to think of it as a transfer or flow of entropy; it is known as the *thermal entropy flux* and has the same direction as the heat transfer with which it is associated. Thermal entropy flux is calculated at the temperature at the point on the boundary of the system where the heat transfer is taking place.

In general there may be a number of TERs interacting with the system so the second term in the middle of (1.53) must be the sum of all the thermal entropy fluxes. Thus:

$$\Pi = (S_2 - S_1) - \sum_r \frac{Q_r}{T_r} \tag{1.54}$$

Using the concept of thermal entropy flux greatly simplifies calculation of entropy production for a closed system, as given by (1.54). The practical significance and applications of this equation should become more apparent in Chapter 3 and Appendix B.

1.10 Entropy production rate in a control region

The concept of entropy production, and its expression in terms of thermal entropy flux, can be extended to control region analysis, a type of analysis which is of particular importance in engineering applications. A new concept, the *matter reservoir*, is required. A matter reservoir is a large, rigid, adiabatic vessel containing matter of prescribed chemical composition and thermodynamic properties. Because of its large size, a matter reservoir is capable of delivering or receiving any quantity of matter without a change in its intensive properties. The process of delivery or storage of matter will be assumed to be quasi-static. Here, two such reservoirs will be used, an inlet matter reservoir, IMR, and an exit matter reservoir, EMR, although in general a number of IMRs and EMRs may be required.

Consider the control region shown in Fig 1.8 with its fictitious surroundings consisting of a TER, a MER, an IMR and an EMR. The control region with the surroundings

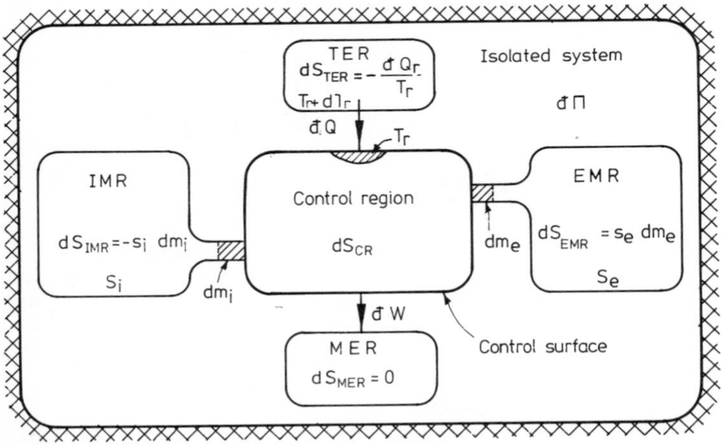

Fig 1.8 Entropy production in a control region.

form an isolated system ('the universe'). It must be clearly understood that the role of the fictitious surroundings is only auxiliary as an aid to the application of the entropy postulate. The control region is assumed to undergo an infinitesimal irreversible non-steady flow process. Since the system and the IMR and the EMR undergo processes in which the quantities of matter inside them change, absolute entropies must be used. During the process which occurs in a time interval dt, the control region receives mass dm_i with specific entropy s_i and rejects mass dm_e with specific entropy s_e. At the same time, the control region receives thermal energy $đQ_r$ from the TER and delivers mechanical energy $đW$ to the MER.

Applying the entropy postulate to the isolated system, its entropy production $đ\Pi$ during the process is equal to the sum of the entropy changes of its constituent parts. Thus:

$$đ\Pi = dS_{CR} + dS_{IMR} + dS_{EMR} + dS_{TER} + dS_{MER} \geqslant 0 \tag{1.55}$$

The entropy terms in (1.55) can be written in terms of the various process parameters as follows:

$$
\begin{aligned}
dS_{IMR} &= -s_i dm_i \\
dS_{EMR} &= s_e dm_e \\
dS_{TER} &= -\frac{đQ_r}{T_r} \\
dS_{MER} &= 0
\end{aligned}
\tag{1.56}
$$

Substituting from (1.56) in (1.55):

$$
đ\Pi = dS_{CR} + s_e dm_e - s_i dm_i - \frac{đQ_r}{T_r} \geqslant 0 \tag{1.57}
$$

In general there may be several inlet and outlet streams and the control region may be interacting with a number of reservoirs. Introducing these generalisations and dividing (1.57) by the time interval dt and expressing on a rate basis:

$$
\dot{\Pi} = \frac{dS_{CR}}{dt} + \sum_{OUT} (s_e \dot{m}_e) - \sum_{IN} (s_i \dot{m}_i) - \sum_{r} \frac{\dot{Q}_r}{T_r} \geqslant 0 \tag{1.58}
$$

In the special case when the control region operates under steady conditions:

$$
\frac{dS_{CR}}{dt} = 0 \tag{1.59}
$$

and (1.58) can be written conveniently as:

$$
\dot{\Pi} = (\dot{S}_{OUT} - \dot{S}_{IN}) - \sum_{r} \frac{\dot{Q}_r}{T_r} \geqslant 0 \tag{1.60}
$$

where

$$
\begin{aligned}
\dot{S}_{OUT} &= \sum_{OUT} s_e \dot{m}_e \\
\dot{S}_{IN} &= \sum_{IN} s_i \dot{m}_i
\end{aligned}
\tag{1.61}
$$

Since under steady conditions:

$$
\sum_{OUT} \dot{m}_e = \sum_{IN} \dot{m}_i \tag{1.62}
$$

the values of specific entropies in (1.61) need not be absolute, provided the process does not involve chemical reactions, ie the proportions of the different chemical species in the process remain unchanged. For a steady internally reversible, adiabatic flow, Eq (1.60) reduces to

$$
\dot{S}_{OUT} = \dot{S}_{IN} \tag{1.63}
$$

1.11 Maximum work obtainable from a system in combination with a thermal energy reservoir

We shall now prove the theorem[1.3] that the total work obtainable from a *combined system*, consisting of a system and a thermal energy reservoir, with a heat engine

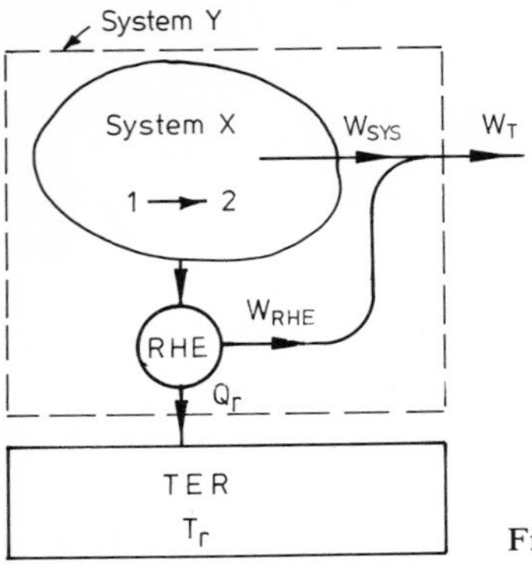

Fig. 1.9 Work from a system in combination with a TER.

operating between them (Fig 1.9) for a process between two equilibrium states, cannot exceed that of a reversible process between the same two states. Furthermore, the total work is the same for all reversible processes occurring between the same two states.

For a process to be fully reversible not only must conditions for reversibility be satisfied within the system itself, but heat transfer between system and TER must also occur reversibly; conditions for both internal and external reversibility must be satisfied. Here external reversibility can be satisfied using a reversible heat engine operating between the system and the TER (Fig 1.9). The total work output of the combined system will consist of W_{SYS} delivered by the system and W_{RHE} obtainable from the heat engine.

Now consider any process, reversible or irreversible, with a total work output $[W_T]_{ANY}$. If, using a reversible process, with the total work input $[W_T]_{REV}$, to restore the system and the TER to their respective initial states, and it is found that $[W_T]_{ANY} > [W_T]_{REV}$, then we could operate System Y, in a cycle delivering a net amount of work in each cycle while interacting with only one TER. This would make System Y a perpetual motion machine of the second kind (PMM2). The impossibility of a PMM2 is one of the different ways in which the Second Law is stated[1.4], originally by Planck[1.5].

1.12 Maximum work of a chemical reaction

Gibbs function of a reaction

We have shown in the previous section that if maximum work is to be obtained, the process must be fully reversible. As no restriction has been imposed on the nature of the process, this also applies to chemical processes.

The work obtainable from a reversible chemical reaction is particularly relevant in the development of the concept of standard chemical exergy in Chapter 2. Thus, consider a chemical reaction in which the reactants enter and the products leave the control region at a steady rate and at standard temperature and pressure T^0, P^0 (Fig 1.10). The device delivers work while interacting with a TER, the conceptual environment, at temperature T^0. An example of an ideal device with an electrical work output is a reversible fuel cell; one with a mechanical work output will be considered below. Note that this type of device is not a heat engine* and therefore its efficiency is not subject to an upper limit given by the Carnot efficiency. Applying the steady flow

* See definition in Section 1.1.

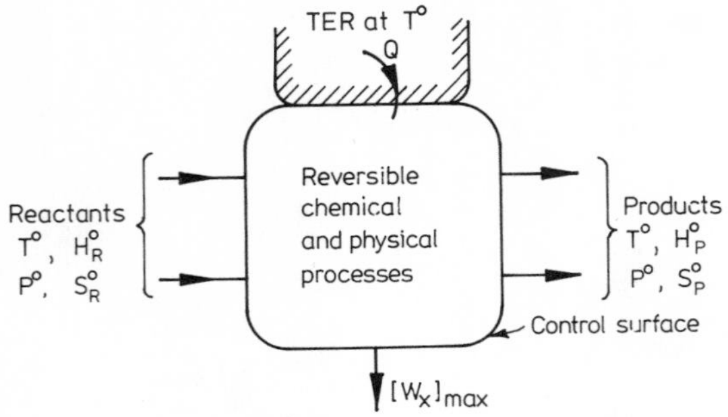

Fig 1.10 Reversible chemical reaction at the reference temperature T^0.

energy equation to the control region, with $\Delta \mathbf{E}_k = 0$ and $\Delta \mathbf{E}_p = 0$ and assuming the process to be internally and externally reversible:

$$Q - [W_x]_{MAX} = H_P^0 - H_R^0 \tag{1.64}$$

and:

$$Q = T^0(S_P^0 - S_R^0) \tag{1.65}$$

From (1.64) and (1.65):

$$[W_x]_{MAX} = -(H_P^0 - T^0 S_P^0) + (H_R^0 - T^0 S_R^0) \tag{1.66}$$

Defining the Gibbs function:

$$G = H - TS \tag{1.67}$$

and the Gibbs function of the reaction:

$$\Delta G = G_P - G_R \tag{1.68}$$

The expression for the maximum work of a chemical reaction can be put as:

$$[W_x]_{MAX} = -\Delta G^0 \tag{1.69}$$

where ΔG^0 is the standard value of the Gibbs function of the reaction. The relevance of (1.69) is not limited to chemical processes. In fact, in any reversible isothermal steady-flow process, the work is equal to the decrease of the Gibbs function of the stream.

Van't Hoff equilibrium box

The derivation of (1.69) gives the maximum work of a chemical reaction in terms of the parameters of the streams entering and leaving the control region; however, the considerations used are of *external* type and make no contribution towards the understanding of the *internal* phenomena which accompany a reversible chemical reaction. Therefore, to gain a better understanding of the physical phenomena which accompany a reversible chemical reaction it may be helpful to consider an ideal, reversible device known as the van't Hoff equilibrium box (Fig 1.11). The processes throughout the control region are maintained at temperature T^0 by reversible heat transfer with the surrounding TER. For the sake of simplicity, both the reactants and the products are assumed here to be ideal gases. Operation of the device depends on the fact that under equilibrium conditions, a reaction in the reaction chamber, or equilibrium box, proceeds simultaneously at equal rates in both directions. Thus, with two

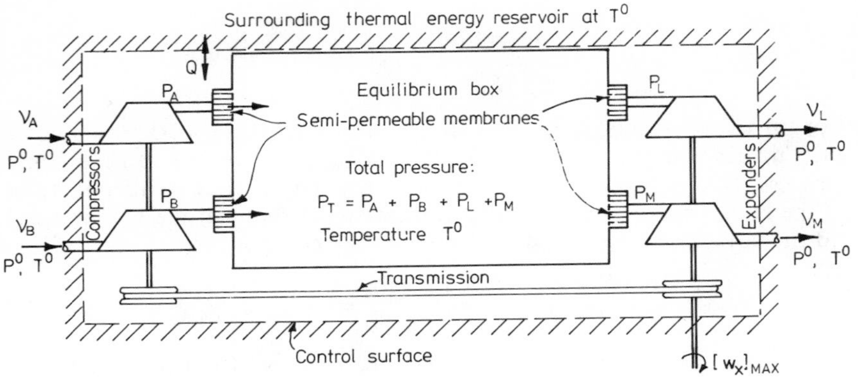

Fig 1.11 Van't Hoff equilibrium box.

reactants A and B, and two products L and M:

$$\nu_A A + \nu_B B \rightleftarrows \nu_L L + \nu_M M \tag{1.70}$$

where ν_A, ν_B, ν_L and ν_M are the stoichiometric coefficients of the reaction. If the reactants are delivered to the equilibrium box and the products are withdrawn from it at steady rates and in stoichiometric quantities, the mole fractions and the partial pressures of each constituent inside the box remain unchanged. The reversible supply of the reactants and the withdrawal of the products, through the semi-permeable membranes*, is by reversible isothermal compressors and expanders which operate between the standard pressure, P^0, of the gases outside the device and the partial pressure of each component of the mixture inside the equilibrium box. Clearly, if each of the machines is to perform as a compressor or an expander, as designated in Fig. 1.11, P^0 must be smaller than any of the partial pressures of the component of the mixture. The compressors and expanders are mechanically coupled and the net shaft work is delivered outside the control region.

The assumption of reversible operation of the device implies that only an infinitesimal change in the external conditions is required to reverse all flow directions and the directions of the work and heat interaction and make the reaction proceed in the opposite direction.

The work per mole delivered by an open system when an ideal gas undergoes a reversible isothermal process is:

$$\tilde{w}_x = \tilde{R}T \ln \frac{P_1}{P_2} \tag{1.71}$$

P_1 being the pressure at entry and P_2 the pressure at exit.

Using (1.71), the net work delivered by the control region shown in Fig 1.11, taking the stoichiometric coefficients as the numbers of moles of the substances involved in the reaction, is:

$$[W_x]_{MAX} = \tilde{R}T^0\left[\nu_A \ln \frac{P^0}{P_A} + \nu_B \ln \frac{P^0}{P_B} + \nu_L \ln \frac{P_L}{P^0} + \nu_M \ln \frac{P_M}{P^0}\right] \tag{1.72}$$

* A semi-permeable membrane allows the passage of only one molecular species. This is an idealised device of which there are very few known examples; red hot palladium is permeable only to hydrogen.

or:

$$[W_x]_{MAX} = \tilde{R}T^0\left[\ln\frac{P_L^{\nu_L}P_M^{\nu_M}}{P_A^{\nu_A}P_B^{\nu_B}} + \ln(P^0)^{(\nu_A+\nu_B-\nu_L-\nu_M)}\right] \tag{1.73}$$

The subscript 'MAX' is justified here, according to the 'maximum work' theorem of Section 1.11. Further Section 1.11 showed that $[W_x]_{MAX}$ will be the same for any reversible process. Now consider the operation of the device at two different total pressures of the equilibrium mixture. Using a prime to distinguish the second total pressure and the corresponding partial pressures of the components:

$$\left.\begin{aligned} P_T &= P_A + P_B + P_L + P_M \\ P'_T &= P'_A + P'_B + P'_L + P'_M \end{aligned}\right\} \tag{1.74}$$

Since $[W_x]_{MAX}$ is the same regardless of the pressure, it follows from (1.73) that:

$$\frac{P_L^{\nu_L}P_M^{\nu_M}}{P_A^{\nu_A}P_B^{\nu_B}} = \frac{P_L'^{\nu_L}P_M'^{\nu_M}}{P_A'^{\nu_A}P_B'^{\nu_B}} = K_p \tag{1.75}$$

The value of K_p, the *equilibrium constant*, is independent of pressure but a function of the temperature for any particular reaction. Despite its appearance in the generally adopted form as given by (1.75), K_p is a dimensionless quantity even when the number of moles taking part on each side is not equal, ie when $\nu_A + \nu_B \neq \nu_L + \nu_M$. This is because the standard pressure, P^0, in (1.73) has been taken, for convenience, as unity in the same units as the partial pressures making up K_p. Thus each partial pressure is implicitly divided by P^0. In tables of chemical thermodynamic data, the standard reference pressure adopted is 1 atm. Hence, from (1.73), the maximum work of a chemical reaction of an ideal gas at the standard temperature T^0 and pressure P^0 can be written as:

$$[W_x]_{MAX} = \tilde{R}T^0 \ln K_p \tag{1.76}$$

Expressing molar Gibbs function of the reaction:

$$\Delta G^0 = \nu_L \tilde{g}_L^0 + \nu_M \tilde{g}_M^0 - \nu_A \tilde{g}_A^0 - \nu_B \tilde{g}_B^0 \tag{1.77}$$

(1.69) and (1.76) give:

$$\ln K_p = -\frac{\Delta G^0}{\tilde{R}T^0} \tag{1.78}$$

As can be seen from (1.69) and (1.76), the maximum work of a chemical reaction involving ideal gases can be determined either from the equilibrium constant or from the Gibbs function of the reaction; the two can be calculated from each other using (1.78).

1.13 Chemical potential

The concept of chemical potential of a substance is very important in chemical thermodynamics. Here it is applied to the concepts of chemical equilibrium and chemical exergy.

To understand these concepts, consider a process in which an open system containing a mixture of gases simultaneously receives and rejects *reversibly* pure substances of which it comprises. Figure 1.12 shows an open system which com-

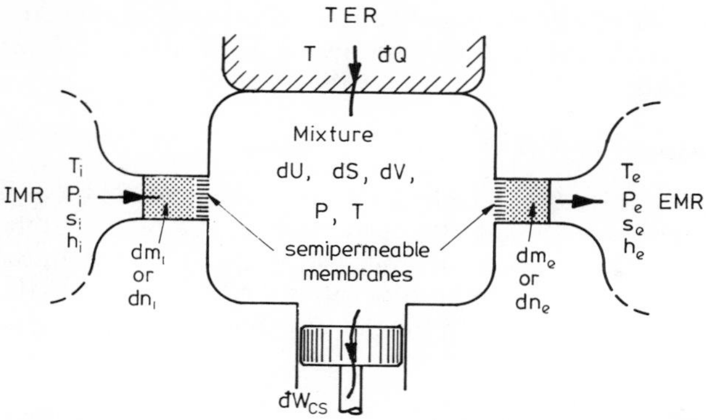

Fig 1.12 An open system undergoing an infinitesimal reversible process.

municates through suitable semi-permeable membranes with two matter reservoirs* containing components of the mixture at the partial pressures P_i and P_e at which they exist in the mixture and at temperature T of the mixture. Under these conditions the pure components outside the open system (on the outer sides of the semi-permeable membranes) are *in membrane equilibrium* with the mixture[1.8]. Let us assume the open system to undergo an infinitesimal reversible process involving heat transfer $đQ$, work due to volume change $đW_{CS}$ and the transfer of quantities dm_i and dm_e of the two components through the semi-permeable membranes in the directions indicated in the figure.

Using (1.26) with (1.27), in which changes in kinetic and potential energy will be neglected, and with $đW_x$10, $đW_{CS} = PdV$ and $dE_{CR} = dU$:

$$đQ - PdV = dU + h_e dm_e - h_i dm_i \tag{1.79}$$

The heat transfer, under reversible conditions, can be expressed from (1.57) as:

$$đQ = T(dS + s_e dm_e - s_i dm_i) \tag{1.80}$$

Substituting $đQ$ from (1.80) in (1.79) and rearranging:

$$dU = TdS - PdV + g_i dm_i - g_e dm_e \tag{1.81}$$

Where, with $T_i = T_e = T$:

$$\left.\begin{aligned} g_i &= h_i - T_i s_i \\ g_e &= h_e - T_e s_e \end{aligned}\right\} \tag{1.82}$$

are the specific Gibbs functions of the pure substances entering and leaving the open system.

Using a sign convention in which dm is positive when entering the open system, (1.81) can be rewritten for the general case of a process with N pure substances entering or leaving the mixture. All the N products of the type $g_j dm_j$ are summed to take account of the mass transfers, so:

$$dU = TdS - PdV + \sum_j g_j dm_j \tag{1.83}$$

* Only one IMR and one EMR are considered for simplicity.

which is the Gibbs equation applicable to systems with variable composition and corresponds to (1.35) for systems with fixed composition. Using molar quantities, the products in the last term on the RHS of (1.83) can be written:

$$g_j \mathrm{d}m_j = \mu_j \mathrm{d}n_j \tag{1.84}$$

where n_j represents the number of moles and μ_j is the *molar Gibbs function* of the j-th single component reversibly entering or leaving the open system. The molar Gibbs function of a single component system is known as its *chemical potential*. The chemical potential of a single component in membrane equilibrium with a mixture is equal to the partial molar Gibbs function of that component in the mixture. Substituting (1.84) in (1.83):

$$\mathrm{d}U = T\mathrm{d}S - P\mathrm{d}V + \sum_j \mu_j \mathrm{d}n_j \tag{1.85}$$

Although obtained, for illustrative purposes, by considering an infinitesimal reversible process, (1.85) is a fundamental relation between thermodynamic properties for a system with variable composition. It can also be derived by a route similar to that used for systems of fixed composition in Section 1.5. Introducing into (1.32) the effect of changes in composition on the internal energy of a multi-component system:

$$U = U(S, V, n_1, n_2, \ldots, n_N) \tag{1.86}$$

From (1.85):

$$\mathrm{d}U = \left(\frac{\partial U}{\partial S}\right)_{V,n_j} \mathrm{d}S + \left(\frac{\partial U}{\partial V}\right)_{S,n_j} \mathrm{d}V + \sum_j \left(\frac{\partial U}{\partial n_j}\right)_{S,V,n_l \neq j} \mathrm{d}n_j \tag{1.87}$$

As in (1.33), the coefficients of dS and dV are defined as thermodynamic temperature and (minus) pressure. The coefficient of dn_j is defined as chemical potential, ie:

$$\mu_j \equiv \left(\frac{\partial U}{\partial n_j}\right)_{S,V,n_l \neq j} \tag{1.88}$$

Thus, an expression identical to that given by (1.85) can be derived without reference to any particular type of process.

The terms in (1.85) show the effect of the various types of reversible interactions on the internal energy of a system:

(i) $T\mathrm{d}S$—the effect of heat transfer
(ii) $P\mathrm{d}V$—the effect of work due to change in volume
(iii) $\sum_j \mu_j \mathrm{d}n_j$—the effect of reversible mass transfers, or change in composition.

The nature of the last type of interaction can be understood[1.7] by regarding (1.85) as an energy equation for an infinitesimal reversible process in a closed system when the terms $P\mathrm{d}V - \sum_j \mu_j \mathrm{d}n_j$ represent the total work done. The second of the two terms is a form of work which the system can do by virtue of its change of composition, known as *chemical work* and demonstrated in Section 1.12 by the van't Hoff equilibrium box.

Thus it should be clear that μ, being an intensive property, plays a similar part in the energy transfer processes to that played by the other intensive properties in this expression, T and P. Differences in P, T or μ within a system are known as generalised *driving forces*, since a difference in pressure will tend to produce a volume change*, a difference in temperature will give rise to heat transfer and a difference in chemical potential may give rise to diffusion from a region of high chemical potential to one of low chemical potential. These driving forces, when they are not infinitesimal, give rise to processes which are inherently irreversible and called *spontaneous non-equilibrium processes* in Section 1.7. The greater the driving forces, the greater will be the *process rates* caused by them; when all the driving forces are zero the process rate is also zero. This clearly corresponds to a condition of thermodynamic equilibrium.

By adding to either side of (1.85) the quantity $\mathrm{d}(PV-TS)$ it can be transformed to give a change in the Gibbs function of a system:

$$\mathrm{d}G = -S\mathrm{d}T + V\mathrm{d}P + \sum_j \mu_j \mathrm{d}n_j \tag{1.89}$$

giving an alternative definition of chemical potential:

$$\mu_j \equiv \left(\frac{\partial G}{\partial n_j}\right)_{T,P,n_{l \neq j}} \tag{1.90}$$

From the form of (1.90), the chemical potential of the j-th constituent can be defined as a measure of the effect on the Gibbs function of the system when the content of this constituent in the mixture is increased.

A more complete treatment of the subject of chemical potential is given in Refs [1.1, 1.6, 1.7, 1.9].

Example 1.1

A lump of ice with a mass of 1.5 kg, at an initial temperature of $T_1 = 260$ K melts at the pressure of 1 bar as a result of heat transfer from the environment. After some time has elapsed the resulting water attains the temperature of the environment, ie $T_2 = T_0 = 293$ K. Calculate the entropy production associated with this process. Take the enthalpy of fusion for water $h_{sf} = 333.4$ kJ/kg and the isobaric specific heat capacities of ice and water as $c_{P,\mathrm{ice}} = 2.07$ kJ/kgK and $c_{P,\mathrm{w}} = 4.20$ kJ/kgK respectively. Ice melts at $T_{\mathrm{m}} = 273.16$ K.

Solution: The entropy production in this process can be calculated from (1.54). As shown in Fig 1.13 the TER in this case is the environment at temperature T_0, the corresponding heat transfer being denoted by Q_0. The boundary of the system is drawn far away from the bowl to make sure that the heat interaction at the boundary takes place at T_0 so that the temperature gradients and the associated entropy production are confined to the system. Hence, from (1.54):

$$\Pi = (S_2 - S_1) - \frac{Q_0}{T_0} \tag{w}$$

---

* In open systems pressure difference will give rise to flow of matter.

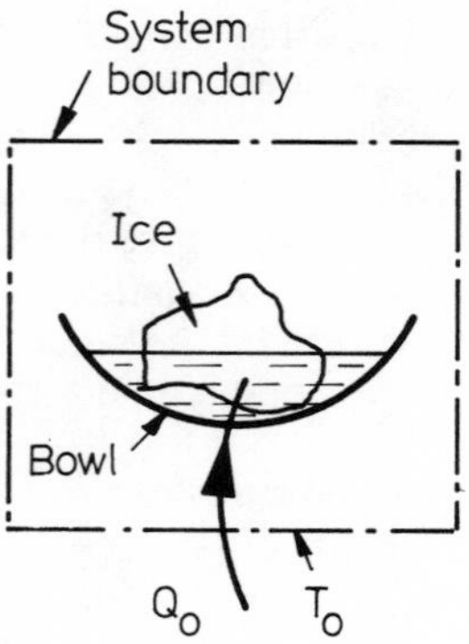

Fig 1.13 Ice melting process.

The heat transfer is obtained from the energy balance for the system:

$$\begin{aligned} Q_0 &= m(h_2 - h_1) \\ &= m[c_{P,w}(T_2 - T_m) + h_{sf} + c_{P,ice}(T_m - T_1)] \\ &= 1.5[4.20(293 - 273.16) + 333.4 + 2.07(273.16 - 260)] \\ &= 665.95 \text{ kJ} \end{aligned}$$

The entropy change at constant pressure between the initial state of ice and the final state of water is given by

$$\begin{aligned} S_2 - S_1 &= m\left(c_{P,w} \ln \frac{T_2}{T_m} + \frac{h_{sf}}{T_m} + c_{P,ice} \ln \frac{T_m}{T_1}\right) \\ &= 1.5\left(4.20 \ln \frac{293}{273.16} + \frac{333.4}{273.16} + 2.07 \ln \frac{273.16}{260}\right) \\ &= 2.426 \text{ kJ/K} \end{aligned}$$

Substituting the numerical values in (w):

$$\begin{aligned} \Pi &= \left(2.426 - \frac{665.95}{293}\right) \\ &= \underline{0.153 \text{ kJ/K}} \end{aligned}$$

Chapter 2 Basic exergy concepts

This chapter introduces the concepts on which the Exergy Method is based. First the characteristics of the various forms of energy are discussed and the need for a universal standard of quality of energy is established. From a basic definition of exergy, expressions for exergy corresponding to different energy forms are derived and their characteristics discussed.

Most classical thermodynamics text books first define quantities relevant to closed system analysis and then derive from them quantities which are applicable to the analysis of open systems. Here, because open system analysis is much more relevant to the analysis of thermal plants or chemical systems than closed system analysis, the exergy of a steady stream of substance is covered first. Non-flow exergy, the exergy function which is applied in closed system analysis, is then derived from steady flow exergy.

Because the concept of chemical exergy will be of greater interest to chemical engineers than to, say, designers of refrigeration plants, some of the more specialist chemical exergy topics appear in Appendix A.

2.1 Classification of forms of energy

Energy manifests itself in many forms, each with its own characteristics and its quality. *Quality of energy* is synonymous with its *capacity to cause change.* For example, the capacity to effect some desired change (eg heat a room, compress gas, or promote an endothermic reaction) of 100 J of electric energy is greater than that of 100 J of thermal energy available temperature of 1000 K and even greater than the same amount of thermal energy available at 400 K, when the temperature of the environment is, say, 300 K. Unless they are taken into account in the analysis, these differences in quality may lead to misleading results when analysing the performance of thermal processes. It may therefore be helpful to examine the characteristics of the different forms of energy and to classify them with a view to establishing a suitable standard of quality of energy.

The quality of a given form of energy depends on its mode of storage. This may be either ordered or disordered (random) in which there are various degrees of randomness. As mentioned in Section 1.5, entropy is a measure of microscopic randomness and the resulting uncertainty about the microscopic state of a system. It is also a measure of the 'unavailability' of a given disordered form of energy for conversion to the ordered form.

Ordered energy

The energy forms in this category are of two kinds:

(a) Potential energy, which may be stored in a gravitational, electrical or magnetic force field. Strain energy stored in a perfectly elastic spring also belongs in this category.
(b) Organised kinetic energy, for example in a spinning flywheel or a jet of an ideal fluid. Here the paths of the particles of the system in motion in which the energy is stored are parallel to one another. Under ideal conditions, ordered kinetic energy, unlike that associated with turbulent motion, can be fully converted to shaft work.

Figure 2.1 shows a number of different devices in which ordered energy undergoes a chain of transformations. Under ideal conditions, ie with no dissipative effects such as

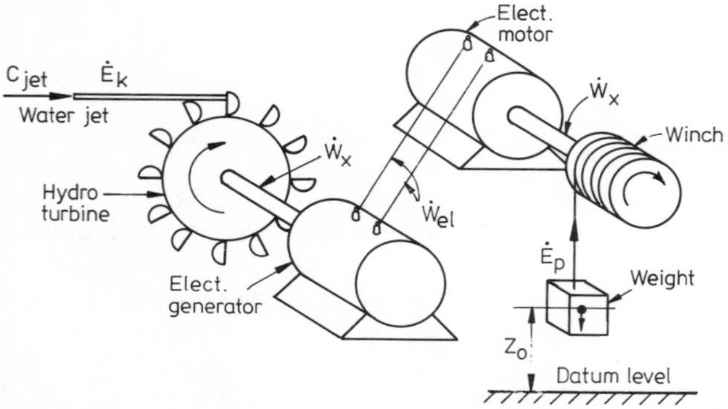

Fig 2.1 Examples of transformation of ordered energy.

friction, electrical resistance, hysteresis, *etc* each device will operate with 100% efficiency so that even after a number of transformations, the rate of kinetic energy $\mathbf{E}_k$ of the water jet entering the turbine would be equal to the rate of increase of the potential energy $\mathbf{E}_p$ of the weight.

Ordered energy has the following characteristics:

(i) Conversion of one form of ordered energy into another form, if carried out reversibly, will take place in full.
(ii) The transfer of ordered energy between two systems manifests itself as a work interaction (not heat!) at the boundary which separates them, ie *work is ordered energy in transit.*
(iii) Reversible transfer of ordered energy occurs without changes in the entropies of the interacting systems and can be analysed with the aid of the First Law only, ie without the use of the Second Law.
(iv) The parameters of the environment are not required for calculating the transfer of high grade energy between two systems.

Disordered energy

Internal energy of matter, thermal radiation*, and chemical energy are different forms of disordered energy. Energy associated with the turbulent motion of a fluid is also of

* Consideration of thermal radiation is outside the scope of this book. The exergy of thermal radiation is discussed in Ref. [2.1].

this form, although it differs from the other members of this group in the important respect that it is a transient form of energy through which ordered energy is ultimately converted to the energy associated with the random molecular motion (disordered energy). Three examples of transformation of disordered energy to ordered energy are shown in Fig 2.2.

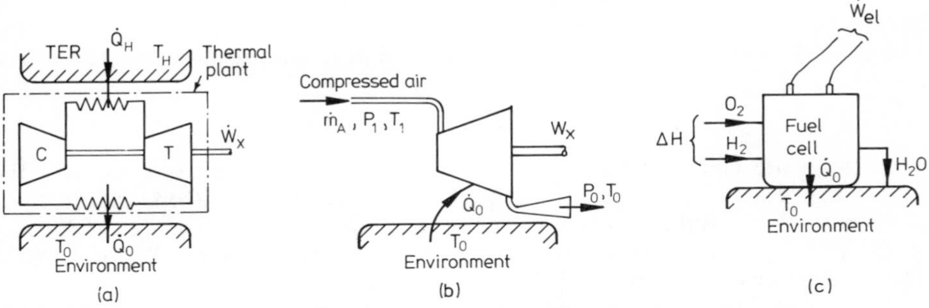

Fig 2.2 Examples of transformation of disordered energy to ordered energy.

In Fig 2.2(a) thermal energy delivered at the rate $\dot{Q}_H$ by a TER at temperature T_H is converted to shaft power by a heat engine using the environment at temperature T_0 as a heat sink. The upper limit of convertibility is given by the Carnot efficiency:

$$[\dot{W}_x]_{REV} = \dot{Q}_H \eta_{CARNOT}$$

where:

$$\eta_{CARNOT} = \frac{T_H - T_0}{T_H}$$

Thus, the quality of $\dot{Q}_H$ as expressed by $[\dot{W}_x]_{REV}$ depends, for a particular value of T_0, on the temperature T_H at which it is available. While the energy transformation is taking place, the two TERs at temperatures T_H and T_0 interacting with the reversible heat engine will undergo entropy changes at the rate of $\dot{Q}_H/T_H$ and $\dot{Q}_0/T_0$ respectively. If the engine operates reversibly:

$$\dot{\Pi} = \frac{\dot{Q}_H}{T_H} - \frac{\dot{Q}_0}{T_0} = 0$$

Figure 2.2(b) shows an open system in which a stream of compressed air initially at pressure P_1 and temperature T_1 is expanded to the environmental pressure P_0 and temperature T_0. The change in the enthalpy of the stream as it passes through the expander bears no relation to the quality of the stream. In fact, in the special case when $T_1 = T_0$, the change in enthalpy assuming ideal gas behaviour, is zero and yet the quality of the energy associated with air stream is not zero since, as it expands reversibly and isothermally the stream delivers an amount of power given by (neglecting $\Delta \mathbf{E}_k$ and $\Delta \mathbf{E}_p$):

$$(\dot{W}_x)_{T_0} = \dot{m}_A R T_0 \ln \frac{P_1}{P_0}$$

During the reversible isothermal expansion, the stream will undergo a change of entropy at the rate:

$$\Delta \dot{S}_A = \dot{m}_A R \ln \frac{P_1}{P_0}$$

which is equal to the rate of decrease in entropy suffered by the environment.

Figure 2.2(c) shows a fuel cell in which the chemical energy of the reactants, H_2 and O_2, is transformed to electric power, $\dot{W}_{el}$. The maximum work of a chemical reaction (Section 1.12) is given by the reduction in the Gibbs function from the reactants to the products. In the special case when the reactants and products are at the environmental temperature T_0:

$$(\dot{W}_{el})_{T_0} = -\Delta \dot{H}_0 + T_0 \Delta \dot{S}_0$$

The entropy rate corresponds to the difference between the entropy rates of the products and reactants, ie $\Delta \dot{S}_0 = \dot{S}_P - \dot{S}_R$. The corresponding entropy rate for the environment, $\dot{Q}_0/T_0$, is the same numerically but of opposite sign. The sign of $\Delta \dot{S}$ may be positive or negative so that the power output may be greater or smaller than $|\Delta \dot{H}_0|$.

This discussions leads to the following statements about processes which are intended to yield the maximum conversion of disordered energy to ordered energy:

(i) The processes used must be fully reversible.
(ii) The upper limit of conversion depends on the thermodynamic parameters of the system (in which the energy is stored) and those of the environment.
(iii) The analysis of the processes must involve the use of the Second Law.
(iv) The conversion is in general accompanied by changes in the entropies of the interacting systems.

2.2 The concept of exergy

Section 2.1 established that the quality (capacity to cause change) of disordered energy forms, characterised by entropy, is variable and depends both on the form of energy (chemical, thermal, *etc*) and on the parameters of the energy carrier and of the environment. On the other hand, ordered forms of energy, which are not characterised by entropy, have invariant quality and are fully convertible, through work interaction, to other forms of energy.

To account for the variable quality of different disordered energy forms in the analysis of thermal and chemical plants, a universal standard of quality is needed. It follows from Section 2.1 that the most natural and convenient standard is the maximum work which can be obtained from a given form of energy using the environmental parameters as the reference state. This standard of energy quality is called *exergy*.

One of the main uses of this concept is in an exergy balance in the analysis of thermal systems. The exergy balance is similar to an energy balance but has the fundamental difference that, while the energy balance is a statement of the law of *conservation of energy*, the exergy balance may be looked upon as a statement of the law of *degradation of energy*. Degradation of energy is equivalent to the irretrievable loss of exergy due to all real processes being irreversible. The use of the exergy balance and the calculation of process irreversibility will be dealt with in Chapter 3. This chapter defines and discusses various exergy terms which go into an exergy balance.

2.3 Exergy concepts for control region analysis

In control region analysis, three types of energy transfer across the control surface have to be considered, in general, in an energy balance:

1. Work transfer.
2. Heat transfer.
3. Energy transfer associated with mass transfer.

In the Exergy Method of control region analysis we shall be using an appropriate form of exergy balance with exergy transfer terms which correspond to the energy transfer forms listed above. Before considering the exergy form of these terms, it is helpful to discuss two basic concepts used in defining the exergy terms.

Environment

The environment, as a concept peculiar to the Exergy Method, is a very large body or medium in the state of perfect thermodynamic equilibrium. Thus, this conceptual environment has no gradients or differences involving pressure, temperature, chemical potential, kinetic or potential energy and, therefore, there is no possibility of producing work from any form of interaction between parts of the environment. Any system outside the environment which has one or more parameters, such as pressure, temperature, or chemical potential, which differs from the corresponding environmental parameter has a work potential in relation to the environment. The environment, therefore, is a natural reference medium for assessing the work potential of different kinds of systems.

For practical purposes, in terrestrial applications, the environment consists of the atmosphere, the seas and the oceans, and the earth's crust. The environment may interact with systems in three different ways:

(a) Through thermal interaction as a reservoir (source or sink) of thermal energy at the temperature T_0: because of the enormous heat capacity of the environment it is able to exchange heat with any man-made system without suffering a significant change in its temperature.

(b) Through mechanical interaction as a reservoir of unusable work: this form of interaction occurs only in systems which undergo a change in volume during the process under consideration, for example, a closed system undergoing an expansion process. It has no relevance in the case of steady flow processes. As terrestrial systems are usually immersed in the atmosphere at pressure P_0, any increase in volume, ΔV, of a system results, assuming the expansion process to be quasi-static, in the work $P_0 \Delta V$ being done on the atmosphere. This part of the work done by the system is, of course, unavailable for use in the technical application for which the system's work is intended but it can be recovered when the volume of the system returns quasi-statically to its original value.

(c) Through chemical interaction as a reservoir of a substance of low chemical potential in stable equilibrium: this type of interaction occurs whenever an open system rejects matter to the environment or draws from it substances of low chemical potential. The environment is assumed to be made up of such substances which are in chemical equilibrium with one another.

Through these types of interactions the environment determines, for the purpose of evaluating exergy, the zero reference levels for pressure, temperature and chemical potential.

Equilibrium

The environmental state

This chapter is concerned with two types of equilibria. Restricted equilibrium, where the conditions of mechanical and thermal equilibrium between the system and the environment are satisfied, requires the pressure and the temperature of the system and environment to be equal. The adjective *restricted* indicates that, under these conditions, the substances of the system are restrained by a physical barrier that prevents the exchange of matter between system and environment. Hence, under conditions of restricted equilibrium there is, in general, no chemical equilibrium between system and environment. The state of restricted equilibrium with the environment will be referred to as the *environmental state.*

The dead state

In unrestricted equilibrium the conditions of mechanical, thermal and chemical equilibrium between the system and the environment are satisfied. Thus, in addition to pressures and temperatures, the chemical potentials of the substances of the system and environment must be equal. Under these conditions of full thermodynamic equilibrium between system and environment, the system cannot undergo any changes of state through any form of interaction with the environment. We shall call this the *dead state.*

To illustrate these equilibrium states, consider a simple substance, CO_2. When pure CO_2 is contained in a vessel or is flowing in a pipe, and its pressure and temperature are those of the environment, P_0 and T_0, it is in the environmental state. However, when CO_2 is at T_0 but its pressure is $(P_{00})_{CO_2}$, ie the partial pressure at which it is normally found in atmospheric air, then it is in unrestricted equilibrium with the environment, the dead state. In this state the gas could also be described as being in membrane equilibrium with the environment since it could pass *reversibly* to the environment through a suitable semi-permeable membrane. Clearly, therefore, all the substances which form the conceptual environment are there in their respective dead states. A more general criterion of unrestricted equilibrium (Section 1.13) than equality of partial pressures is equality of chemical potentials (Section 1.13).

Exergy associated with a work transfer

Since we have defined the work equivalent of a given form of energy as a measure of its exergy, clearly work is equivalent to exergy in every respect. Thus exergy transfer can be specified both in magnitude and direction by the work transfer to which it corresponds, using the same symbol, W, for both. Similarly, $\dot{W}_x$ will be used for both the work transfer rate, or power, and for the associated exergy transfer rate.

Exergy associated with a heat transfer

The exergy of a heat transfer at the control surface is determined from the maximum work that could be obtained from it using the environment as a reservoir of zero-grade thermal energy. For a heat transfer rate $\dot{Q}_r$ and a temperature at the control surface where the heat transfer is taking place T_r, the maximum rate of conversion from thermal energy to work is:

$$\dot{W}_{MAX} = \dot{E}^Q = \dot{Q}_r \tau \qquad (2.1)$$

where:

$$\tau = 1 - T_0/T_r \qquad (2.2)$$

τ is called *dimensionless exergetic temperature*[2.2] and is equal to the Carnot efficiency for the special case when the environment, at temperature T_0, is used as the other TER. The exergy associated with a heat transfer rate is called *thermal exergy flow* and is denoted in open system analysis, by $\dot{E}^Q$. In Eq (2.1), in accordance with the sign convention adopted, $\dot{Q}_r$ is positive when the transfer is to the system. If the heat transfer takes place at a temperature higher than the environmental temperature, τ is positive and hence a heat transfer to the control region results in a gain of exergy by the control region.

A reversible model for determining the exergy associated with a heat transfer rate $\dot{Q}_r$, taking place at T_r, with $T_r > T_0$, is shown in Fig 2.3(a). The thermal energy is shown as being available from a TER at T_r. The power output obtainable from a reversible heat engine RHE operating between this TER and the environment at T_0 is a measure of the thermal exergy associated with the heat transfer rate $\dot{Q}_r$. The T–s diagram represents the relationship between $\dot{E}^Q$ and the corresponding $\dot{Q}_r$ at T_r.

A similar ideal model is shown for $T_r < T_0$, in Fig 2.3(b). Here the natural direction of the heat transfer is from the environment to the TER. The power output obtainable from a RHE operating between T_0 and T_r is a measure of the thermal exergy associated with the heat transfer rate $\dot{Q}_r$ at T_r.

Thermal exergy flux at temperatures $T_r < T_0$ may also be looked upon as the minimum power necessary to maintain a *cooling rate* equal to $\dot{Q}_r$ in a system at T_r. As shown on the T–s diagram in Fig 2.3(b), for two different temperatures (T_r' and T_r''), the lower the temperature of the system which is being cooled the greater the minimum power necessary to maintain a given cooling rate $\dot{Q}_r$. Using the sign convention adopted, positive thermal exergy flow (from (2.1) and (2.2)) corresponds to negative $\dot{Q}_r$

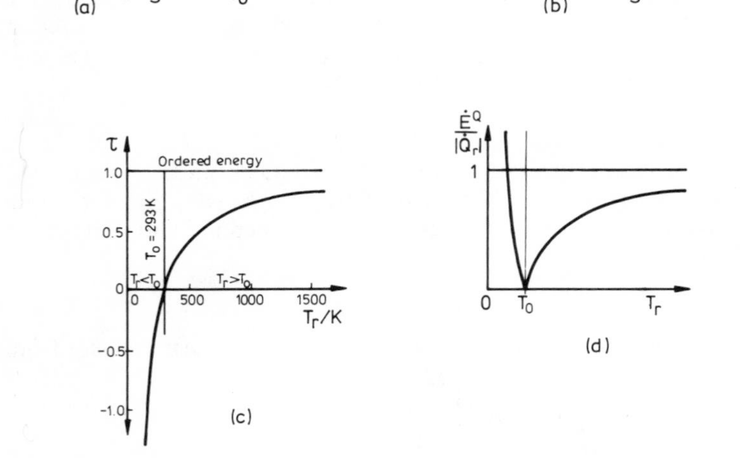

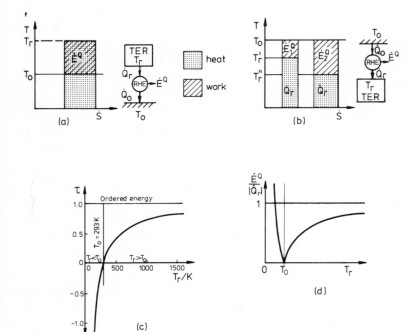

Fig 2.3 Thermal exergy.

since when $T_r < T_0$, $\tau < 0$. Hence, in the sub-environmental temperature range, the exergy of a system increases when it is being cooled, but decreases when it is being heated.

The relationship between τ and T_r ((2.2)) is shown in Fig 2.2(c) for a fixed value of T_0. At $T_r = T_0$, τ is zero, ie thermal energy transferred at T_0 is associated with zero thermal exergy flow. As T_r increases in relation to T_0 the value of both τ and the corresponding thermal exergy increases. In the limit, as $T_r \to \infty$, the quality of thermal energy tends to that of ordered energy, ie the numerical value of $\dot{E}^Q$ tends to that of $\dot{Q}_r$.

In the sub-environmental temperature range τ has negative values and its magnitude exceeds unity at lower temperatures. The interpretation of this is that at low temperatures, the minimum power input to a refrigerator (ie $\dot{E}^Q$) may be greater than the refrigeration duty (ie $\dot{Q}_r$). Note, from the above discussion and in particular from Figs 2.3(a) and (b), that the work potential of any system due to its temperature difference with the environment is always positive regardless of whether $T_r > T_0$ or $T_r < T_0$. In the first case the system would act as a heat source and in the second case as a heat sink. The ratio of the thermal exergy flux to the corresponding absolute value of the heat transfer rate is shown as a function of the temperature in Fig 2.3(d).

There are occasionally circumstances when the heat transfer occurs at a temperature which varies from point to point on the control surface. If the heat flux $\dot{Q}_A$ (heat transfer rate per area) distribution at the various temperatures, T, is known, the associated thermal exergy flux can be determined from:

$$\dot{E}^Q = \int_A \left(\frac{T - T_0}{T}\right)\dot{Q}_A \, dA \qquad (2.3)$$

where A is the heat transfer area. Although (2.3) is a more general definition of $\dot{E}^Q$ than (2.1), there are few practical applications where it might be found useful. Clearly, in order to simplify the calculations whenever possible, the control surface should be selected to give uniform temperature distribution so (2.1) can be used.

$\dot{E}^Q$ gives the exergy transfer rate corresponding to a heat transfer rate $\dot{Q}_r$ when the temperature at the control surface at the points where the heat transfer is occurring is T_r. Since heat transfer takes place in the direction of decreasing temperature the last point is crucial. Consider, for example, heat conduction along a thermally insulated bar; the rate of heat conduction at all normal sections of the bar will be the same but the associated values of $\dot{E}^Q$ will decrease as the temperature decreases. Care is needed in selecting the position of the control surface since this will, in general, affect the value of thermal exergy flux associated with a given heat transfer rate.

Example 2.1

A heat exchanger which acts as a mercury condenser and a H_2O evaporator is shown in Fig 2.4. The heat transfer rate, $\dot{Q}$, is 500 MW, with zero losses to the environment. Using data given in the figure calculate the thermal exergy flow from mercury and the thermal exergy flow to H_2O.

Solution

Using (2.1) and (2.2), the thermal exergy flow associated with the heat transfer from mercury is:

$$\dot{E}^Q_M = 500\,\frac{590 - 290}{590} \qquad \left(\frac{T_m - T_0}{T_m}\right)Q$$

$$= \underline{254.2 \text{ MW}}$$

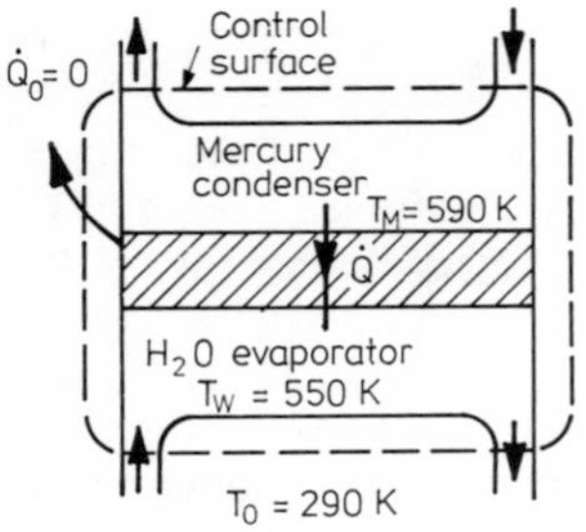

Fig 2.4 Heat exchanger.

Similarly, the thermal exergy flow associated with the heat transfer to water substance is

$$\dot{E}^Q_W = 500\,\frac{550-290}{550}$$

$$= \underline{236.4\ \text{MW}}$$

Note that, although all the thermal energy lost by mercury is transferred to the water substance, some of the thermal exergy flow (17.8 MW) is lost in the heat transfer process. This phenomenon will be discussed in Chapter 3.

Exergy associated with a steady stream of matter

Definition

Exergy of a steady stream of matter is equal to the maximum amount of work obtainable when the stream is brought from its initial state to the dead state by processes during which the stream may interact only with the environment.

Thus, exergy of a stream of matter is a property of two states, the state of the stream and the state of the environment.

Exergy components

As with energy, exergy of a stream of matter, $\dot{E}$, can be divided into distinct components. In the absence of nuclear effects, magnetism, electricity and surface tension, $\dot{E}$ is:

$$\dot{E} = \dot{E}_k + \dot{E}_p + \dot{E}_{ph} + \dot{E}_0 \tag{2.4}$$

where $\dot{E}_k$ is kinetic exergy, $\dot{E}_p$ potential exergy, $\dot{E}_{ph}$ physical exergy, and $\dot{E}_0$ chemical exergy. $\dot{E}_k$ and $\dot{E}_p$ are associated with high grade energy and $\dot{E}_{ph}$ and E_0 with low grade energy. (2.4) may also be written in a 'specific' form. Introducing specific exergy $\varepsilon = \dot{E}/\dot{m}$:

$$\varepsilon = \varepsilon_k + \varepsilon_p + \varepsilon_{ph} + \varepsilon_0 \tag{2.5}$$

Kinetic and potential components of exergy

The kinetic and potential energies of a stream of substance are ordered forms of energy and thus fully convertible to work. Therefore, when evaluated in relation to the environmental reference datum levels, they are equal to kinetic and potential exergy respectively. Thus:

$$\dot{E}_k = \dot{m}\,\frac{C_0^2}{2} \tag{2.6a}$$

and:

$$\dot{E}_p = \dot{m} g_E Z_0 \tag{2.6b}$$

where $\dot{m}$ is the mass flow rate of the fluid stream, C_0 bulk velocity of the fluid stream relative to the surface of the earth, Z_0 altitude of the stream above the sea level, and g_E gravitational acceleration (specific gravitational force), considered a constant.

Use of environmental reference states for C and Z is only important where there is direct interaction of the stream with the environment, for example when evaluating kinetic exergy of the exhaust gases of a jet engine. In most other cases, only changes in kinetic and potential exergies are considered and, therefore, any inertial reference frame may be used.

Physical and chemical components of exergy

Because of the disordered, entropy dependent nature of these forms of energy, the corresponding exergy components can only be determined by considering a composite, two-part system, the stream under consideration and the environment.

In principle one could determine the total exergy derived from disordered energy forms in one idealised device where the stream would undergo physical and chemical processes while interacting with the environment. It is convenient, however, to separate physical exergy $\dot{E}_{ph}$ and chemical exergy $\dot{E}_0$, enabling calculation of exergy values using standard chemical exergy tables. The dividing state in the processes which are used to determine physical and chemical exergy is the environmental state (T_0, P_0).

2.4 Physical exergy

The first component into which exergy derived from disordered energy forms is divided is physical exergy, formally defined as:

Physical exergy is equal to the maximum amount of work obtainable when the stream of substance is brought from its initial state to the environmental state defined by P_0 and T_0, by physical processes involving only thermal interaction with the environment.

An expression for physical exergy can be derived using this definition in connection with Fig 2.5 where Module X represents an ideal device in which the stream will undergo some reversible processes. Assuming the stream to have negligible kinetic and potential energy, the state of the stream under consideration at the entrance to the module is defined by P_1 and T_1. The exit state corresponds to the environmental state, ie the pressure and temperature of the stream are P_0 and T_0. The only interaction associated with the processes in the module is reversible heat transfer with the

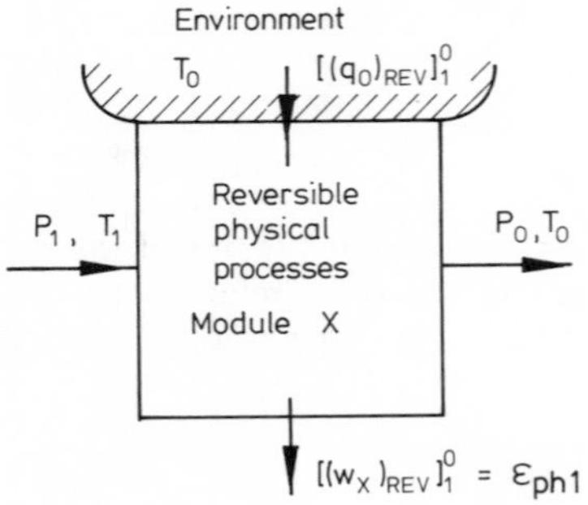

Fig 2.5 A reversible module for determining the physical exergy of a steady stream of matter.

environment which, per mass, is:

$$[(q_0)_{REV}]_1^0 = T_0(s_0 - s_1) \tag{2.7}$$

The steady flow energy equation for the module, per mass, is:

$$[(q_0)_{REV}]_1^0 - [(w_x)_{REV}]_1^0 = (h_0 - h_1) \tag{2.8}$$

As follows from the definition above, the reversible work delivered by the module is equal to the specific physical exergy of the stream. Thus, from (2.7) and (2.8):

$$\varepsilon_{ph1} = (h_1 - T_0 s_1) - (h_0 - T_0 s_0) \tag{2.9}$$

ie ε_{ph1} is equal to the difference in the value of $(h - T_0 s)$ at the initial state 1 and the environmental state. $(h - T_0 s)$ is the *specific exergy function*, denoted by β, ie:

$$\beta = h - T_0 s \tag{2.10}$$

β, in its general form, is similar to the Gibbs function. h and s are evaluated at the state of the substance and T_0 is the temperature of the environment. In the final state, as shown by the second term on the RHS in (2.9), this function is identical with the Gibbs function of the substance in the environmental state. Hence from (2.9) and (2.10):

$$\varepsilon_{ph1} = \beta_1 - \beta_0 \tag{2.11}$$

The hypothetical reversible physical processes which could be used to determine physical exergy must satisfy conditions of both internal and external reversibility; many different combinations of processes could satisfy these conditions. One particular combination of reversible processes, an adiabatic process followed by an isothermal process, is shown in Fig 2.6(a). The condition of external reversibility is satisfied in this case by arranging the heat transfer to occur at $T = T_0 = \text{const}$. The physical exergy of the stream of a homogeneous simple fluid in state 1 is given by the sum of the cross-hatched areas shown on the diagram. The hypothetical device used to determine the physical exergy of the stream would, in this case, consist of an adiabatic expander followed by an isothermal one.

Usually, analysis of a physical process requires the difference in ε_{ph} for two states rather than the separate values. From (2.9):

$$\varepsilon_{ph1} - \varepsilon_{ph2} = (h_1 - h_2) - T_0(s_1 - s_2) \tag{2.12}$$

Using the same type of representation as in Fig 2.6(a), this difference in ε_{ph} is shown by the sum of the two cross-hatched areas in Fig 2.6(b). It is obvious from (2.12) that T_0 is the only relevant environmental parameter when changes in ε_{ph} are evaluated.

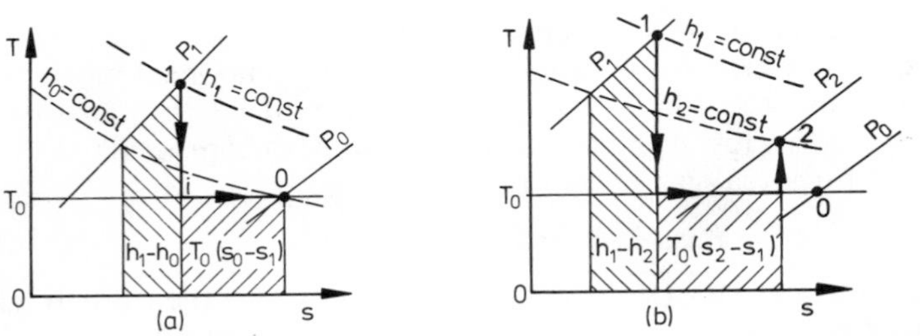

Fig 2.6 Reversible processes for determining (a) physical exergy (b) a change in physical exergy, of a homogenous simple fluid.

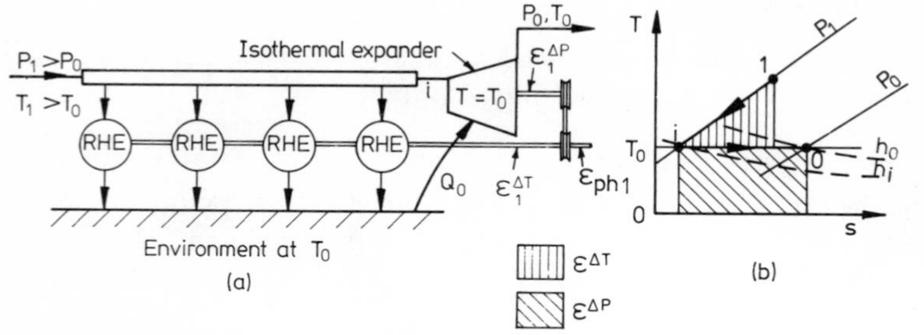

Fig 2.7 Determination of ε_{ph} for a stream of a homogenous simple fluid when $P_1 > P_0$ and $T_1 > T_0$.

Components of physical exergy

The physical exergy of a stream of substance is naturally divisible into two components. Consider a pair of hypothetical, reversible processes (which may be used for determining ε_{ph}); an isobaric process taking place at the initial pressure P_1 followed by an isothermal process corresponding to the environmental temperature T_0. The reversible ideal device which could be used for determining ε_{ph} is shown in Fig 2.7(a) and the processes represented in T–s co-ordinates in Fig 2.7(b). In this case we have selected $P_1 > P_0$ and $T_1 > T_0$ for the initial state of the stream.

The reversible isobaric process is shown (Fig 2.7(b)) in its idealised form as a cooling process in which the thermal energy rejected by the stream is used in a series of reversible heat engines, operating between the temperature of the gas at the different stages of the cooling process and the environmental temperature T_0. The component of physical exergy corresponding to this reversible work term will be denoted by $\varepsilon^{\Delta T}$. Its magnitude, represented as a triangular area in Fig 2.7(b), is given by:

$$\varepsilon_1^{\Delta T} = \left[-\int_{T_1}^{T_0} \frac{T - T_0}{T} \, dh \right]_{P_1} \tag{2.13}$$

This component resulting from the temperature difference between the stream and the environment is termed the *thermal component* of physical exergy. Note that this component has some features in common with the concept of thermal exergy, introduced in Section 2.3, but thermal exergy is associated with *heat transfer* at the control surface while the thermal component of physical exergy is associated with a *stream of matter* crossing the control surface. These two concepts correspond to two different forms of exergy and must not be confused.

The second process takes place reversibly in the isothermal expander shown in Fig 2.7(a). The reversible work obtained from the expansion process is equal to the second component of physical exergy of the stream, denoted by $\varepsilon^{\Delta P}$. This component resulting from the pressure difference between the stream and the environment is termed the *pressure component* of physical exergy. Applying (2.9) to the isothermal process i–0, the pressure component of ε_{ph} for a real-gas stream is:

$$\varepsilon_1^{\Delta P} = T_0(s_0 - s_i) - (h_0 - h_i) \tag{2.14}$$

(2.14) may be looked upon as the SFEE (with $\Delta \mathbf{E}_k = 0$ and $\Delta \mathbf{E}_p = 0$) for the isothermal expander. In such a case $\varepsilon^{\Delta P}$ would correspond to the reversible work output, $T_0(s_0 - s_i)$ the reversible heat transfer and $(h_0 - h_i)$ the enthalpy change due to the non-ideal behaviour of the fluid.

For an ideal gas stream, the second term on the RHS of (2.14) is zero. For this particular case, $\varepsilon_1^{\Delta P}$ is represented by the rectangular, cross-hatched area under the process line i–0 shown on the *T–s* diagram in Fig 2.7(b).

Clearly, physical exergy of a stream of substance can be expressed as a two-component function:

$$\varepsilon_{ph} = \varepsilon^{\Delta T} + \varepsilon^{\Delta P} \tag{2.15}$$

The equivalence of (2.9) and (2.15) can be proved by substituting (2.13) and (2.14) in (2.15) and integrating (with $dh/T = đq/T = ds$).

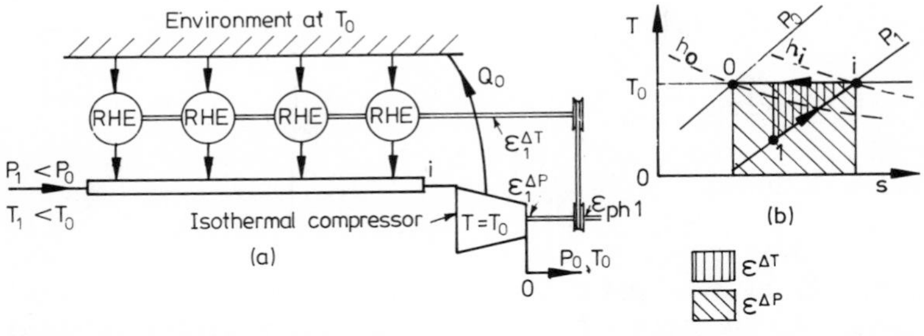

Fig 2.8 Determination of ε_{ph} for a stream of a homogenous simple fluid when $P_1 < P_0$ and $T_1 < T_0$.

To illustrate further the nature of the two components, the ideal device for determining ε_{ph} when $P_1 < P_0$ and $T_1 < T_0$ is shown in Fig 2.8. Note that, although $T_1 < T_0$, the sign of the thermal component is positive; $\varepsilon^{\Delta T}$ is always positive, or zero. This is not true of the pressure component, $\varepsilon^{\Delta P}$, which is negative when $P_1 < P_0$. For the case illustrated in Fig. 2.8(b), $|\varepsilon^{\Delta P}| > |\varepsilon^{\Delta T}|$ so the physical exergy corresponding to state 1 is negative.

Physical exergy of a perfect gas

Introducing perfect gas relations into (2.9) gives an expression for the specific physical exergy of a perfect gas:

$$\varepsilon_{ph} = c_P(T - T_0) - T_0\left(c_P \ln \frac{T}{T_0} - R \ln \frac{P}{P_0}\right) \tag{2.16}$$

Fig 2.9 shows the variation of ε_{ph} of a perfect gas with pressure and temperature. When $P \geqslant P_0$, ε_{ph} is positive for the whole range of temperatures, except in the environmental state where $\varepsilon_{ph} = 0$. At pressures below P_0, the range of temperatures, both below and

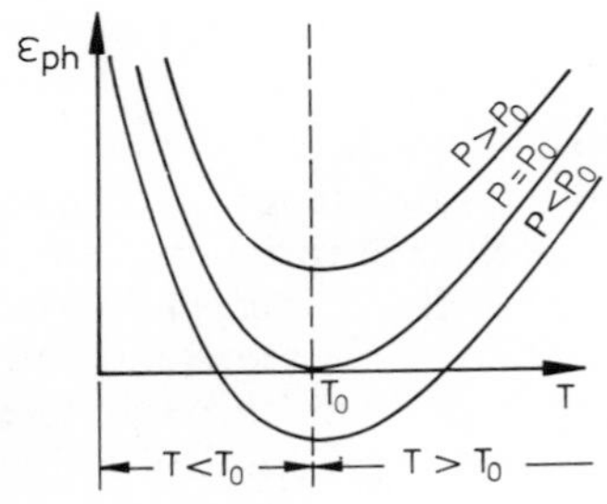

Fig 2.9 Variation of the physical exergy of an ideal-gas stream with its pressure and temperature.

above T_0 for which ε_{ph} assumes negative values increases with the pressure difference $(P_0 - P)$. The significance of negative values of ε_{ph} can be seen from Fig 2.8.

Example 2.2

Calculate the specific physical exergy of air for a state defined by $P_1 = 2$ bar and $T_1 = 393.15$ K when the environmental parameters are $P_0 = 1.0$ bar, $T_0 = 293.15$ K.

Solution: Assuming air to be a perfect gas and taking $c_P = 1.0$ kJ/kgK and $R = 0.2871$ kJ/kgK, (2.16) gives:

$$\varepsilon_{ph1} = c_P\left(T_1 - T_0 - T_0 \ln \frac{T_1}{T_0}\right) + RT_0 \ln \frac{P_1}{P_0}$$

$$= 1.0\left(100 - 293.15 \ln \frac{393.15}{293.15}\right) + 0.2871 \times 293.15 \ln 2$$

$$= 13.97 + 58.34$$

$$= \underline{72.31 \text{ kJ/kg}}$$

ε_{ph1} comprises two terms, one involving temperature and the other pressure, which correspond to $\varepsilon^{\Delta T}$ and $\varepsilon^{\Delta P}$ (see Figs 2.7 and 2.8). Here, both are positive. Note that air is a unique substances for which the environmental state is the same as the dead state. Consequently, the values of ε_{ph1} calculated is equal to ε_1, ie $\varepsilon_0 = 0$.

Example 2.3

Calculate the specific physical exergy of CO_2 ($c_p = 0.8659$ kJ/kgK, $R = 0.1889$ kJ/kgK) for a state defined by $P_1 = 0.7$ bar and $T_1 = 268.15$ K and for the environment $P_0 = 1.0$ bar, $T_0 = 293.15$ K.

Solution: Using the general expression for the physical exergy of a perfect gas:

$$\varepsilon_{ph1} = c_P\left(T_1 - T_0 - T_0 \ln \frac{T_1}{T_0}\right) + RT_0 \ln \frac{P_1}{P_0}$$

$$= 0.8659\left(-25 - 293.15 \ln \frac{268.15}{293.15}\right) + 0.1889 \times 293.15 \ln 0.7$$

$$= 0.9791 - 19.751$$

$$= \underline{-18.772 \text{ kJ/kg}}$$

In this case $\varepsilon^{\Delta T}$ has a small positive value whilst, $\varepsilon^{\Delta P}$ has a large negative value giving overall a negative ε_{ph}. This corresponds to the case illustrated in Fig 2.8.

Physical exergy near the absolute zero

Figure 2.10 shows[2.3] two reversible processes, 1–i and i–0 which can be used to determine the physical exergy of a substance in state 1, near the absolute zero. ε_{ph} can be considered as an algebraic sum of $\varepsilon^{\Delta P}$ and $\varepsilon^{\Delta T}$. To consider the effect of variation of state 1 along the isobar $P_1 = \text{const}$, we will take $\varepsilon^{\Delta P}$, which corresponds to process i–0 to be fixed, and investigate the effect of the changes on the magnitude of $\varepsilon^{\Delta T}$, represented by the area between the isotherm $T = T_0$, the isobar $P = P_1$ and the ordinate through state 1. As $T_1 \rightarrow 0$, the magnitude of $\varepsilon^{\Delta T}$ tends to a finite value represented in Fig 2.10 by the

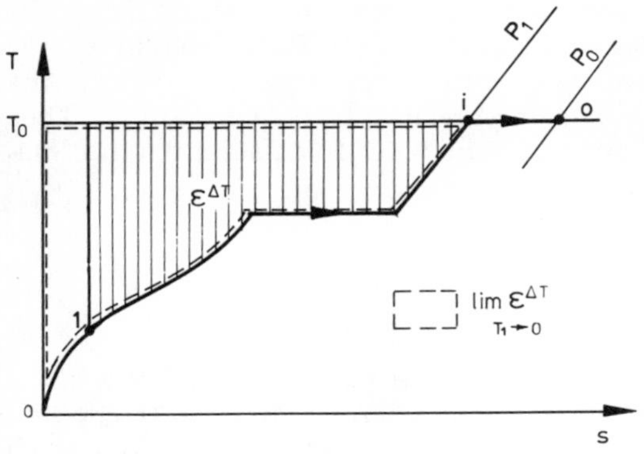

Fig 2.10 Physical exergy for states near the absolute zero temperature (reproduced from Ref [2.3], by permission).

area enclosed by the dotted line. This does not contradict the principle of the unattainability of the absolute zero of temperature. It only shows that it is possible to reach states near the absolute zero, as closely as may be desired, using a finite quantity of work. According to the Third Law, however, one cannot reach the absolute zero by a finite number of repeated processes[1.1].

Physical exergy values from *h–s* charts

For substances for which ε–h charts are not available but h–s charts exist, the construction shown in Fig 2.11 will give values of physical exergy. Since $T=(\partial h/\partial s)_P$, a tangent to the constant pressure line P_0=const, drawn at the points where it is intersected by the isotherm T_0=const, is inclined at an angle $\lambda=\tan^{-1} T_0$. The rest of

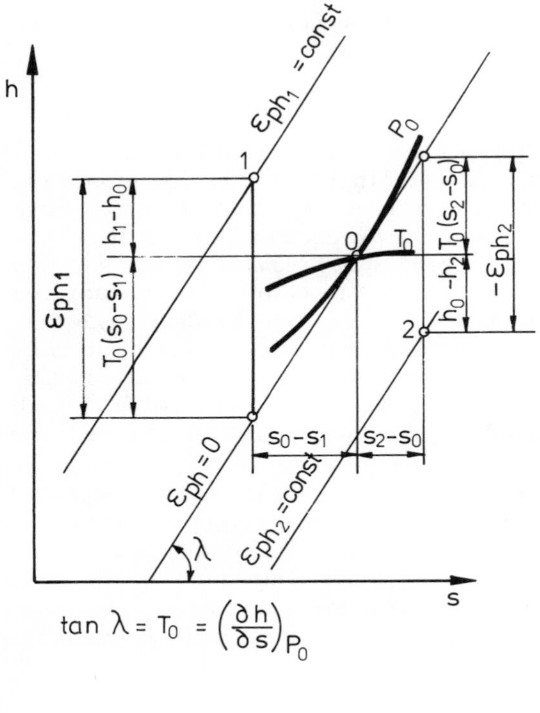

Fig 2.11 Determination of physical exergy from an enthalpy–entropy diagram.

the construction, shown for two states, should be clear from the diagram. Now, the tangent to the curve P_0=const coincides with the line $\varepsilon_{ph}=0$. States lying above the tangent have positive ε_{ph} values while those lying below it, have negative ε_{ph} values. Any line on the *h–s* chart, parallel to the $\varepsilon_{ph}=0$ line represents a line of constant ε_{ph}. Thus simply superimposing on an existing *h–s* chart a number of equidistant parallel lines would give ε_{ph} values at convenient intervals.

2.5 Chemical exergy

Definitions

In determining physical exergy, the final state of the stream (see Fig 2.5) was the environmental state defined by P_0 and T_0 and characterised by zero kinetic and potential exergy. This state will now be the initial state in the reversible processes which will determine the chemical exergy of the stream of substance. The final state to which the substance will be reduced according to the definition of exergy (Section 2.3) is one of unrestricted equilibrium with the environment, the dead state. Chemical exergy may be defined formally, thus:

Chemical exergy is equal to the maximum amount of work obtainable when the substance under consideration is brought from the environmental state to the dead state by processes involving heat transfer and exchange of substances only with the environment.

Comparing this definition with a similar definition given earlier for the exergy of a stream of substance shows that *chemical exergy is the exergy of a stream of substance when the state of the substance corresponds to the environmental state.*

The concept of reversibility, that all flow directions and interactions can be reversed, gives the alternative definition:

Chemical exergy is equal to the minimum amount of work necessary to synthesise, and to deliver in the environmental state, the substance under consideration from environmental substances by means of processes involving heat transfer and exchange of substances only with the environment.

Reference substances

To assess the work potential (ie exergy) of a stream of substance by virtue of the difference between its chemical potential and that of the environment, the properties of the chemical elements comprising the stream must be referred to the properties of some corresponding suitably selected substances in the environment. One essential characteristic of these *reference substances* is that they must be *in equilibrium with the rest of the environment.* Thus CO_2 is a suitable reference substance for C but not unoxidised carbon which exists in fossil fuel deposits in the Earth's crust, or CO which is only rarely found in the environment. Clearly, among different environmental substances containing a particular chemical element the one with the *lowest chemical potential* is most suitable as a reference substance for the chemical element in question. Also, the concentration of a proposed reference substance in the environment must be known with adequate precision.

When the system under consideration consists of reference substances in some arbitrary state, its reference substances are the corresponding substances in their dead states. For example, if the substance in the system is N_2 its reference substance would be atmospheric N_2 in the state defined by the environmental temperature and the partial pressure at which N_2 exists in the atmosphere. When the system under consideration does not comprise a reference substance, then its reference substances must contain the constituent elements of the substance. For example, the reference substances for CH_4 are CO_2 (for carbon) and H_2O (for hydrogen) in their dead states. A general scheme of *standard reference substances*, one for each chemical element, is described in Appendix A. Tables of standard chemical exergies based on this system of reference substances are also presented, simplifying the calculation involved in the exergy balances of chemical systems. A technique for calculating the standard chemical exergy of industrial fuels is presented in Appendix C and simple numerical examples given here.

Chemical exergy of reference substances

CO_2 and O_2 which are common constituents of the atmosphere and are in a state of thermodynamic equilibrium with other constituents of the environment. Therefore, these substances have been adopted as references substances for carbon and oxygen respectively.

The first definition of chemical exergy given above indicates that, to get the maximum amount of work, the processes used must be fully reversible. The initial state in this process is the environmental state, defined by P_0 and T_0, and the final state is the dead state, defined by T_0 and the partial pressure, P_{00} of the gaseous reference substances under consideration as components of the atmosphere. Since both the initial and the final state are characterised by the same temperature, T_0, a reversible isothermal process can be used to reduce the substance from T_0, P_0 to T_0, P_{00}. Such a process satisfies the condition of reversible 'heat transfer only with the environment' stipulated in the definition of ε_0. At the end of the expansion process, when the gas pressure is reduced to its partial pressure in the atmospheric air, the gas can be discharged reversibly (through a semi-permeable membrane) into the atmosphere. This process corresponds to the 'exchange of substances only with the environment' referred to in the definition. The work obtained from such a process per mole of substance will equal the molar chemical exergy, $\tilde{\varepsilon}_0$. For an ideal gas:

$$\tilde{\varepsilon}_0 = \tilde{R}T_0 \ln \frac{P_0}{P_{00}} \tag{2.17}$$

Example 2.4

Calculate the molar chemical exergy of CO_2 and O_2. Take $P_0 = 1$ bar and $T_0 = 298.15$ K and the partial pressures of CO_2 and O_2 in the atmosphere as $(P_{00})_{CO_2} = 0.0003$ bar and $(P_{00})_{O_2} = 0.2040$ bar. The molar ideal-gas constant is 8.3144 kJ/kmolK.

Solution: Substituting the given data in (2.17) for CO_2:

$$(\tilde{\varepsilon}_0)_{CO_2} = 8.3144 \times 298.15 \ln \frac{1}{0.0003}$$

$$= \underline{20\,108 \text{ kJ/kmol}}$$

For O_2:

$$(\tilde{\varepsilon}_0)_{O_2} = 8.3144 \times 298.15 \ln \frac{1}{0.2040}$$

$$= \underline{3\,941 \text{ kJ/kmol}}$$

Chemical exergy of a gaseous fuel

Calculating chemical exergy for a gaseous fuel involves an additional complication because fuels do not form part of the system of common environmental substances of

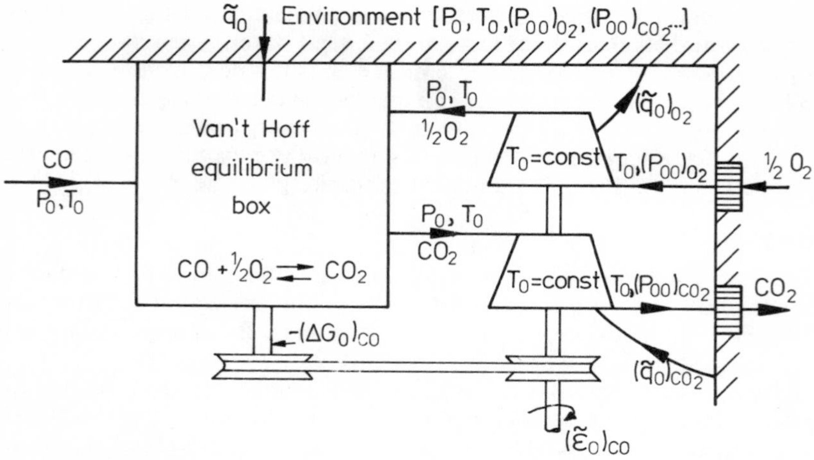

Fig 2.12 A reversible device for determining the chemical exergy of carbon monoxide.

low Gibbs function, ie reference substances, of which the standard environment is assumed to comprise. To overcome this difficulty, the reversible processes selected for determining the chemical exergy of the fuel must include a reversible chemical reaction to change the fuel, with the aid of oxygen brought from the environment, into one or more reference substances. As reaction would take place reversibly, it would deliver the maximum work of the reaction. A reversible device of this type for determining the chemical exergy of carbon monoxide (Fig 2.12) consists of a van't Hoff equilibrium box with the compressors and expanders necessary to deal with the reversible reaction of CO with O_2 (see Section 1.12). The two reactants are delivered to the equilibrium box and the product, CO_2, removed from it in the environmental state (P_0, T_0). The net work output from the equilibrium box is $-(\Delta G_0)_{CO}$. The reversible isothermal compressor delivers atmospheric oxygen to the equilibrium box, while that of the reversible isothermal expander discharges CO_2 reversibly to the atmosphere. The work done in the compressor per mole of CO is $\frac{1}{2}(\tilde{\varepsilon}_0)_{O_2}$ and the corresponding work output from the expander is $(\tilde{\varepsilon}_0)_{CO_2}$, as shown above. The work from these two machines is combined with the work output from the equilibrium box. The net work output from these reversible processes per mole of CO is equal to the molar chemical exergy of CO. Thus the molar chemical exergy of CO can be expressed as:

$$(\tilde{\varepsilon}_0)_{CO} = -(\Delta G_0)_{CO} + (\tilde{\varepsilon}_0)_{CO_2} - \tfrac{1}{2}(\tilde{\varepsilon}_0)_{O_2} \qquad (2.18)$$

Example 2.5

Calculate the molar chemical exergy of CO. The molar Gibbs function of formation at 25°C of CO is −137 160 kJ/kmol and that of CO_2 is −394 390 kJ/kmol when carbon is in graphite form[2.6]. Use the values of molar chemical exergy of CO_2 and O_2 obtained in Example 2.4.

Solution: The molar Gibbs function of the reaction $(\Delta G_0)_{CO}$ can be calculated from the following general relation:

$$\Delta G_0 = \sum_{PROD} \nu_k \Delta\tilde{g}_{fk} - \sum_{REACT} \nu_j \Delta\tilde{g}_{fj} \qquad (2.19)$$

where ν_k, ν_j and $\Delta\tilde{g}_{fk}, \Delta\tilde{g}_{fj}$ are the stoichiometric coefficients and the molar Gibbs functions of formation of the products and the reactants, respectively. Hence, since $(\Delta\tilde{g}_f)_{O_2}=0$:

$$(\Delta G_0)_{CO} = -394\,390 + 137\,160 = -257\,230 \text{ kJ/kmol}$$

Using the values of molar chemical exergy of the two reference substances from Example 2.4:

$$(\tilde{\varepsilon}_0)_{CO} = 257\,230 + 20\,108 - \tfrac{1}{2}\times 3\,941$$
$$= \underline{275\,368 \text{ kJ/kmol}}$$

Chemical exergy of a mixture

In many important applications the working medium consists of a mixture of ideal gases, for example gaseous fuels, combustion products, *etc.* To derive an expression for the chemical exergy of the mixture consider the reversible device shown in Fig 2.13. The mixture is supplied at a steady rate at P_0, T_0. Each of the N components is separated in turn by a semi-permeable membrane, and then compressed, reversibly and isothermally from its partial pressure in the mixture P_i to P_0. The total work of compression per mole of the gas mixture is:

$$\sum_i [\tilde{w}_{xi}]_{REV} = \tilde{R}T_0 \sum_i x_i \ln x_i \tag{2.20}$$

where x_i is the mole fraction of the i-th component in the mixture. Each separated component emerging from its compressor at P_0, T_0 has at this point molar exergy $\tilde{\varepsilon}_{0i}$ while its flow rate per mole of mixture is x_i. As all the processes taking place in the device are reversible, the exergy of the gas mixture is the sum of the exergies of its components minus the work of compression. Hence:

$$\tilde{\varepsilon}_{0M} = \sum_i x_i\tilde{\varepsilon}_{0i} + \tilde{R}T_0 \sum_i x_i \ln x_i \tag{2.21a}$$

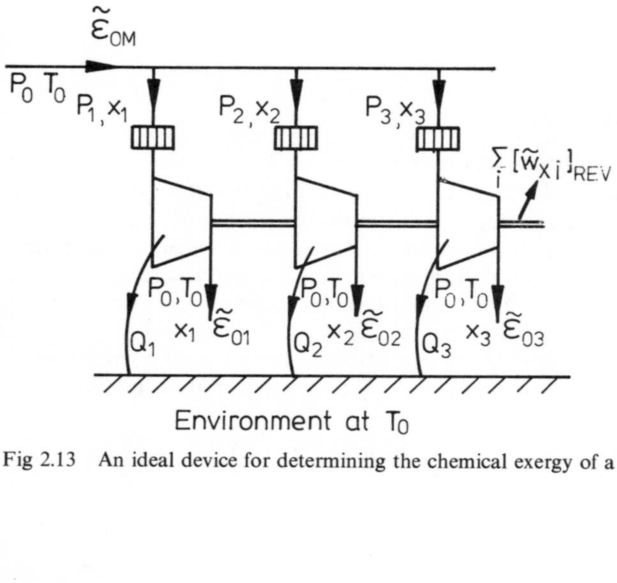

Fig 2.13 An ideal device for determining the chemical exergy of a mixture of ideal gases.

Since the second term on the RHS of (2.21) is always negative, the exergy of the mixture is always less than the sum of the exergies of its components at the pressure and temperature of the mixture. Eqs (2.20) and (2.21a) are also applicable to *ideal solutions* of liquids, ie solutions produced without a change in volume or enthalpy. In the case of *real solutions* the exergy of a mixture can be obtained if the activity coefficients, γ_i, for the components are known. The more general version of (2.21a) is:

$$\tilde{\varepsilon}_{0M} = \sum_i x_i \tilde{\varepsilon}_{0i} + \tilde{R} T_0 \sum_i x_i \ln \gamma_i x_i \tag{2.21b}$$

Activity coefficients may be greater or smaller than unity, and are unity for ideal solutions. The chemical exergy of real solutions can also be obtained from exergy-concentration diagrams which will be introduced in Section 3.6.

The work of compression, the second term on the RHS of (2.21a) and (2.21b), can be interpreted as the minimum work necessary for separating the components of the mixture and delivering them at the environmental pressure and temperature. This interpretation will be useful in the analysis of distillation processes.

Example 2.6

Calculate the molar chemical exergy of a mixture of gases with the following composition by volume (or by mole): CO—0.15, air—0.85. Take $T_0 = 298.15$ K and $P_0 = 1$ bar.

Solution: Using the value of $(\tilde{\varepsilon}_0)_{CO}$ calculated in Example 2.5, for CO:

$$(\tilde{\varepsilon}_0)_{CO} = 275\,368 \text{ kJ/kmol} \qquad x_{CO} = 0.15$$

and for air:

$$(\tilde{\varepsilon}_0)_{air} = 0 \qquad x_{air} = 0.85$$

From Eq (2.21):

$$\begin{aligned}\tilde{\varepsilon}_{0M} &= (0.15 \times 275\,368 + 0.85 \times 0) + 8.3144 \times 298.15(0.15 \ln 0.15 + 0.85 \ln 0.85)\\ &= 41\,305.1 - 1047.9\\ &= \underline{40\,257.2 \text{ kJ/kmol}}\end{aligned}$$

Chemical exergy in the Exergy Method

Clearly, chemical exergy is an integral component of exergy for the latter is to be evaluated in relation to its natural reference state offered by the environment. This type of reference state is essential in any calculations involving a chemical reaction*. If, however, processes being analysed do not involve chemical reactions and, hence, there are no changes in chemical composition of the streams crossing the control surface, and if there is no exchange of substances with the environment, the chemical component of exergy will cancel out when exergy of the incoming and outgoing streams are subtracted in an exergy balance. Under these conditions, the values of exergy may be given in a table or a chart as relative values. These values are computed relative to any

* Alternative reference datum states for evaluating exergy are discussed in Appendix A.

convenient reference state a defined by properties h_a and s_a. This leads to:

$$\varepsilon = h - T_0 s - \beta_a \tag{2.22}$$

where:

$$\beta_a = h_a - T_0 s_a$$

One possible choice for the reference state is the environmental state P_0, T_0, when the 'relative' value of exergy becomes the physical component of exergy. It is generally advantageous to select a reference state so that the values of exergy over the useful range of states are all positive. To achieve this, one could take $T_a = T_0$, the value of P_a corresponding to the lowest pressure in the useful range of values.

As in (2.22), ε will be also used for the relative values of exergy since the nature of the quantity should be generally clear from the context.

The examples of calculation of chemical exergy (Examples 2.5 and 2.6) given above introduced the concepts involved, and illustrated the method for simple substances. For practical exergy analysis, values of standard chemical exergy for a variety of chemical substances are given in Tables A3 and A4. For substances not listed there, a calculation method using the Gibbs function of formation and the values of chemical exergy of the constituent elements of the compound is given in Appendix A. A method for calculating chemical exergy of industrial fuels from empirical expressions is shown in Appendix C.

Example 2.7

Ammonia vapour enters a refrigeration plant condenser at 0.05 kg/s at a pressure of 11.67 bar and a temperature of 80°C. It leaves the condenser in the saturated liquid state at the same pressure. Calculate the change in the exergy rate of the stream. Take the environmental temperature $T_0 = 298.15$ K.

Solution: As the stream does not undergo a change in its chemical composition, its chemical exergy at the inlet and outlet is the same. Therefore, the change in the exergy of the stream is equal to the change in the physical exergy as given by (2.12). Values of specific enthalpy and specific entropy can be taken relative to any convenient reference state; for ammonia, these properties are usually tabulated for a reference state corresponding to the saturated liquid state at -40°C. The values for the inlet and outlet states are[2.4]:

$$h_1 = 1\,610.5 \text{ kJ/kg} \qquad s_1 = 5.417 \text{ kJ/kgK}$$

$$h_2 = 323.1 \text{ kJ/kg} \qquad s_1 = 1.204 \text{ kJ/kgK}$$

Neglecting changes in kinetic energy and potential energy, the change in the exergy rate of the stream is:

$$\begin{aligned}\dot{E}_1 - \dot{E}_2 &= \dot{m}[(h_1 - h_2) - T_0(s_1 - s_2)] \\ &= 0.05[(1\,610.5 - 323.1) - 293.15(5.417 - 1.204)] \\ &= \underline{1.78 \text{ kW}}\end{aligned}$$

Example 2.8

A combustible mixture of CO and air containing 15% CO by volume enters a combustion chamber at 0.5 kg/s. The pressure, temperature and the mean velocity of the stream are 2.1 bar, 125°C and 120 m/s while the environmental pressure and temperature are 1 bar and 25°C respectively. Calculate the exergy rate of the stream.

Solution: In gaseous streams the potential exergy component can be usually neglected. Consequently, the expression for the exergy rate of a stream of a substance as given by (2.4) reduces to:

$$\dot{E}_1 = \dot{E}_{k1} + \dot{E}_{ph1} + \dot{E}_0$$

or

$$\dot{E}_1 = \dot{m}[\varepsilon_{k1} + \varepsilon_{ph1} + \varepsilon_0]$$

Kinetic exergy

The specific kinetic exergy follows from (2.6a):

$$\varepsilon_{k1} = \frac{\dot{E}_{k1}}{\dot{m}} = \frac{C_0^2}{2}$$

Substituting:

$$\varepsilon_{k1} = \frac{120^2}{2} \times 10^{-3} = 7.2 \text{ kJ/kg}$$

Physical exergy

For the two components of the mixture[2.4]:

	CO	Air
Molar mass, $\tilde{m}$/[kg/kmol]	28	28.96
Specific heat capacity (mean value), c_P/[kJ/kgK]	1.043	1.008

Hence, for the mixture:

$$\tilde{m} = 0.15 \times 28 + 0.85 \times 28.96 = 28.82 \text{ kg/kmol}$$

and:

$$c_P = 0.15 \times 1.043\left(\frac{28}{28.82}\right) + 0.85 \times 1.008\left(\frac{28.96}{28.82}\right) = 1.013 \text{ kJ/kgK}$$

The specific gas constant for the mixture can be obtained from the molar value:

$$R = \frac{\tilde{R}}{\tilde{m}} = \frac{8.3144}{28.82} = 0.2885 \text{ kJ/kgK}$$

Using (2.16):

$$\varepsilon_{ph1} = c_P(T_1 - T_0) - T_0\left(c_P \ln \frac{T_1}{T_0} - R \ln \frac{P_1}{P_0}\right)$$

$$= 1.013 \times 100 - 298.15\left(1.013 \ln \frac{398.15}{298.15} - 0.2885 \ln \frac{2.1}{1.0}\right)$$

$$= 77.76 \text{ kJ/kg}$$

An alternative way of calculating $\tilde{\varepsilon}_{ph}$ for ideal gases is described in Section 4.5.

Chemical exergy

The molar chemical exergy of this gaseous mixture was calculated in Example 2.6. The specific value is:

$$\varepsilon_0 = \frac{\tilde{\varepsilon}_0}{\tilde{m}} = \frac{40\ 257.2}{28.82} = 1\ 396.85\ \text{kJ/kg}$$

From the three specific values of exergy components we get the total exergy rate of the stream:

$$\dot{E}_1 = 0.5(7.2 + 77.76 + 1\ 396.85)$$
$$= \underline{740.9\ \text{kW}}$$

2.6 Exergy concepts for closed system analysis

When analysing a closed system using an energy equation, two forms of interaction, work and heat, and the energy of the control mass must be considered. The exergy forms associated with the two interactions were discussed in connection with control region analysis in Section 2.3, and all the conclusions reached there apply here with minor modifications.

Thus, work transfer is equivalent both in direction and magnitude to the exergy transfer associated with it; both will be denoted by the symbol W.

The exergy associated with heat transfer in closed system analysis is called *thermal exergy* and denoted by Ξ^Q. The only difference between this concept and thermal exergy rate is that the latter is defined for a steady heat transfer rate whereas the former corresponds to the transfer of a fixed amount of thermal energy through heat interaction. Adapting (2.1) accordingly:

$$\Xi^Q = \tau Q_r \tag{2.23}$$

where τ has the same meaning as in (2.2).

2.7 Non-flow exergy

When the definition of exergy is applied to a control mass or fixed quantity of substances, ie a closed system identified by a system boundary, we have *non-flow exergy*, denoted by Ξ with the corresponding specific and molar quantities denoted by ξ and $\tilde{\xi}$ respectively. As stated earlier, the exergy of a fluid in steady flow has been taken as the fundamental quantity because of the practical importance of steady-flow processes; consequently, all the considerations so far have been limited to this. Deriving a relationship between exergy of a fluid in steady flow and exergy in a closed system will extend the applicability of these considerations to the latter. This relationship will be derived by using an ideal device consisting of a horizontal frictionless, leakproof, and adiabatic piston and cylinder assembly (Fig 2.14). The fluid enters the cylinder through a reversible diffuser so that the fluid velocity inside the cylinder can be made arbitrarily small by choosing an appropriate cylinder-to-pipe diameter ratio. The piston, which separates the fluid at pressure P from the environmental medium at pressure P_0, moves at a small steady velocity as the system is charged with the fluid. Subsequently, the supply is cut off and the fluid contained within the system boundary can be regarded as

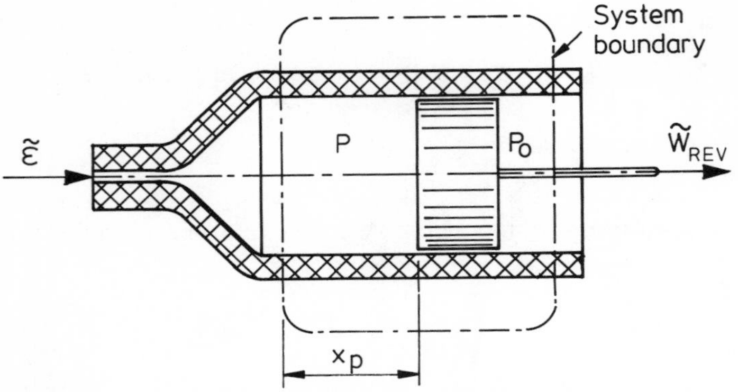

Fig 2.14 An ideal device for determining the non-flow exergy of a fluid.

a closed system. If x_p is the displacement of the piston when a mole of the fluid enters the system, the corresponding reversible net work delivered by the system is:

$$\tilde{w}_{REV} = (P - P_0)A_p x_p$$
$$= (P - P_0)\tilde{v} \tag{2.24}$$

where A_p is the piston area and $\tilde{v}$ is the molar volume. Note that the environmental medium displaced by the piston has zero exergy. Hence, in the absence of any irreversibilities, the exergy of the fluid inside the system boundary is equal to the exergy of the fluid delivered to the cylinder minus the work done by the system. Thus the exergy of a homogenous substance in a closed system is:

$$\tilde{\xi} = \tilde{\varepsilon} - \tilde{w}_{REV} \tag{2.25}$$

or, using (2.24):

$$\tilde{\xi} = \tilde{\varepsilon} - (P - P_0)\tilde{v} \tag{2.26}$$

Combining expressions for physical exergy from (2.9) with chemical exergy, all in molar quantities, and neglecting E_k and E_p:

$$\tilde{\varepsilon} = (\tilde{h} - T_0\tilde{s}) - (\tilde{h}_0 - T_0\tilde{s}_0) + \tilde{\varepsilon}_0 \tag{2.27}$$

Substituting (2.27) in (2.26), and using the relation for enthalpy:

$$\tilde{\xi} = (\tilde{u} + P_0\tilde{v} - T_0\tilde{s}) - (\tilde{u}_0 + P_0\tilde{v}_0 - T_0\tilde{s}_0) + \tilde{\varepsilon}_0 \tag{2.28}$$

Now, non-flow exergy can be divided conveniently into two components, physical non-flow exergy and chemical non-flow exergy. The physical component of non-flow exergy is:

$$\tilde{\xi}_{ph} = (\tilde{u} + P_0\tilde{v} - T_0\tilde{s}) - (\tilde{u}_0 + P_0\tilde{v}_0 - T_0\tilde{s}_0) \tag{2.29}$$

Note that $\tilde{\xi}_{ph}$ is equal to the difference in the value of a function corresponding to two different states. This function will be called *specific non-flow exergy function* denoted by α. In molar form:

$$\tilde{\alpha} = \tilde{u} + P_0\tilde{v} - T_0\tilde{s} \tag{2.30}$$

Hence Eq (2.29) can be re-written:

$$\tilde{\xi}_{ph} = \tilde{\alpha} - \tilde{\alpha}_0 \tag{2.31}$$

Molar non-flow exergy function is made up of three properties of the system, $\tilde{u}$, $\tilde{v}$ and $\tilde{s}$, and the properties of the environment, P_0 and T_0. Clearly, in the case of non-flow exergy, the environment plays its part as a reservoir of unusable mechanical energy and hence we must have its pressure, P_0, in the function representing ξ_{ph}. In the second term on the RHS of (2.29) the three properties of the system are evaluated at the environmental state.

The chemical component of non-flow exergy is given by the third term of (2.28). It is therefore clear that chemical exergy is identical to chemical non-flow exergy, ie:

$$\tilde{\varepsilon}_0 \equiv \tilde{\xi}_0 \tag{2.32}$$

This follows from (2.26) by considering the case when the substance is in the environmental state, and therefore under pressure $P = P_0$. Chemical non-flow exergy is relavant to the analysis of a number of non-flow combustion processes, such as combustion in reciprocating spark ignition or compression ignition engines, bomb calorimeter process, *etc.*

The concept of chemical exergy is superfluous for non-reacting substances in a closed system. Here the environmental state is the appropriate reference datum state for evaluating non-flow exergy which then becomes identical with physical non-flow exergy.

Physical non-flow exergy of an ideal gas system

Consider, by way of illustration, a closed system consisting of a non-reacting ideal gas undergoing some selected reversible processes with a view to determining its physical non-flow exergy. The gas, as shown in Fig 2.15(a), is contained in a frictionless, leakproof piston and cylinder assembly which may have an adiabatic boundary or, when required, a diathermal one for exchanging heat reversibly with the environment at T_0. Note also that the environmental pressure P_0 acts on the outside surface of the piston. The following processes, shown in Fig 2.15(b), bring the system from its initial state 1 to the environmental state:

(i) Reversible adiabatic (isentropic) expansion from state 1 to state i where $T_i = T_0$ and, since it is an ideal gas, $U_i = U_0$. During this process the reversible work done by the gas on the piston is:

$$[W_{REV}]_1^i = U_1 - U_0 \tag{a}$$

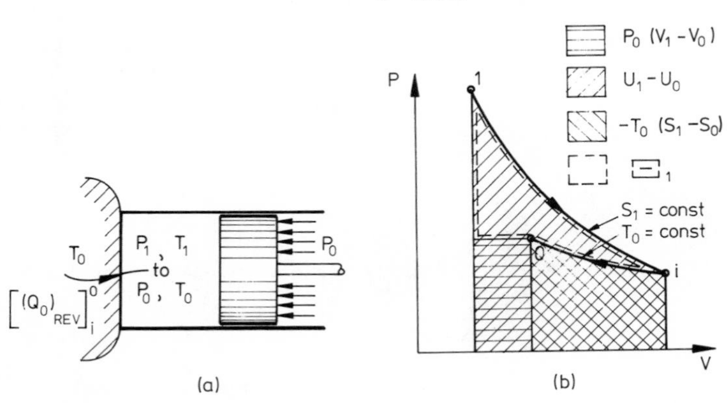

Fig 2.15 Determination of the physical non-flow exergy of an ideal gas.

(ii) Reversible isothermal compression from P_i to P_0 at T_0. Since for the previous process $S_1 = S_i$, the work in this process can be expressed as:

$$[W_{REV}]_i^0 = -T_0(S_1 - S_0) \tag{b}$$

Adding (a) and (b), the total work done by the gas is:

$$[W_{REV}]_1^0 = U_1 - U_0 - T_0(S_1 - S_0) \tag{c}$$

However, to get the *net useful work* which is delivered by the system we must subtract from $[W_{REV}]_1^0$ any work which the system does on the environment. This is:

$$[W_{REV}]_0 = P_0(V_0 - V_1) \tag{d}$$

Hence the net useful work, which is equal to the non-flow exergy of the system, is:

$$[(W_{REV})_{NET}]_1^0 = \Xi_{ph1} = (U_1 + P_0V_1 - T_0S_1) - (U_0 + P_0V_0 - T_0S_0) \tag{e}$$

The various work quantities corresponding to the processes described above and the net useful work delivered by the system are shown as areas in the P–V coordinates in Fig 2.15(b).

Characteristics of non-flow exergy

Non-flow exergy is a fundamental function with important practical properties, particularly in connection with the determination of equilibrium states. Consider, for example, a perfect gas assumed to be inert so that $\xi = \xi_{ph}$ and with $\gamma = 1.4$. Figure 2.16 has been constructed[2.5] by substituting in (2.29) perfect gas relations for the gas properties and non-dimensionalising the resulting relation using $\bar{P} = P/P_0$, $\bar{T} = T/T_0$ and $\bar{\xi} = \xi/RT_0$. The lines of constant non-flow exergy form a family of closed, non-intersecting curves around the point corresponding to the state for which $T = T_0$ and $P = P_0$. The values of $\bar{\xi}$ are positive for the whole range of possible states except for T_0, P_0. In contrast (see Fig 2.9), ε_{ph} could have negative values for some states for which $P < P_0$. Any line representing some constant property will intersect all the $\bar{\xi} = \text{const}$ curves at two points, except the one curve to which it is tangent. For example in case of the $V = \text{const}$ line (which is a straight line for an ideal gas) shown on the diagram, the points of intersection with the $\bar{\xi} = 0.3$ curve are at 2 and 3, and the point of tangency occurs with the curve $\bar{\xi} = 0.2$ at 1. Quite clearly, the points of intersection always

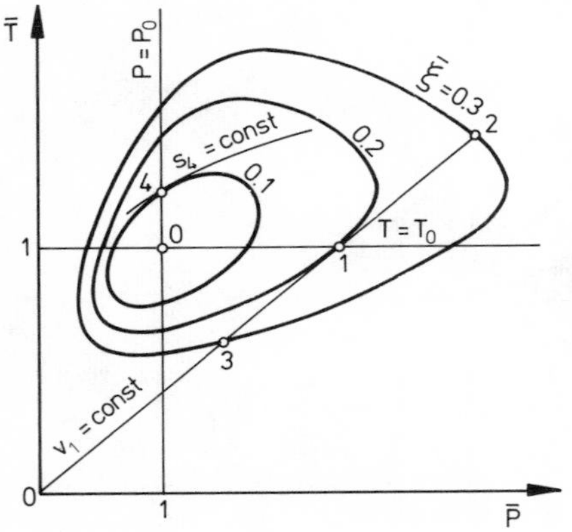

Fig 2.16 Lines of constant physical non-flow exergy in dimensionless co-ordinates for a perfect gas with $\gamma = 1.4$.

correspond to higher values of $\bar{\xi}$ than the point of tangency, so that the latter marks the point of *minimum non-flow exergy* on a given constant property line.

Stable equilibrium states

Criteria of stable equilibrium are discussed in Ref [1.3, p. 438] where it is shown that for a closed system with an environment at T_0 and P_0, stable equilibrium corresponds to the condition of minimum non-flow exergy within the constraints imposed on the system. This can be demonstrated by the following arguments:

Any interaction between the system and the environment must lead inevitably to a reduction in the non-flow exergy of the system. A contrary assumption would violate the concept of the environment as a reservoir of zero grade thermal energy. In other words there is no exergy available within the environment which could benefit the system. Therefore, any spontaneous change in the state of the system will lead to a reduction in the non-flow exergy of the system until the minimum value possible, under given constraints, is reached when no further change of state is possible and the system is in a state of equilibrium.

In the absence of any constraints on the system, such as an adiabatic boundary or an undeformable boundary, the system will tend to the state of restricted equilibrium with the environment (when $\bar{\xi}=0$) at the centre of the family of $\bar{\xi}=$const curves, where $\bar{P}=1$ and $\bar{T}=1$. Constraints will prevent the system from reaching this particular state of equilibrium. If the system is constrained from changing its volume, no work interaction is possible and equality of pressure between the system and the environment cannot be achieved ($P \neq P_0$). However, if the system is not, at the same time, constrained from exchanging thermal energy, ie the boundary is diathermal, an equality of temperature with the environment can be achieved ($T=T_0$). The corresponding equilibrium state is indicated on Fig 2.16 at point 1 where the constant volume line $V_1=$const has a point of tangency with a $\bar{\xi}=$const curve (in this case $\bar{\xi}=0.2$) at $\bar{T}=1$.

When the constraint is imposed by an adiabatic boundary equality of temperature with the environment cannot be achieved ($T \neq T_0$). If the process is also reversible, the system is constrained to follow an isentropic process arriving at a state of equilibrium for which (in the absence of a rigid boundary) $P=P_0$. This case is illustrated by point 4 in Fig 2.16 at which the curve $S_4=$const is a tangent to a $\bar{\xi}=$const curve at $\bar{P}=1$.

Example 2.9

A quantity of air is contained inside an *adiabatic* frictionless and leakproof cylinder and piston assembly. The initial state of the air is $P_1=3$ bar, $T_1=400$ K and the environmental state is $P_0=1$ bar, $T_0=300$ K. The properties of the air are $R=0.2871$ kJ/kgK and $\gamma=1.4$. Assuming the processes between the initial state and the equilibrium state take place quasi-statically calculate:

1. The pressure and temperature defining the equilibrium state of the air.
2. The final specific physical non-flow exergy.
3. The reduction in the specific non-flow exergy of the air between the initial and the equilibrium states.

Solution

1. Since the system boundary is adiabatic, the piston is frictionless and the process takes place quasi-statically, the process is therefore describable by $S=$const. The

equilibrium state for this case is given by point 4 in Fig 2.16. Hence the final pressure is $P_2 = P_0 = 1$ bar and the temperature is determined from the isentropic relationship

$$T_2 = 400(\tfrac{1}{3})^{0.4/1.4}$$
$$= \underline{292.24 \text{ K}}$$

2. From (2.29):

$$\xi_{ph2} = (u_2 - u_0) + P_0(v_2 - v_0) - T_0(s_2 - s_0)$$

Using the ideal gas equation, perfect gas relations and $P_2 = P_0$:

$$\xi_{ph2} = \frac{\gamma R}{\gamma - 1}\left[(T_0 - T_2) - T_0 \ln \frac{T_2}{T_0}\right]$$
$$= \frac{1.4 \times 0.2871}{0.4}\left[(300 - 292.24) - 300 \ln \frac{292.24}{300}\right]$$
$$= \underline{15.70 \text{ kJ/kg}}$$

3. The reduction in the physical non-flow exergy of the system during the process is:

$$\xi_1 - \xi_2 = (u_1 - u_2) + P_0(v_1 - v_2)$$

Substituting, as above:

$$\xi_1 - \xi_2 = R\left[\frac{T_1 - T_2}{\gamma - 1} + P_0\left(\frac{T_1}{P_1} - \frac{T_2}{P_2}\right)\right]$$
$$= 0.2871\left[\frac{400 - 292.24}{0.4} + 1\left(\frac{400}{3} - \frac{292.24}{1}\right)\right]$$
$$= \underline{31.72 \text{ kJ/kg}}$$

Chapter 3 Elements of plant analysis

Previous chapters have considered the concepts on which the Exergy Method is based. This chapter introduces techniques which use these concepts in plant analysis.

Application of the Exergy Method to control mass analysis will be dealt with only very briefly here. The rest of the chapter will be devoted to the analysis of steadily operating open systems, reflecting their predominance among industrial energy systems.

The techniques which will be discussed here will involve only thermodynamic concepts. Thermoeconomic applications of the exergy concepts will be discussed in Chapter 6.

3.1 Control mass analysis

The non-flow exergy balance

(B.17) derived in Appendix B for a control mass, m, can be written for convenience in the compact form:

$$\Xi_1 + \Xi^Q = \Xi_2 + W_{NET} + I \tag{3.1}$$

where:

$$\Xi_1 = m\xi_1 \tag{3.2}$$

$$\Xi_2 = m\xi_2 \tag{3.3}$$

$$\Xi^Q = \sum_r \left[Q_r \frac{T_r - T_0}{T_r} \right] \tag{3.4}$$

$$W_{NET} = W_x + W_{CSN} \tag{3.5}$$

The expression for specific non-flow exergy is (see Section 2.9):

$$\xi = (u + P_0 v - T_0 s) - (u_0 + P_0 v_0 - T_0 s_0) + \xi_0 \tag{3.6}$$

The effect of changes of kinetic energy and potential energy is neglected. Also, as will be recalled from Section 2.9 ((2.35)) chemical non-flow exergy is identical to chemical exergy, and can be obtained from tables of standard chemical exergy in Appendix A. In physical processes, when the chemical composition of the system remains unaltered, the value of ξ_0 need not be obtained since it will cancel out in the non-flow exergy balance.

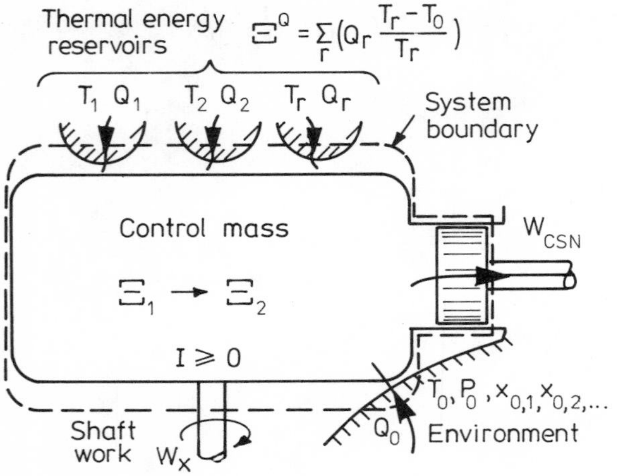

Fig 3.1 Non-flow exergy balance.

The terms used in the non-flow exergy balance (3.1) are indicated against the background of a closed system in Fig 3.1. If we consider *an extended system* consisting of the control mass, all the TERS interacting with it and the MER in which the work is stored as an isolated system, then the law of degradation of energy can be stated in exergy terms as:

The non-flow exergy of an isolated system at the beginning of a real process is always greater than that at the end of the process and the difference is equal to the irreversibility, I, of the process. Since all the terms, except I, which appear in the non-flow exergy balance (3.1) have already been defined and discussed, one can write the non-flow exergy balance directly as a form of statement of the law of degradation of energy.

The Gouy–Stodola relation for a control mass

The other expression from which the irreversibility of a process can be calculated has been derived in Appendix B ((B.19)). In its more general form it can be written as:

$$I = T_0 \Pi \tag{3.7}$$

This relation states that the irreversibility of a process is equal to the product of the temperature of the environment and the entropy production which results from the process in all the systems participating in the process. Section 1.8 showed that this relation has been found to hold for three particular processes. In those cases the product $T_0(\Delta S)_{ISOL}$, was equal to the restoring work which is now called the irreversibility of the process.

The version of (B.19) from which the irreversibility of a process can be calculated is:

$$I = T_0 \left[(S_2 - S_1) - \sum_r \frac{Q_r}{T_r} \right] \tag{3.8}$$

Comparison of (3.7) with (3.8) shows that the expression in the square brackets in the latter equation is equal to the entropy Π produced during the process. The first term, $(S_2 - S_1)$ is the increase in the entropy of the system and the second term, $\sum_r (Q_r/T_r)$ is the total thermal entropy flux into the system.

The expressions for the irreversibility in the forms given by (3.7) and (3.8) are known as the Gouy–Stodola relations in honour of two distinguished contributors to the development of Second Law Analysis.

(3.1) and (3.7) are *not independent*; either can be used to calculate the irreversibility of a process in a closed system. Since process irreversibility and the associated entropy production occur within a certain physical space, the extent of the system must be clearly specified by means of a system boundary.

Evaluation of irreversibility in closed systems

To illustrate the use of the non-flow exergy balance and the Gouy–Stodola relation and to demonstrate the nature of some forms of irreversibility, the two relations will now be applied to some simple systems.

Uncontrolled expansion in an adiabatic container

In this process a quantity of gas, initially contained in a compartment separated from an evacuated space by means of a membrane, expands in an uncontrolled way when the membrane is burst (Fig 3.2(a)). With $Q=0$ the irreversibility of the process from the Gouy–Stodola relation is

$$I=T_0\Pi=T_0(S_2-S_1) \qquad \text{(a)}$$

Thus in the absence of any heat transfer, the entropy change of the control mass is equal to the entropy produced in the process. From (a), the irreversibility is represented in Fig 3.2(b) in the T–S co-ordinates as the cross-hatched area.

Frictionless isobaric expansion

In this process a quantity of gas in a piston-cylinder assembly shown in Fig 3.3(a) expands without friction at constant pressure while being heated from a TER at temperature T_r. The finite temperature difference between the TER and the gas is the cause of the irreversibility of the process. By extending the system boundary to include the region of temperature gradients between the TER and the piston, this form of irreversibility is included in the irreversibility of the system. Applying the Gouy–Stodola relation to the system so defined:

$$I=T_0\Pi=T_0\left[(S_2-S_1)-\frac{Q_r}{T_r}\right] \qquad \text{(b)}$$

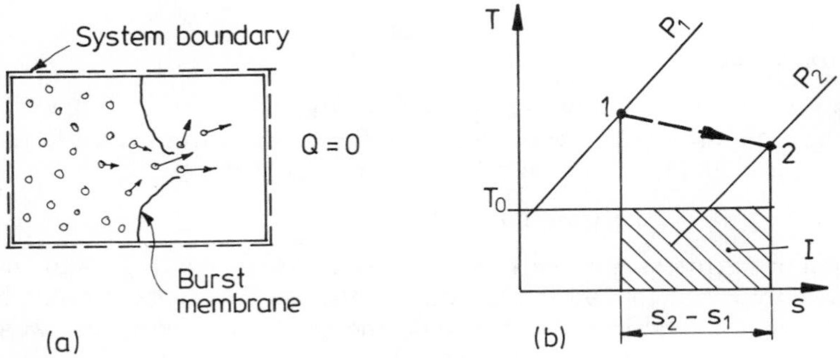

Fig 3.2 Uncontrolled expansion.

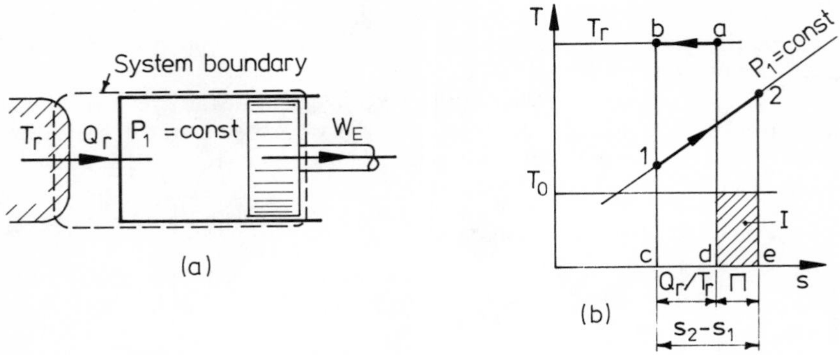

Fig 3.3 Frictionless, isobaric expansion.

During the process the system undergoes a change of entropy $(S_2 - S_1)$ while the thermal entropy flux is Q_r/T_r. Since the 'heat loss' of the TER is equal to the 'heat gain' by the system, the two areas a, b, c, d, a and 1, 2, e, c, 1, on the *T–S* diagram in Fig 3.3(b), must be equal. This enables the entropy production, Π, for the process to be represented as a change in the abscissa on the *T–S* diagram. Hence, from (b), the irreversibility of the process is represented by the cross-hatched area. Note that the cross-hatched area, and thus also the irreversibility of the process, becomes smaller as the temperature difference between the gas and T_r decreases.

It may be instructive to apply the non-flow exergy balance to this case. From (3.5), for the present case $W_{NET} = W_E$ and from (3.4):

$$\Xi^Q = Q_r \frac{T_r - T_0}{T_r} \tag{c}$$

the increase in non-flow exergy of the system is:

$$\Xi_2 - \Xi_1 = Q_r \frac{T_r - T_0}{T_r} - W_E - I \tag{d}$$

Clearly, since the thermal energy is supplied to the system at T_r and not at the instantaneous temperature of the system, the original thermal exergy, $Q_r[(T_r - T_0/T_r]$ at the system boundary is reduced at the point of acceptance by the gas to $\{Q_r[(T_r - T_0)/T_r] - I\}$.

Frictionless isothermal compression

A quantity of gas is compressed without friction and isothermally (Fig 3.4(a)), while simultaneously a heat quantity Q_0, necessary to keep T_1 = const, is transferred to the environment. Applying the Gouy–Stodola relation to this process:

$$I = Q_0 - T_0(S_1 - S_2) \tag{e}$$

An interpretation of this expression in the *T–S* coordinates is shown in Fig 3.4(b), the cross-hatched area representing the irreversibility of the process. This area corresponds to the work which could be obtained if the thermal energy rejected by the gas were utilised in a reversible engine. Thus, just as in the previous case, the irreversibility is due to heat transfer over a finite temperature difference.

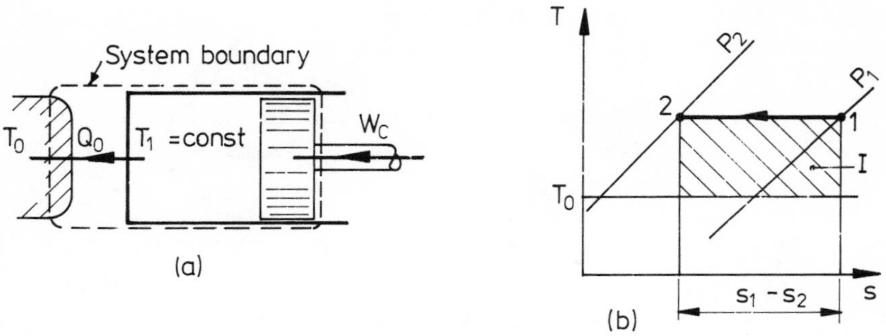

Fig 3.4 Frictionless, isothermal compression.

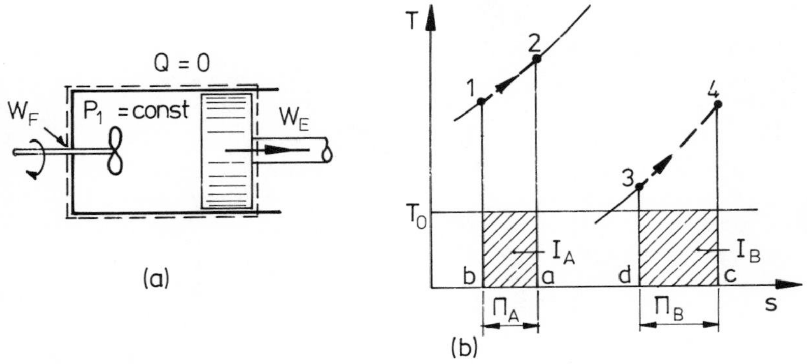

Fig 3.5 Isobaric expansion with viscous dissipation.

Isobaric expansion with stirring

The system shown in Fig 3.5(a) represents the unlikely, though perhaps interesting, case of energy being supplied to the gas in the cylinder by dissipating mechanical work through stirring. As the system is adiabatic ($Q=0$) the application of the Gouy–Stodola relation yields:

$$I = T_0(S_2 - S_1) \qquad \text{(f)}$$

A graphical representation of this process in T–S co-ordinates is shown in Fig 3.5(b) for two cases, the mechanical work dissipated being the same, ie area 1, 2, a, b, 1 is equal to area 3, 4, c, d, 3. Note that the higher the average temperature of the process, the lower the irreversibility. Thus in systems designed to operate at low temperatures, greater care must be taken to prevent dissipation of mechanical energy. This is particularly true when the average temperature is less than T_0, eg cryogenic systems, since then $I > W_F$.

Example 3.1

0.1 kg of steam contained in a cylinder fitted with a frictionless, leak-proof piston is heated at a constant pressure of 10 bar from a TER at a constant temperature T_r (see Fig 3.3). The initial steam temperature is 200°C and the final temperature 300°C. Calculate the irreversibility of the process from the Gouy–Stodola relation and, independently, from the non-flow exergy balance for the cases when $T_r = 620$ K and 700 K. Take $P_0 = 1$ bar and $T_0 = 290$ K.

Solution: The following values of properties of steam at 10 bar have been extracted from steam tables[2.4]:

Temperature	u/[kJ/kg]	h/[kJ/kg]	s/[kJ/kgK]	v/[m³/s]
200°C	2623	2829	6.695	0.2061
300°C	2794	3052	7.124	0.2580

Irreversibilities from the non-flow energy balance: For the process:

$$\Xi_2 - \Xi_1 = m[(u_2 - u_1) + P_0(v_2 - v_1) - T_0(s_2 - s_1)]$$

$$= 0.1[2794 - 2623) + (0.2580 - 0.2061) \times 10^2 - 290(7.124 - 6.695)]$$

$$= 5.18 \text{ kJ}$$

Useful work done during the process:

$$W_{NET} = m(P - P_0)(v_2 - v_1)$$

$$= 0.1(10 - 1)(0.2580 - 0.2061) \times 10^2$$

$$= 4.67 \text{ kJ}$$

Heat transfer during the process:

$$Q_r = m(h_2 - h_1)$$

$$= 0.1(3\,052 - 2\,829) = 22.3 \text{ kJ}$$

Thermal energy for $T_r = 620$ K:

$$\Xi_a^Q = 22.3 \frac{620 - 290}{620} = \underline{11.87 \text{ kJ}}$$

for $T_r = 700$ K:

$$\Xi_b^Q = 22.3 \frac{700 - 290}{700} = \underline{13.06 \text{ kJ}}$$

Irreversibility for $T_r = 620$ K:

$$I_a \Xi_a^Q - (\Xi_2 - \Xi_1) - W_{NET}$$

$$= 11.87 - 5.18 - 4.67 = \underline{2.02 \text{ kJ}}$$

Similarly for $T_r = 700$ K

$$I_b = 13.06 - 5.18 - 4.67 = \underline{3.21 \text{ kJ}}$$

Irreversibility from the Gouy–Stodola relation

$$I = T_0\left[m(s_2 - s_1) - \frac{Q_r}{T_r}\right]$$

For $T_r = 620$ K

$$I_a = 290\left[0.1(7.124 - 6.695) - \frac{22.3}{620}\right]$$

$$= \underline{2.01 \text{ kJ}}$$

For $T_r = 700$ K

$$I_b = 290\left[0.1(7.124 - 6.695) - \frac{22.3}{700}\right]$$
$$= \underline{3.20 \text{ kJ}}$$

As expected, except for rounding errors, the same values of I were obtained from the two relations. A higher value of I was obtained for $T_r = 700$ K than for $T_r = 620$ K because the temperature differences between T_r and the temperature of steam during the process is greater. The effect of T_r on I can be seen from Fig 3.3(b).

3.2 Control region analysis

The exergy balance

The exergy balance for a control region undergoing a steady-state process was obtained in Appendix B and is given by (B.21). For convenience, exergy balance may be written:

$$\dot{E}_i + \dot{E}^Q_e = \dot{E}_e + \dot{W}_x + \dot{I} \tag{3.9}$$

where:

$$\dot{E}_i = \sum_{IN} \dot{m}\varepsilon \tag{3.10}$$

$$\dot{E}_e = \sum_{OUT} \dot{m}\varepsilon \tag{3.11}$$

$$\dot{E}^Q = \sum_r \left[\dot{Q}_r \frac{T_r - T_0}{T_r}\right] \tag{3.12}$$

Also, (from Chapter 2) the expression for specific exergy may be written as:

$$\varepsilon = (h - T_0 s) - (h_0 - T_0 s_0) + \varepsilon_0 + \frac{C_0^2}{2} + g_E Z_0 \tag{3.13}$$

Here the velocity C_0 and the altitude Z_0 are evaluated relative to the environmental datum levels.

The chemical exergy, ε_0, may be evaluated separately (Section 2.6) or obtained from tables of standard chemical exergy (Appendix A). In the case of physical processes which do not involve exchange of matter between the control region and the environment, chemical exergy and the term $(h_0 - T_0 s_0)$ will cancel out in the exergy balance.

Exergy flows in relation to a control region are indicated in Fig 3.6. The various terms of which the exergy balance is made up, as given in (3.9), are indicated in the figure. The flow of exergy to or from the control region is associated with the inflow or outflow of matter, heat transfer and work transfer. In general each of these different modes of exergy transfer may have a positive or a negative sign, ie it may contribute to either inflow or outflow of exergy. Note from (3.9) and from Fig 3.6 that the exergy balance expresses *the law of degradation of energy* which can be stated in exergy terms for a control region undergoing a steady-state process as:

The exergy flow to the control region is always greater than that from the control region. The difference between the two, the rate of loss of exergy, is called the irreversibility rate. This statement is applicable to all real processes. In the ideal case of a

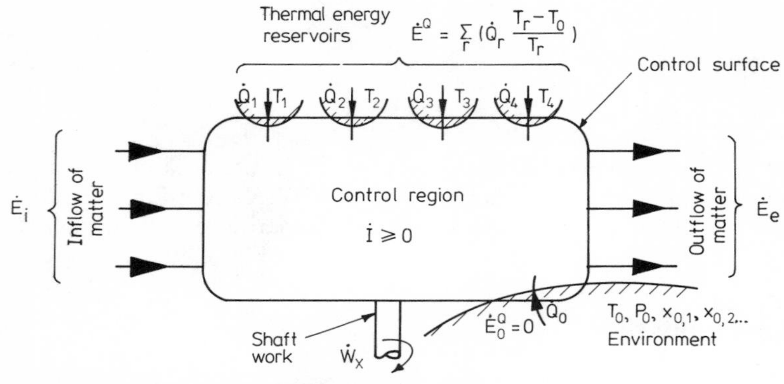

Fig 3.6 Steady-state process in a control region.

reversible process, the inflow and outflow of exergy are equal and the irreversibility rate is zero.

Recall that all the terms of the exergy balance, except $\dot{I}$, have already been defined (Chapter 2) in terms of the maximum work potential of a given form of energy. Also Chapter 1 discussed the nature of irreversible processes and the associated degradation of energy. This provides sufficient basis for writing (rather than deriving) the exergy balance by inspection from Fig 3.6.

The Gouy–Stodola relation for a control region

Also derived in Appendix B is an expression (B.22) for the irreversibility rate in an open system; its more general form is:

$$\dot{I} = T_0 \dot{\Pi} \tag{3.14}$$

which states that the irreversibility rate of a process is the product of the entropy production rate for all systems participating in the process and the temperature of the environment.

A more detailed version of (B.22):

$$\dot{I} = T_0 \left[\sum_{\text{OUT}} \dot{m}_e s_e - \sum_{\text{IN}} \dot{m}_i s_i - \sum_r \frac{\dot{Q}_r}{T_r} \right] \tag{3.15}$$

can be used to calculate irreversibility rate from quantities at various points of the control surface. (3.14) and (3.15) are versions of the Gouy–Stodola relation applicable to a control region undergoing a steady flow process. Clearly, the expression in the square brackets in (3.15) is equal to entropy production rate for a steady flow process. The first two terms give the entropy flux associated with the flow of matter out of and into the control region. The third term gives the sum of thermal entropy fluxes.

Note that (3.9) and (3.15) are not independent; either can be used to calculate the irreversibility rate for a given control region. (3.15) may be more convenient for physical processes while, for chemical processes or processes involving exchange of matter with the environment, (3.9) may be more convenient, particularly if values of chemical exergy are obtained as outlined in Appendix A or, for fuels, in Appendix C.

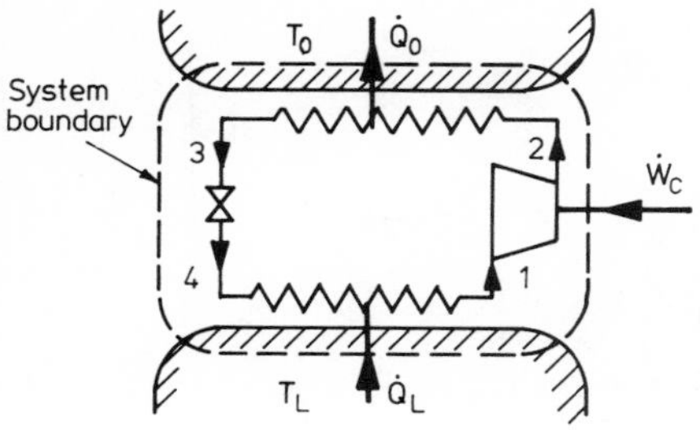

Fig 3.7 Simple vapour-compression refrigerator.

Application to closed, steadily operating systems

Although (3.9) and (3.15) are applicable to steadily operating open systems, they can also be applied to closed systems, provided these operate under steady-state conditions, eg heat engines and reversed heat engines. When such thermal plants are fully enclosed by the system boundary there are no streams of matter crossing it, only heat and work. Figure 3.7 shows such a system boundary, drawn around a simple vapour-compression refrigerator. For this type of system the exergy terms associated with the flow of matter are zero, ie:

$$\dot{E}_i = \dot{E}_e = 0 \tag{3.16}$$

and hence the exergy balance becomes:

$$\dot{E}^Q = \dot{W}_x + \dot{I} \tag{3.17}$$

Similarly the entropy flux associated with the flow of matter in (3.15) is zero, ie

$$\sum_{\text{OUT}} \dot{m}_e s_e = \sum_{\text{IN}} \dot{m}_i s_i = 0 \tag{3.18}$$

Hence the irreversibility rate is given by:

$$\dot{I} = -T_0 \sum_r \frac{\dot{Q}_r}{T_r} \tag{3.19}$$

Despite the negative sign in (3.19), the irreversibility rate, as follows from the Second Law, is always a positive quantity. The change of sign will come about because the net value of $\sum_r (\dot{Q}_r/T_r)$ is always negative.

Exergy balance for a hydraulic system

A hydraulic system of the type shown in Fig 3.8 is a special case of an open system which can be described in terms of the following characteristics:

(i) The fluid is incompressible with coefficients of thermal expansion and isothermal compressibility which may be taken to be zero, ie:

$$\alpha_P = \frac{1}{V}\left(\frac{\partial V}{\partial T}\right)_P = 0 \tag{3.20}$$

$$\kappa_T = -\frac{1}{V}\left(\frac{\partial V}{\partial P}\right)_T = 0 \tag{3.21}$$

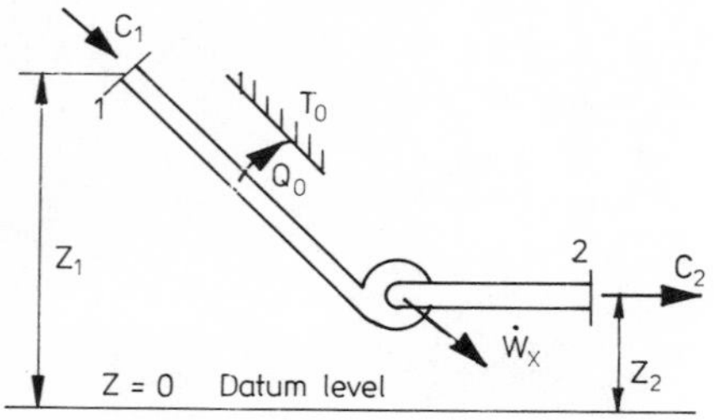

Fig 3.8 Example of a hydraulic system.

(ii) The fluid temperature is constant throughout the system, ie:

$$T = \text{const} \tag{3.22}$$

(iii) The fluid exchanges heat only with the environment.
(iv) The fluid undergoes no change in its chemical composition.

For a general system the exergy balance can be written in terms of specific quantities as:

$$\varepsilon_1 - \varepsilon_2 = -\varepsilon^Q + w_x + i \tag{3.23}$$

The LHS of (3.23) will now be written in a detailed form:

$$\varepsilon_1 - \varepsilon_2 = (h_1 - h_2) - T_0(s_1 - s_2) + \frac{C_1^2 - C_2^2}{2} + g_E(Z_1 - Z_2) \tag{3.24}$$

For the case of the fluid which conforms to (3.20) and (3.21)[3.9]:

$$dh = c_p dT + \frac{dP}{\rho} \tag{3.25}$$

and:

$$ds = c_p \frac{dT}{T} \tag{3.26}$$

But, since the process is isothermal, from (3.25) and (3.26):

$$h_1 - h_2 = \frac{P_1 - P_2}{\rho} \tag{3.27}$$

$$s_1 - s_2 = 0 \tag{3.28}$$

Using (3.27) and (3.28), (3.24) assumes the following special form applicable to a hydraulic system:

$$(\varepsilon_1 - \varepsilon_2)_H = \frac{P_1 - P_2}{\rho} + \frac{C_1^2 - C_2^2}{2} + g_E(Z_1 - Z_2) \tag{3.29}$$

Also, because of the restriction expressed by (iii):

$$\varepsilon^Q = \varepsilon_0^Q \equiv 0 \tag{3.30}$$

Substituting (3.29) and (3.30) in (3.23) and dividing throughout by g_E:

$$h_{H1} - h_{H2} = \frac{w_x}{g_E} + \Delta h_{HL} \tag{3.31}$$

where:

$$h_{H1} - h_{H2} = \frac{P_1 - P_2}{\rho g_E} + \frac{C_1^2 - C_2^2}{2g_E} + Z_1 - Z_2 \tag{3.32a}$$

$$\Delta h_{HL} = \frac{i}{g_E} \tag{3.32b}$$

(3.32a) gives the change of total head $(h_{H1} - h_{H2})$ of the hydraulic fluid, while (3.32b) gives the head loss in terms of the specific irreversibility. Hence, it follows that expression (3.31), known in hydraulics as the work–mechanical energy balance, is a special form of the exergy balance. The total head, h_H, is a restricted form of exergy for this type of system and fluid.

Irreversibility rate and the control surface

One particular advantage of the Exergy Method is that it allows calculation of numerical values of process irreversibilities. As shown by the Gouy–Stodola theorem, the irreversibility rate in control region is given by the product of entropy production rate and the temperature of the environment. As entropy production occurs within a physical space, the limits of that space must be defined clearly by a control surface.

It is sometimes convenient to distinguish between internal irreversibilities, which occur within the space in which the principal process takes place, and external irreversibilities in the immediate environment. For example, taking an internal combustion engine as the open system under consideration, internal irreversibilities occur within the engine due to such irreversible phenomena as uncontrolled combustion, heat transfer over temperature gradients, irreversible mixing, and friction. External irreversibilities occur in the immediate environment, outside the engine, and are associated with degradation of thermal energy ('heat loss' from the engine) in the environment, dissipation of the kinetic energy of the exhaust gases, and irreversibility due to the mixing of the exhaust gases with atmospheric air. Such division of irreversibilities can involve some degree of arbitrariness, differentiation is not always easy and is not always necessary. It is important, however, not to overlook related irreversibilities occurring in the immediate environment and to draw the overall system boundary so that it includes all irreversible phenomena which have their origin in the principal process under consideration.

In some applications it is helpful to divide the control region into a number of sub-regions corresponding to distinct process or plant components (Fig 3.9). The spaces in the immediate environment corresponding to the production of external irreversibilities may then be designated as separate sub-regions. This is illustrated by sub-region IV in Fig 3.9 where the cold air rejected by the heat pump mixes with the atmospheric

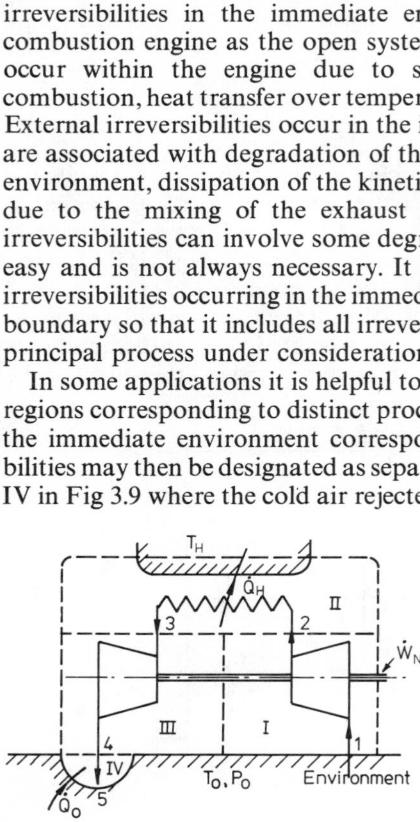

Fig 3.9 Open circuit, air heat pump.

air. Entropy is an extensive property so the sum of irreversibility rates evaluated for all the sub-regions is equal to the irreversibility rate of the whole control region. Thus:

$$\dot{I} = \sum_i \dot{I}_i \tag{3.33}$$

where the subscript i denotes the index number of the sub-region.

In some cases the control region may interact with one or more TER including the environment (eg cold chambers, heated spaces, geothermal sources, *etc*). To ensure that any irreversibilities due to the heat transfer between the reservoir and the control region are assigned to the latter, the control surface must envelop those parts of the thermal energy reservoir in which temperature gradients associated with the heat transfer exist, eg thermal boundary layers. For example, the part of the TER with temperature gradients resulting from the heat transfer is included in sub-region II in Fig 3.9. Thus the control surface at the point where heat transfer occurs is at temperature T_H.

Evaluation of irreversibility rate in some simple steady-flow processes

To demonstrate the use of the exergy balance and the Gouy–Stodola relation in the calculation of irreversibility rates, consider the following simple steady-flow processes.

Frictionless isobaric heating: A fluid in steady flow is heated at constant pressure by heat transfer from a constant temperature thermal energy reservoir at T_r (Fig 3.10(a)). From (3.14) and (3.15):

$$\dot{I} = T_0 \dot{\Pi} = T_0 \left[(\dot{S}_2 - \dot{S}_1) - \frac{\dot{Q}_r}{T_r} \right] \tag{g}$$

During this process the fluid undergoes a rate of change of entropy $(\dot{S}_2 - \dot{S}_1)$ while the corresponding thermal exergy rate is $(\dot{Q}_r/T_r)$. Since the 'heat loss' of the TER is equal to the 'heat gain' by the fluid, the two areas a, b, c, d, a and 1, 2, e, c, 1 in Fig 3.10(b) must be equal. Thus $\dot{\Pi}$ can be represented as a change in the abscissa in Fig 3.10(b). Hence, from the Gouy–Stodola relation, the irreversibility rate is given as the cross-hatched area in the T–$\dot{S}$ co-ordinates. Note that in this process all the irreversibility is due to heat transfer over a finite temperature difference. The associated temperature gradients occur within the fluid itself and in the space between the pipe carrying the fluid and the

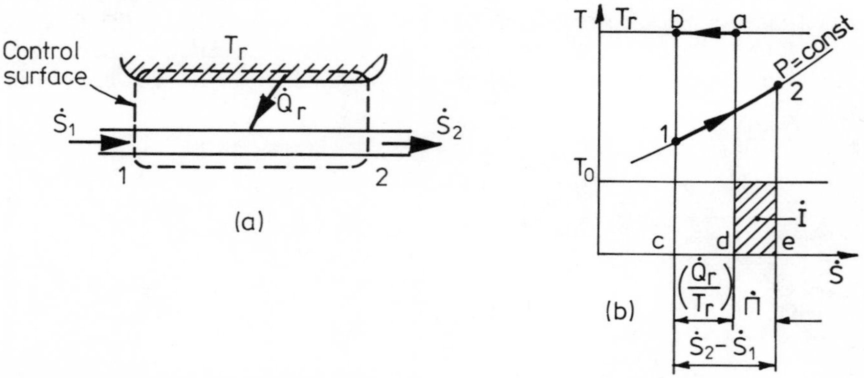

Fig 3.10 Frictionless isobaric heating.

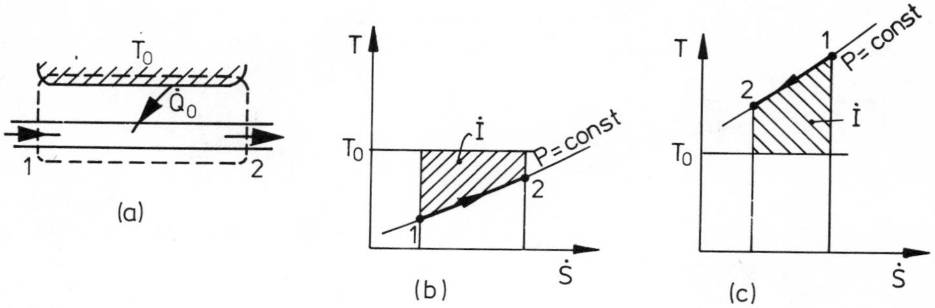

Fig 3.11 Heat exchange with the environment. Frictionless flow.

TER. Consequently the control surface must enclose the region corresponding to temperature gradients as shown in Fig 3.10(a).

Heat exchange with the environment: Here the TER is the environment at T_0. The thermal interaction may correspond either to heat transfer to or from the fluid as illustrated in the T–$\dot{S}$ co-ordinates in Fig 3.11(b) and Fig 3.11(c). Considering the first of the two cases, (3.15) can be adapted as:

$$\dot{I} = T_0(\dot{S}_2 - \dot{S}_1) - \dot{Q}_0 \tag{h}$$

From (h) the irreversibility rate, when the flow is frictionless and isobaric, is represented in Fig 3.11(b) by the cross-hatched area.

When the heat transfer takes place from the fluid to the environment, $\dot{Q}_0$ becomes negative ($\dot{Q}_0 = -\dot{Q}_0'$) and from (3.15):

$$\dot{I} = \dot{Q}_0' - T_0(\dot{S}_1 - \dot{S}_2) \tag{i}$$

From (i) the irreversibility rate is represented by the cross-hatched area in Fig 3.11(c). In the case illustrated, the cross-hatched areas are equal to the reversible work which could be obtained by using the temperature differences between the fluid flowing in the pipe and the environment.

Adiabatic expansion in a nozzle with frictional dissipation: It is instructive to analyse this type of process using an exergy balance. Expressing (3.9) in the specific form:

$$\varepsilon_1 + \varepsilon^Q = \varepsilon_2 + w_x + i \tag{j}$$

Since in this case (see also Fig 3.12(a)) $q=0$, $w_x=0$, $C_1=0$ and $\Delta Z=0$:

$$(h_1 - T_0 s_1) - (h_2 - T_0 s_2) = \frac{C_2^2}{2} + i \tag{k}$$

This shows that the change in the exergy function is equal to the specific kinetic exergy at the exit from the nozzle (the useful effect) plus the specific irreversibility. As the latter decreases, the useful effect gets larger for given initial and final states.

From the steady flow energy equation, for this process:

$$h_1 - h_2 = \frac{C_2^2}{2} \tag{l}$$

Subtracting (l) from (k):

$$i = T_0(s_2 - s_1) \tag{m}$$

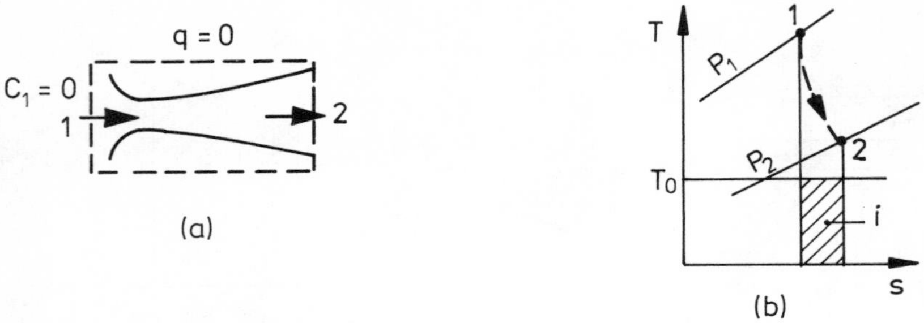

Fig 3.12 Expansion in an adiabatic nozzle with frictional dissipation.

The specific irreversibility as given by (m) is represented by the cross-hatched area in Fig 3.12(b). Note that (m) can be obtained easily and directly from the Gouy–Stodola relation.

Adiabatic flow with fluid friction: This process represents the common case of adiabatic pipe flow (Fig 3.13(a)). The irreversibility rate follows directly from the Gouy–Stodola relation:

$$\dot{I} = T_0(\dot{S}_2 - \dot{S}_1) \tag{n}$$

and is represented graphically in Fig 3.13(b). Note that both expression (n) and its graphical interpretation (Fig 3.13(b)) are also applicable to adiabatic throttling.

Example 3.2

A pipe carries a stream of brine with a mass flow rate of 5 kg/s. Because of poor thermal insulation the brine temperature increases from 250 K at the pipe inlet to 253 K at the exit. Neglecting pressure losses and taking the isobaric specific heat capacity for the brine to be 2.85 kJ/kgK calculate the irreversibility rate associated with the 'heat leak'. Take $T_0 = 293$ K.

Solution: The system and the process described in this question are shown in Figs 3.11(a) and (b). Using the Gouy–Stodola relation:

$$\dot{I} = T_0\left[(\dot{S}_2 - \dot{S}_1) - \frac{\dot{Q}_0}{T_0}\right]$$
$$= \dot{m}T_0(s_2 - s_1) - \dot{Q}_0$$

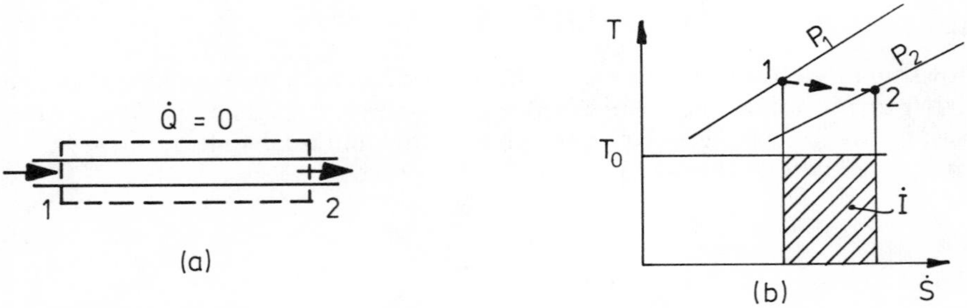

Fig 3.13 Flow with frictional dissipation in an adiabatic pipe.

Now:

$$\dot{Q}_0 = \dot{m}(h_2 - h_1)$$

hence, substituting for $\dot{Q}_0$ in $\dot{I}$:

$$\dot{I} = \dot{m}[(h_1 - h_2) - T_0(s_1 - s_2)]$$

As will be realized (see (2.12)) this represents the rate of loss of exergy due to heat transfer to the brine. Hence for a fluid with $c_p = \text{const}$ undergoing an isobaric process:

$$\dot{I} = \dot{m}c_P\left[T_1 - T_2 - T_0 \ln \frac{T_1}{T_2}\right]$$

Substituting the numerical values:

$$\dot{I} = 5 \times 2.85\left[250 - 253 - 293 \ln \frac{250}{253}\right]$$

$$= \underline{7.05\ \text{kW}}$$

This value is the minimum power, delivered to a reversible refrigeration plant, needed to restore the brine stream to its inlet temperature. This is clearly much more useful as a measure of the effect of the 'heat leak' than the value of $\dot{Q}_0$ which would be obtained from an energy balance for the pipe. $\dot{Q}_0$ does not take into account either the brine temperature or the environmental temperature.

3.3 Avoidable and intrinsic irreversibilities

The irreversibility rate is a yardstick, unique to the Exergy Method, by which losses in the plant and its components (or sub-regions into which the plant is divided) can be quantified and compared on a rational basis. By comparing the magnitudes of the irreversibility rates for the various plant components, one can see at a glance where the greatest losses occur, ie which plant components contribute most to plant inefficiency. However, such a list of irreversibility rates of plant components does not offer on its own a pointer to the components with the greatest *potential for improvement* in performance. The potential for improvement in a given component is determined by its irreversibility rate under a given set of conditions in relation to the minimum irreversibility rate possible within the limits imposed by physical, technological, economic and other constraints. This type of minimum irreversibility rate is called the *intrinsic irreversibility rate*. The difference between the actual and intrinsic irreversibility rates is the *avoidable irreversibility rate*. Thus:

$$\dot{I} = \dot{I}_{\text{INTRINSIC}} + \dot{I}_{\text{AVOIDABLE}} \tag{3.34}$$

Constraints which may contribute to the intrinsic irreversibility include:

Mismatching of heat capacities of heat transfer media: Consider heat transfer taking place in a parallel-flow mode or when the heat capacities of the streams are mismatched in a counter-flow heat exchanger (Fig 3.14). Even when the temperature difference is very small at one end of the heat exchanger, there will still be appreciable irreversibility rate due to heat transfer over a finite temperature difference at other points in the heat exchanger. This type of intrinsic irreversibility is associated with the particular physical configuration of the plant.

Uncontrolled chemical reaction: Except when a chemical reaction takes place under controlled conditions, eg in a reversible fuel cell or a galvanic cell, all uncontrolled

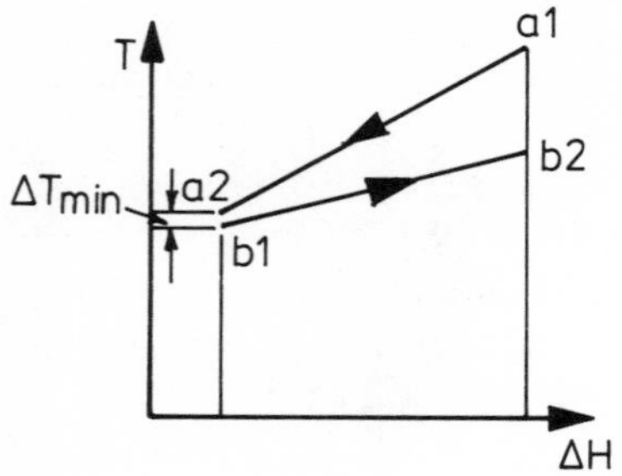

Fig 3.14 Counter-flow heat transfer with mis-matched stream heat capacities, $(mc_p)_a < (mc_p)_b$.

chemical reactions are inherently irreversible. Thus, in every plant component such as a boiler, combustion chamber or a chemical reactor there is some intrinsic irreversibility which cannot, owing to the present state of technological development, be eliminated.

Economic constraints: Efficiency of some plant components is improved by increasing their size. For example, heat exchangers of a given design perform better when the heat transfer areas are increased. However, this involves extra cost and hence there is a limiting size beyond which further increase would not be justified economically.

A more complex plant might be more efficient, but generally will have a higher capital cost and lower reliability, leading to higher maintenance costs and reduced plant availability. Use of expensive materials with better refractory properties, higher strength, or greater corrosion resistance might lead to a more efficient operation of the plant. An extensive use of such materials may not be justified economically.

The effect of some types of constraints can be evaluated easily. For example, with mismatched heat capacities of streams in a counter-flow heat exchanger, the minimum irreversibility corresponds to $\Delta T_{min} \rightarrow 0$ (Fig 3.14) and negligibly small pressure losses. The size, and hence cost, of such a heat exchanger would be very large indeed. However, the irreversibility rate so calculated would be a very useful first approximation, a target figure which would put attempts to improve performance into a more realistic perspective. The effect of the economic constraint on the size and hence the irreversibility rate of the heat exchanger involves generally rather more complex calculations. The thermoeconomic optimisation of plant components and in particular heat exchangers will be discussed in Chapter 6.

3.4 Criteria of performance

Rational efficiency

There are two main types of traditional performance criteria for thermal plant; energy conversion ratios, eg overall thermal efficiency and coefficients of performance, which are used for assessing the performance of complete power plants, and ratios involving actual output and ideal outputs, usually applied to plant components. Examples are isentropic efficiencies, thermal ratio, mechanical efficiency, *etc*. In formulating these criteria, all forms of energy are taken as equivalent, ie no reference is made to the Second Law.

Recent developments in exergy analysis allow definition of new performance criteria which offer some advantages over the traditional ones. Several workers including Bošnjaković[3.1] and Fratzscher[3.2] have proposed general criteria based on the concept of exergy. Szargut and Petela[3.3] defined exergy efficiencies for a number of particular thermal plants while Horlock[3.4] covered mainly combustion power plants.

Here a general technique based on the concept of exergy for formulating criteria of performance for a variety of thermal plants is presented[3.5]. Consider a system which undergoes a process under steady or quasi-steady conditions. Unless the process is purely dissipative, the exergy transfers can generally be grouped into those which represent the *desired output* of the process and those which represent the *necessary input*. The exergy inputs and outputs may take different forms such as work, exergy associated with heat transfer, exergy transfer associated with the flow of matter in or out of the control region or *change* of exergy of a stream of matter passing through a control region, eg in heat exchangers. If the exergy input and output are correctly identified, in relation to a control surface which encloses *all* irreversibilities related to the process under consideration, then, together, they account for all exergy transfers. In other words, there can be no exergy transfer terms in the exergy balance which cannot be included in either exergy input or exergy output; if there were such transfers, then they would correspond to *external* irreversibilities which are excluded, a priori. Hence:

$$\sum \Delta \dot{E}_{IN} = \sum \Delta \dot{E}_{OUT} + \dot{I} \tag{3.35}$$

where $\sum \Delta \dot{E}_{IN}$ is the sum of all exergy transfers making up the input and $\sum \Delta \dot{E}_{OUT}$ is the sum of all exergy transfers making up the output. Since, by the Second Law:

$$\dot{I} \geqslant 0 \tag{3.36}$$

then:

$$\frac{\sum \Delta \dot{E}_{OUT}}{\sum \Delta \dot{E}_{IN}} \leqslant 1 \tag{3.37}$$

The ratio of exergy output to exergy input is less than one, the difference depending on the degree of irreversibility of the process; for full reversibility, the ratio is one. This feature of the ratio makes it particularly suitable as a criterion of the degree of thermodynamic perfection of a process. In accordance with established practice[3.4] it will be called *rational efficiency* and denoted by ψ. From (3.35), two equivalent expressions for the rational efficiency are:

$$\psi = \frac{\sum \Delta \dot{E}_{OUT}}{\sum \Delta \dot{E}_{IN}} \tag{3.38}$$

$$\psi = 1 - \frac{\dot{I}}{\sum \Delta \dot{E}_{IN}} \tag{3.39}$$

Identifying exergy transfers corresponding to the output and the input follows from the purpose of the plant under consideration and its physical characteristics. In certain complex systems, (3.35) can be helpful in checking the choice of quantities or in the identification process, since the input and output quantities must satisfy this equation.

Efficiency defect

When dealing with a multicomponent system in which the control region can be divided into a number, N, of distinct sub-regions, it is advantageous to introduce (3.33) into the second form of rational efficiency as given by (3.39):

$$1 - \psi = \frac{\sum_i \dot{I}_i}{\sum \Delta \dot{E}_{IN}} \tag{3.40}$$

The difference $1-\psi$ is clearly that fraction of the input which is lost through irreversibility and will be called[3.5] *efficiency defect*. Its physical significance is clear from the RHS of (3.40) where it is given as a sum of fractions of the input which are lost through irreversibilities in the different sub-regions. Thus, (3.40) gives a direct causal relationship between component irreversibilities and their effect on the efficiency of the plant. The fraction representing the proportion of the input lost through irreversibility in the i-th sub-region, *or the i-th component efficiency defect*, is denoted conveniently by δ_i:

$$\delta_i = \frac{\dot{I}_i}{\sum \Delta \dot{E}_{IN}} \tag{3.41}$$

Note that the input $\sum \Delta \dot{E}_{IN}$ by which $\dot{I}_i$ is divided corresponds to the whole plant, not to the i-th component. Using this notation, (3.40) can be written:

$$1 = \psi + \delta_1 + \delta_2 + \delta_3 + \cdots + \delta_N \tag{3.42}$$

which is a form of dimensionless exergy balance for a multi-component plant, and will be used in this chapter as a basis for a pictorial representation of the exergy balance.

Relative irreversibilities

Instead of using the efficiency defect, local irreversibilities can be made non-dimensional by dividing them by the total plant irreversibility. Hence, using (3.33):

$$1 = \frac{\dot{I}_1}{\dot{I}} + \frac{\dot{I}_2}{\dot{I}} + \frac{\dot{I}_3}{\dot{I}} + \cdots + \frac{\dot{I}_N}{\dot{I}} \tag{3.43}$$

which is a convenient form for examining the relative contributions different plant components make to total irreversibility. (3.43) is particularly useful when analysing multi-component plants which have no useful output expressible in terms of exergy, eg a drying plant.

Examples of formulation of rational efficiency

To demonstrate the technique of formulating expressions for rational efficiency, a number of examples of application will be given. Further examples will be presented in Chapters 4 and 5.

Back pressure condenser: The arrangement shown in Fig 3.15 is of the type found in back pressure process heating or district heating. The useful heating $\dot{Q}_B$ is done on a

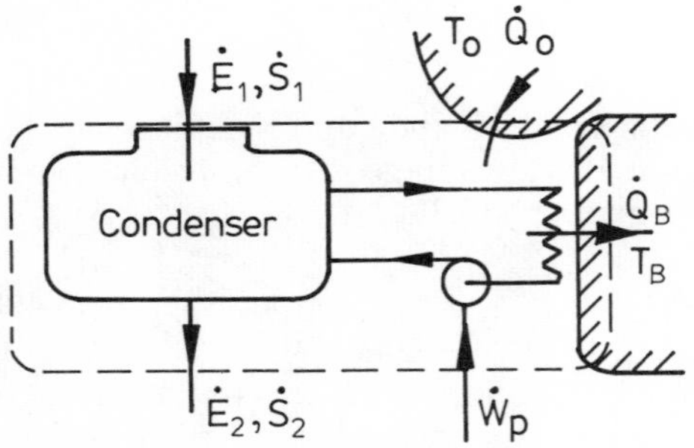

Fig 3.15 Back-pressure heating plant.

TER. There is also some 'heat loss', $\dot{Q}_0$, to the environment. Using the symbols from the diagram:

$$\dot{E}_B^Q = -\dot{Q}_B \frac{T_B - T_0}{T_B}$$

and:

$$\dot{W}_x = -\dot{W}_P$$

Hence the exergy balance can be written:

$$(\dot{E}_1 - \dot{E}_2) + \dot{W}_P - \dot{I} = \dot{Q}_B \frac{T_B - T_0}{T_B}$$

The desired exergy output being:

$$\dot{Q}_B \frac{T_B - T_0}{T_B}$$

the expressions for the rational efficiency of the system are:

$$\psi = \frac{\dot{Q}_B[(T_B - T_0)/T_B]}{(\dot{E}_1 - \dot{E}_2) + \dot{W}_P} \tag{3.44}$$

and:

$$\psi = 1 - \frac{\dot{I}}{(\dot{E}_1 - \dot{E}_2) + \dot{W}_P} \tag{3.45}$$

Using (3.15) the irreversibility rate for the system is:

$$\dot{I} = T_0(\dot{S}_2 - \dot{S}_1) + \frac{T_0}{T_B}\dot{Q}_B + \dot{Q}_0$$

Refrigeration plant evaporator: In a refrigeration plant, the cold chamber may be regarded as a TER at a temperature $T_L < T_0$. Referring to Fig 3.16, the exergy balance for the control region indicated may be written:

$$(\dot{E}_1 - \dot{E}_2) - \dot{I} = \dot{Q}_L \frac{T_0 - T_L}{T_L}$$

The desired output is the increase in the exergy of the cold chamber which, since $T_L < T_0$, is associated with heat transfer from the cold chamber. Thus:

$$\text{Output} = \dot{Q}_L \frac{T_0 - T_L}{T_L}$$

$$\text{Input} = \dot{E}_1 - \dot{E}_2$$

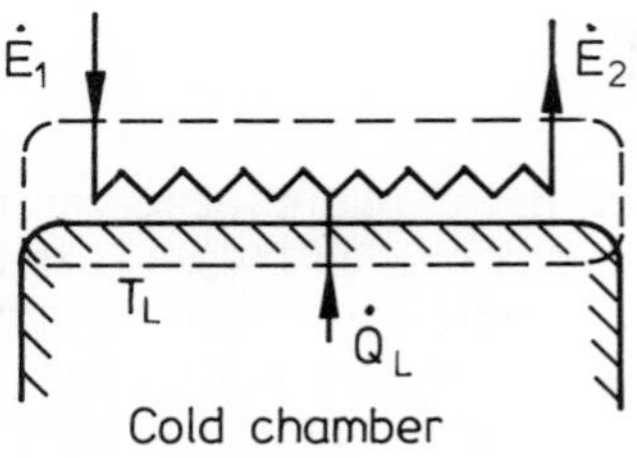

Fig 3.16 Refrigeration plant evaporator.

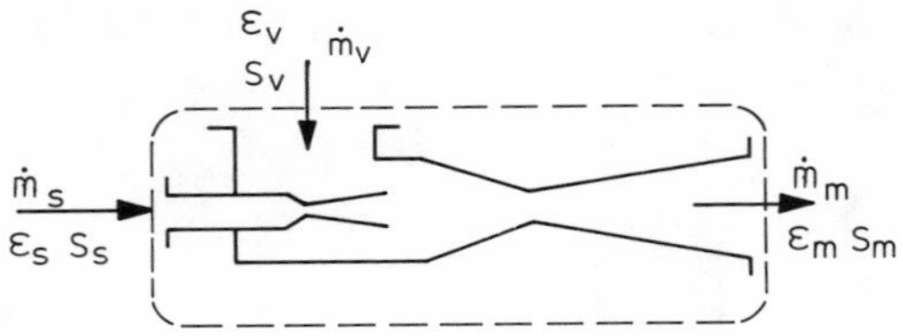

Fig 3.17 Steam ejector.

Using the general expressions, the two forms of the rational efficiency for this system are:

$$\psi = \frac{\dot{Q}_L[(T_0 - T_L)/T_L]}{\dot{E}_1 - \dot{E}_2} \tag{3.46}$$

$$\psi = 1 - \frac{\dot{I}}{\dot{E}_1 - \dot{E}_2} \tag{3.47}$$

and the irreversibility rate is:

$$\dot{I} = T_0\left[(\dot{S}_2 - \dot{S}_1) - \frac{\dot{Q}_L}{T_L}\right]$$

Steam ejector: The steam ejector shown in Fig 3.17 is of the type used for evacuating water vapour from a low pressure region, eg the evaporator of an air-conditioning plant. Assuming the ejector to be adiabatic, the exergy balance may be written:

$$\dot{E}_i = \dot{E}_e + \dot{I}$$

where:

$$\dot{E}_i = \dot{m}_s \varepsilon_s + \dot{m}_v \dot{\varepsilon}_v$$

$$\dot{E}_e = (\dot{m}_s + \dot{m}_v)\varepsilon_m$$

Hence:

$$\dot{m}_s(\varepsilon_s - \varepsilon_m) - \dot{I} = \dot{m}_v(\varepsilon_m - \varepsilon_v)$$

Here the input is the decrease in the exergy of the driving steam and the output is the increase in the exergy of the evacuated vapour. Thus:

$$\psi = \frac{\dot{m}_v(\varepsilon_m - \varepsilon_v)}{\dot{m}_s(\varepsilon_s - \varepsilon_m)} \tag{3.48}$$

$$\psi = 1 - \frac{\dot{I}}{\dot{m}_s(\varepsilon_s - \varepsilon_m)} \tag{3.49}$$

and the irreversibility rate is:

$$\dot{I} = T_0[\dot{m}_m s_m - (\dot{m}_v s_v + \dot{m}_s s_s)]$$

Thermally driven heat pump: For the analysis of a reversed power plant, the relevant forms of the exergy balance and the Gouy–Stodola relation are given by (3.17) and (3.19) respectively. In a thermally driven heat pump the input consists of a high temperature thermal source ($T_s > T_0$) with, in the case of an absorption heat pump, a small amount of mechanical power. The basic elements of the control region are shown in Fig 3.18. Thus in general:

$$\text{Input} = \dot{W}_P + \dot{Q}_s \frac{T_s - T_0}{T_s}$$

$$\text{Output} = \dot{Q}_H \frac{T_H - T_0}{T_H}$$

The two forms of rational efficiency are, therefore:

$$\psi = \frac{\dot{Q}_H[(T_H - T_0)/T_H]}{\dot{W}_P + \dot{Q}_s[(T_s - T_0)/T_s]} \tag{3.50}$$

$$\psi = 1 - \frac{\dot{I}}{\dot{W}_P + \dot{Q}_s[(T_s - T_0)/T_s]} \tag{3.51}$$

and the irreversibility rate is:

$$\dot{I} = T_0\left(\frac{\dot{Q}_H}{T_H} - \frac{\dot{Q}_s}{T_s} - \frac{\dot{Q}_0}{T_0}\right)$$

Consider now the ideal case of a plant operating reversibly ($\psi = 1$) with negligible mechanical power input, ie $\dot{W}_P = 0$. Under these conditions the minimum heat transfer rate $[\dot{Q}_s]_{MIN}$ from the thermal source of temperature T_s to produce the rate of heating $\dot{Q}_H$ at temperature T_H is given from (3.50) by:

$$[\dot{Q}_s]_{MIN} = \dot{Q}_H \frac{1 - T_0/T_H}{1 - T_0/T_s} \tag{3.52}$$

Here $[\dot{Q}_s]_{MIN}$ plays a part similar to $\dot{E}^Q$ in the expression:

$$\dot{E}^Q = \dot{Q}_H(1 - T_0/T_H) \tag{3.53}$$

From this expression $\dot{E}^Q$ may be looked upon as the minimum power necessary to produce the rate of heating $\dot{Q}_H$ at temperature T_H. Thus $[\dot{Q}_s]_{MIN}$ may be considered to be a fundamental quantity, named *thermergy* by Silver[3.6].

Single purpose power plant: Power plants which depend on some form of chemical reaction of either controlled or uncontrolled type for the input energy, are of primary importance. Expressions for the two forms of the rational efficiencies will be derived for

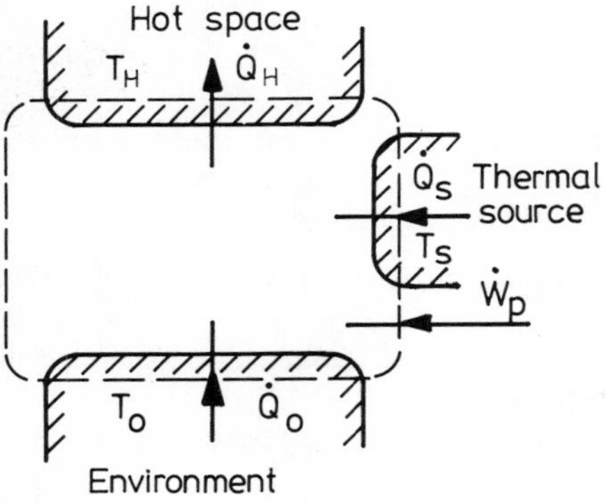

Fig 3.18 Thermally driven heat pump.

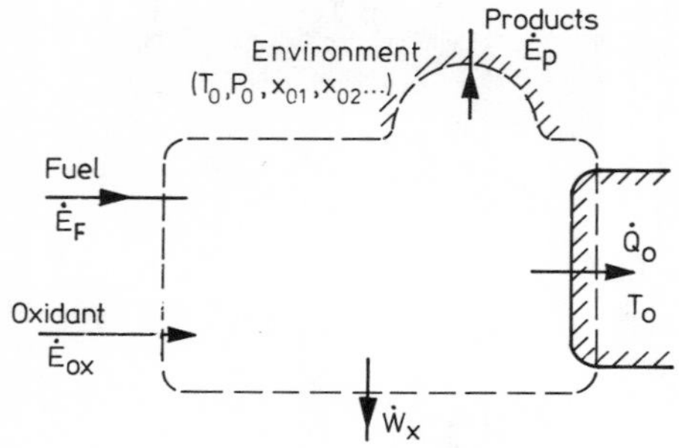

Fig 3.19 Power plant involving chemical reaction.

three common plant configurations. When chemical reactions take place inside the control region, the use of the exergy balance (3.9) is more convenient for the calculation of irreversibility rates rather than (3.15) because of the availability of chemical exergy values from tables (see Appendix A). Figure 3.19 shows the control surface of a single purpose power plant, eg an external combustion closed cycle plant such as a condensing steam plant, when $\dot{Q}_0$ would be principally the heat transfer from the steam condenser. When looked upon as an internal combustion open plant, eg internal combustion engine or gas turbine, $\dot{Q}$ would represent the stray heat transfer losses to the environment. If the control surface is drawn to include the region of mixing of the products with the atmosphere, then $\dot{E}_P = 0$ and exergy balance is:

$$\dot{E}_{ox} + \dot{E}_F = \dot{W}_x + \dot{I}$$

As the input and output for this type of plant are easily identifiable, the two forms of rational efficiency are:

$$\psi = \frac{\dot{W}_x}{\dot{E}_{ox} + \dot{E}_F} \tag{3.53}$$

$$\psi = 1 - \frac{\dot{I}}{\dot{E}_{ox} + \dot{E}_F} \tag{3.54}$$

When the oxidant used is atmospheric air, $\dot{E}_{ox} = 0$. Therefore (3.53) and (3.54) become:

$$\psi = \frac{\dot{W}_x}{\dot{E}_F} \tag{3.55}$$

$$\psi = 1 - \frac{\dot{I}}{\dot{E}_F} \tag{3.56}$$

The general technique presented above can be used in defining a rational criterion of performance for any thermal plant with a useful output expressible in terms of exergy. This criterion, the rational efficiency, has the following advantages:

(1) It gives an indication of the degree of thermodynamic perfection, ie how close the plant is to reversible operation.
(2) It is a univeral criterion applicable to a wide range of thermal plants of open and closed cycle type as well as to plant components.
(3) It gives a rational assessment of the performance of dual purpose plant such as combined heat and power plant.

(4) In multicomponent plant, 'the efficiency defect', $1-\psi$ can be expressed as a linear function of component efficiency defects thus giving an indication of the contribution of each component to plant inefficiency.

Criteria for dissipative systems

Some plants and plant components have no useful output expressible in terms of exergy; some such systems, classified into three groups are[3.5]:

(i) Systems which have as their primary function heat exchange with the environment; the thermal exergy associated with such heat transfer is zero. The most common examples are heat pump evaporators, refrigerator condensers, and natural draught cooling towers.
(ii) Systems designed to accelerate otherwise spontaneous processes, eg drying plants and mechanical separation plant such as dust separators and moisture separators. The streams of matter in these systems undergo a reduction, or no significant change, in exergy and hence the output cannot be expressed in terms of exergy.
(iii) Devices which are dissipative by design. These involve inherently irreversible processes such as throttling and stirring.

Although these devices lack any useful exergy output it may be nevertheless important, at least in the case of Group (i) and (ii) devices, that they perform their function with minimum irreversibility. A number of attempts have been made to formulate criteria for various dissipative devices. Bes[3.7] proposed the ratio of irreversibility rate to the rate of moisture removal as a criterion for assessing drying plants; for cooling towers, Szargut and Petela[3.3] proposed the ratio of irreversibility rate to the enthalpy loss rate by the cooling water. A similar criterion can be applied to heat pump evaporators and refrigeration condensers.

Such special criteria of performance may be useful when dissipative devices are being developed or tested on their own. However, when assessing the performance of a dissipative device on the context of a plant, it may be more appropriate and useful to use the efficiency defect or relative irreversibility as a criterion of performance.

Example 3.3

A vapour-compression refrigerator using Refrigerant 12 as the working fluid operates between condensation and evaporation pressures of 6.516 bar and 2.191 bar respectively, while the temperature of the cold chamber is 0°C and that of the outside air 15°C. The compression process is adiabatic and starts from the saturated vapour state, the isentropic efficiency of compression being 0.75. Expansion of the fluid between the two pressure limits takes place by adiabatic throttling, starting from 5 K below the condenser saturation temperature. Calculate:

(a) Irreversibility of the plant per mass of the working fluid.
(b) The rational efficiency of the refrigerator.

The enthalpy and entropy of compressed liquid may be taken to be equal to the corresponding saturated liquid values at the same temperature. Assume that the thermal energy extracted from the working fluid in the condenser is dissipated in the environment.

Solution: (a) Using (3.19) we can write:

$$i=\dot{I}/\dot{m}=-T_0\sum_r \frac{q_r}{T_r}$$

In terms of the symbols used in Fig 3.7 this becomes:

$$i=T_0=\left[\frac{q_0}{T_0}-\frac{q_L}{T_L}\right]$$

where:

$$-q_0=q_{2,3}=h_3-h_2$$

and:

$$q_L=q_{4,1}=h_1-h_4$$

Under the operating conditions given for the cycle, the specific enthalpy and specific entropy for the four states indicated in Fig 3.7 are[2.4]:

$$h_1=183.19 \text{ kJ/kg} \qquad s_1=0.7020 \text{ kJ/kgK}$$

$$h_2=208.75 \text{ kJ/kg} \qquad s_2=0.7228 \text{ kJ/kgK}$$

$$h_3=54.87 \text{ kJ/kg} \qquad s_3=0.2078 \text{ kJ/kgK}$$

$$h_4=54.87 \text{ kJ/kg} \qquad s_4=0.2144 \text{ kJ/kgK}$$

Hence:

$$q_0=208.75-54.87=153.88 \text{ kJ/kg}$$

$$q_L=183.19-54.87=128.32 \text{ kJ/kg}$$

Substituting the numerical values in the expression for i:

$$i=288.15\left[\frac{153.88}{288.15}-\frac{128.32}{273.15}\right]$$

$$=\underline{18.51 \text{ kJ/kg}}$$

Note that, since the condition $i\geqslant 0$ must always apply, the following must also be true:

$$\frac{q_0}{T_0}\geqslant\frac{q_L}{T_L}$$

The equality sign, above, applies to the reversible mode of operation of the plant.

(b) For the plant operating on a closed cycle the exergy balance given by (3.17) can be written per mass of the working fluid as:

$$\varepsilon^Q=w_x+i$$

where $w_x=-w_c$ and:

$$\varepsilon^Q=q_L\frac{T_L-T_0}{T_L}$$

Now, since $T_0>T_L$, the expression for ε^Q will be rearranged to give:

$$\varepsilon^Q=-q_L\frac{T_0-T_L}{T_L}$$

ie when the temperature of the TER is less than T_0 the direction of thermal exergy transfer is opposite to that of the heat transfer.

Substituting w_x and ε^Q in the expression for i and rearranging:

$$q_L \frac{T_0 - T_L}{T_L} = w_c - i$$

In this exergy balance w_c can be identified easily as the exergy input to the plant while the LHS of this equation is clearly the exergy output. Hence, the two alternative forms of rational efficiency of this plant are:

$$\psi = 1 - \frac{i}{w_c}$$

and:

$$\psi = \frac{q_L}{w_c} \bigg/ \frac{T_L}{T_0 - T_L}$$

$$= \frac{(\mathrm{CP})_{\mathrm{ref}}}{(\mathrm{CP})_{\mathrm{Carnot}}}$$

The specific work input is:

$$w_c = h_2 - h_1 = 208.75 - 183.19$$

$$= 25.56 \text{ kJ/kg}$$

Hence, the rational efficiency, from the first expression, is:

$$\psi = 1 - \frac{18.51}{25.56} = \underline{0.2758}$$

Example 3.4

An experimental domestic heat pump (Fig 3.9) takes in dry atmospheric air at 0.96 bar and 2°C and compresses it in an adiabatic compressor to 2 bar. The compressed air is then passed through a heat exchanger inside the house where it is cooled. At the exit from the heat exchanger the pressure and temperature of the air are 1.99 bar and 30°C, respectively. The air is then expanded in an adiabatic turbine to the atmospheric pressure and discharged back to the atmosphere. The temperature inside the house is 18°C. Assume air to be a perfect gas with $c_P = 1.00$ kJ/kgK and $\gamma = 1.4$ and take the isentropic efficiency of the compressor to be 0.80 and that of the turbine 0.85. Calculate:

(a) The specific irreversibilities and the corresponding component efficiency defects for the sub-regions indicated in Fig 3.9.
(b) The rational efficiency of the plant.

Solution: From the usual cycle analysis:

$$T_2 = 355.4 \text{ K} \qquad T_4 = 254.7 \text{ K}$$

$$q_H = 52.24 \text{ kJ/kg} \qquad w_c = 31.80 \text{ kJ/kg}$$

Sub-region I; compressor: As the compressor is adiabatic, the specific irreversibility for this sub-region, obtained from (3.15), becomes:

$$i_I = T_0(s_2 - s_1)$$

$$= T_0 c_P \left(\ln \frac{T_2}{T_1} - \frac{\gamma - 1}{\gamma} \ln \frac{P_2}{P_1} \right)$$

Substituting the numerical values:

$$i_{\mathrm{I}} = 275.15 \times 1.00\left(\ln \frac{355.4}{275.15} - \frac{0.4}{1.4}\ln \frac{2.00}{0.96}\right)$$

$$= \underline{12.72 \text{ kJ/kg}}$$

Sub-region II; heat exchanger: As this sub-region involves heat transfer, the Gouy–Stodola relation takes the form:

$$i_{\mathrm{I}} = T_0\left[(s_3 - s_2) - \frac{q_{23}}{T_{\mathrm{H}}}\right]$$

where:

$$q_{23} = -q_{\mathrm{H}} = c_P(T_3 - T_2)$$

$$= -52.24 \text{ kJ/kg}$$

and:

$$s_3 - s_2 = c_P\left(\ln \frac{T_3}{T_2} - \frac{\gamma - 1}{\gamma}\ln \frac{P_3}{P_2}\right)$$

$$= 1.00\left(\ln \frac{303.15}{355.4} - \frac{0.4}{1.4}\ln \frac{1.99}{2.00}\right)$$

$$= -0.1576 \text{ kJ/kgK}$$

Hence:

$$i_{\mathrm{II}} = 275.15\left(-0.1576 + \frac{52.24}{291.15}\right)$$

$$= \underline{6.01 \text{ kJ/kg}}$$

Here the irreversibility is due to both heat transfer over a finite temperature difference and the small pressure drop ($P_2 - P_3 = 0.01$ bar) which the air stream suffers during the heat transfer process.

Sub-region III; turbine: The specific irreversibility of the adiabatic turbine is:

$$i_{\mathrm{III}} = T_0(s_4 - s_1)$$

$$= T_0 c_P\left(\ln \frac{T_4}{T_3} - \frac{\gamma - 1}{\gamma}\ln \frac{P_4}{P_3}\right)$$

$$= 275.15 \times 1.00\left(\ln \frac{254.7}{303.15} - \frac{0.4}{1.4}\ln \frac{0.96}{1.99}\right)$$

$$= \underline{9.39 \text{ kJ/kg}}$$

Sub-region IV; mixing region: In this case one non-zero exergy stream crosses the control surface, and the process involves interaction with the environment, so it should prove both convenient and instructive to calculate the specific irreversibility from the exergy balance. Thus, with $\varepsilon_5 = 0$ and $\varepsilon_0^Q = 0$:

$$i_{\mathrm{IV}} = \varepsilon_4$$

As the working medium, air, has zero chemical exergy, its exergy is equal to its physical exergy. Hence:

$$\varepsilon_4 = (h_4 - h_0) - T_0(s_4 - s_0)$$

Since $P_4 = P_0$ and hence $\varepsilon_4 = \varepsilon_4^{\Delta T}$:

$$i_{IV} = \varepsilon_4 = c_P\left[T_4 - T_0 - T_0 \ln \frac{T_4}{T_0}\right]$$
$$= 1.00\left[254.7 - 275.15 - 275.15 \ln \frac{254.7}{275.15}\right]$$
$$= \underline{0.80 \text{ kJ/kg}}$$

Specific irreversibility of the whole plant: This can be calculated for the four sub-regions combined from the Gouy–Stodola relation:

$$i_{plant} = T_0\left[(s_5 - s_1) - \frac{q_0}{T_0} + \frac{q_H}{T_H}\right]$$

Now, since $s_5 = s_1 = s_0$:

$$i_{plant} = T_0\left[\frac{q_H}{T_H} - \frac{q_0}{T_0}\right]$$

The heat transfer, q_0, (Fig 3.9) from the environment to the mixing region and due to the flow of the cold air from the turbine, is:

$$q_0 = c_P(T_0 - T_4)$$
$$= 20.45 \text{ kJ/kg}$$

Substituting now the numerical values:

$$i_{plant} = 275.15\left[\frac{52.24}{291.15} - \frac{20.45}{275.15}\right]$$
$$= \underline{28.92 \text{ kJ/kg}}$$

This is equal to the sum of the specific irreversibilities calculated for the four sub-regions, in agreement with (3.33).

Efficiency defects: As the exergy input to this plant is w_c, the component efficiency defects have the general form:

$$\delta_i = \frac{i_i}{w_c}$$

Substituting the numerical values:

$$\delta_I = \frac{12.72}{31.80} = 0.400$$

$$\delta_{II} = \frac{6.01}{31.80} = 0.189$$

$$\delta_{III} = \frac{9.39}{31.80} = 0.295$$

$$\delta_{IV} = \frac{0.80}{31.80} = 0.025$$

The sum of the component efficiency defects gives the efficiency defect for the plant, ie:

$$\sum_i \delta_i = \delta$$

In this case:

$$\delta = 0.909$$

Hence, from (3.40), the rational efficiency of the plant is:

$$\psi = 1 - \delta$$
$$= 0.091$$

The calculated values of δs show that the mixing region contributes least to the total plant irreversibility. Note, however, that this contribution will increase with the increase in the isentropic efficiency of the turbine since an increase in the latter leads to a reduction in the air exit temperature. A similar type of coupling may be found between the compressor and the heat exchanger. In this case, however, a reduction in the irreversibility of the compressor will lead to a reduction in the temperature difference $(T_2 - T_H)$ and hence to a reduction in the irreversibility of the heat exchanger. Thus an improvement in the performance of one component may affect the performance of other plant components appreciably. This type of coupling between plant components will be discussed under the sub-title 'structural bonds' in Chapter 6.

3.5 Pictorial representation of the exergy balance

Grassmann diagram

Listing component irreversibilities may help in assessing the performance of a simple plant. With more complex plants, however, there is a distinct advantage in presenting the information pictorially. A very useful representation of exergy flows and losses is the *Grassmann diagram* which may be considered as an adaptation of the Sankey diagram used for energy transfers within a plant.

The Grassmann diagram for a one-stage air compressor plant is shown in Fig 3.20. Because of the sequential nature of energy transformations in this plant, the Grassmann diagram takes a linear form. The width of the band is a measure of the magnitude of the exergy flux at the entry or the exit of a particular sub-region. Each sub-region is represented as a rectangular box and the reduction in the width of the band, shown dotted, is a measure of the exergy loss, or irreversibility rate, in a given sub-region.

In more complex plants the Grassmann diagram can be particularly valuable since it shows not only exergy losses but also the splitting of exergy streams and recirculation of exergy. It also shows graphically how a part of the original input of exergy is dissipated in the successive stages of energy transformation. The losses in each sub-region may be further sub-divided into different types where such differentiation can be made. For an electric motor (sub-region A) it might be possible to make an assessment of the proportion of the irreversibility rate which is due to frictional losses, $(\dot{I}_F)_A$ (aerodynamic and in the bearing friction) and electrical dissipation, $(\dot{I}_E)_A$ (electrical resistance and magnetic hysteresis). Similarly, the irreversibility rate in the aftercooler may be divided into two parts, $(\dot{I}^{\Delta T})_D$ due to heat transfer over a finite temperature difference and $(\dot{I}^{\Delta P})_D$ due to frictional dissipation (pressure drop).

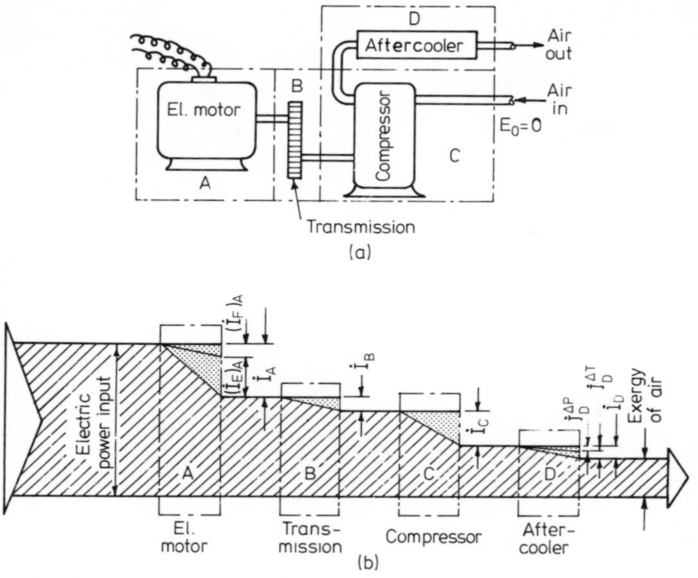

Fig 3.20 Grassmann diagram for a single-stage air compressor plant. An example of application.

The diagram can be presented either in a dimensional or dimensionless form. In the first case, eg Fig 3.20, the values of exergy and irreversibility rate are expressed in units of power, say, kW. In the dimensionless representation the exergy input is unity so that the component losses appear as component efficiency defects, δ_i, and the output as the rational efficiency ψ of the plant. The reason for it should be clear from expression (3.42).

Other examples of the use of Grassmann diagrams in thermal plant analysis are given in Chapters 4 and 5.

Pie diagram

The exergy balance for a multicomponent plant can be represented as a pie diagram, eg for a one-stage air compressor plant (Fig 3.20(a)) in Fig 3.21. Here all the quantities are in dimensionless form, so that the exergy balance is based on (3.42). The magnitudes of the various quantities are given by the *angle* of the sector representing a given quantity. The full 360° of the inner circle represents the input and the angular portions of the outer circle are proportional to the terms on the RHS of (3.42). As with the Grassmann diagram, exergy losses in the electric motor have been split into frictional, $(\delta_F)_A$, and electrical, $(\delta_E)_A$, components.

Similarly, exergy losses in the after cooler have been split into the fractions $(\delta^{\Delta T})_D$ due to heat transfer over a finite temperature difference and $(\delta^{\Delta T})_D$ due to pressure losses. Although the pie diagram does not provide any information about exergy flows, recirculation and stream splitting, it is generally simpler to construct and the relative magnitudes of the various quantities can be shown more accurately.

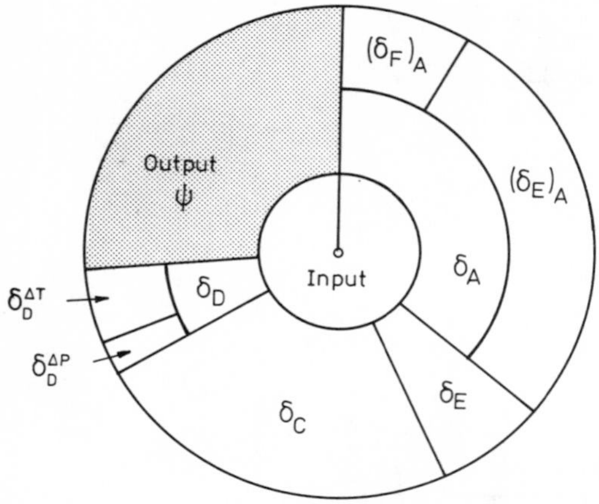

Fig 3.21 Dimensionless exergy balance in the form of a pie diagram.

3.6 Exergy-based property diagrams

Use of T–s and h–s diagrams to obtain quantities such as I and ε_{ph} has already been demonstrated; property diagrams specially devised for use in the analysis of processes by the exergy method are covered here.

Physical exergy–enthalpy diagram

Taking the environmental pressure and temperature to be fixed, the function representing the specific physical exergy can be written as:

$$\varepsilon_{ph} = h - T_0 s + C \tag{3.57}$$

Clearly, then, in ε_{ph}–h co-ordinates the function $s = \text{const}$ is a straight line, inclined at an angle, depending on the relative scales used for ε_{ph} and h, to the co-ordinate axes. In Fig 3.22 the lines $s = \text{const}$ are at 45°, corresponding to the use of the same scales for both the principal co-ordinate axes. Note that an ε_{ph}–h diagram is an oblique version of the h–s diagram.

Figure 3.23(a) is an ε_{ph}–h diagram, typical to a number of substances such as water, ammonia and a number of fluorinated refrigerants. One common characteristic is that their critical temperatures are higher than the environmental temperature, ie $T_C > T_0$. For these substances the environmental state P_0, T_0 is represented in the ε_{ph}–h co-ordinates by a straight horizontal line in the liquid-vapour region which coincides with the lines $P_0 = \text{const}$ and $T_0 = \text{const}$. The value of the Gibbs function of the substance is constant on this line and $\varepsilon_{ph} = 0$.

As will be observed from the diagram, the lines in the two-phase region corresponding to $T = \text{const}$ and $P = \text{const}$ are straight, and their slopes increase with increasing temperature. To prove this, differentiate partially (3.57) with respect to h with $P = \text{const}$, yielding:

$$\left(\frac{\partial \varepsilon_{ph}}{\partial h}\right)_P = 1 - T_0\left(\frac{\partial s}{\partial h}\right)_P$$

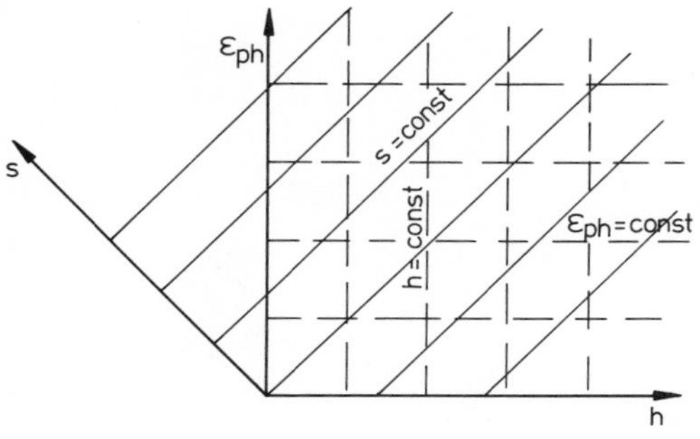

Fig 3.22 Co-ordinates of a ε_{ph}–h diagram[2.2].

Now, since $(\partial s/\partial h)_P = 1/T$:

$$\left(\frac{\partial \varepsilon_{ph}}{\partial h}\right)_P = 1 - \frac{T_0}{T}$$

$$= \tau$$

Hence, the slope of these lines is equal to the dimensionless exergetic temperature obtained from the particular value of T = const. The same applies to the next diagram.

For substances for which $T_C < T_0$ (Fig 3.23(b)), the environmental state lies in the superheated vapour region at the intersection of the lines P_0 = const and T_0 = const. Substances with this characteristic include air, nitrogen, oxygen, helium, neon and a number of working fluids used in cryogenic engineering. A ε_{ph}–h chart for air is given in Appendix E. The bottom right-hand quarter of the property plane shown in Fig 3.23(b) corresponds approximately to the ideal gas region. Such charts can be used for the study and analysis of gas cycles and processes. A number of applications of ε_{ph}–h charts and diagrams will be given in Chapters 4 and 5. Exergy–enthalpy charts are constructed for a particular value of T_0, usually $T_0 = 293$ K. In some application areas and in some climatic regions this value of T_0 may not be representative of the environmental conditions under which the plant operates. It is, however, still possible to apply the chart in such cases by applying a correction to the values read from the chart.

Consider an exergy–enthalpy chart which has been constructed for some specific set of values of T_0, h_0 and s_0. h_0 and s_0 are the values of the specific enthalpy and the specific entropy of the particular fluid under consideration corresponding to the environmental values of pressure and temperature, P_0, T_0. The difference in the value of ε_{ph} between that given by the chart and one corresponding to the set of values T_0', h_0' and s_0' appropriate to the actual environmental state is given by:

$$\Delta\varepsilon = (\varepsilon_{ph} - \varepsilon_{ph}') = (T_0' - T_0)s - A \tag{3.58}$$

where:

$$A = (h_0 - h_0' + T_0' s_0 - T s_0)$$

As follows from (3.57), $A = C' - C$.

Often it is the difference in the values of exergy at two different thermodynamic states which is required. In such cases:

$$\varepsilon_{ph1} - \varepsilon_{ph2} = h_1 - h_2 - T_0(s_1 - s_2)$$

showing that this difference depends on T_0 and not on h_0 or s_0. For the same two

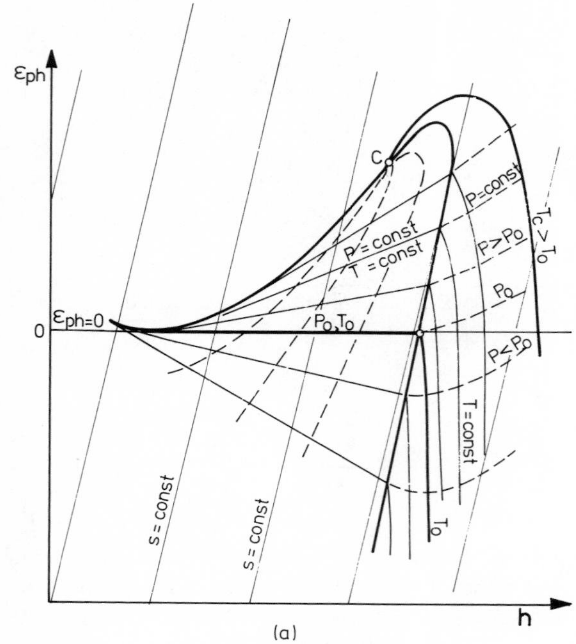

(a)

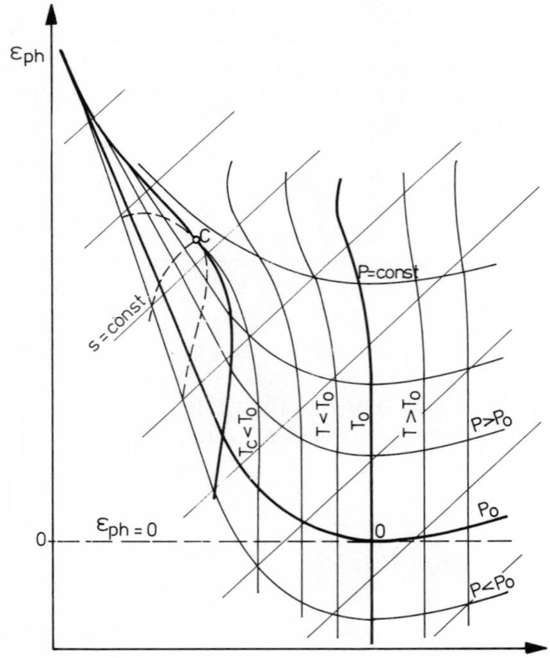

(b)

Fig 3.23 Physical exergy–enthalpy diagram, (a) state (P_0, T_0) as a line in the two-phase region, (b) state (P_0, T_0) in the superheated region.

thermodynamic states 1 and 2 but a different environmental temperature T_0 this exergy difference would be, say, $(\varepsilon_{ph1} - \varepsilon_{ph2})'$, so that:

$$(\varepsilon_{ph1} - \varepsilon_{ph2})' - (\varepsilon_{ph1} - \varepsilon_{ph2}) = (T_0 - T_0')(s_1 - s_2)$$
$$= \Delta T_0 \times \Delta s \quad (3.59)$$

As will be noted from (3.59), in the case when the two states (1 and 2) lie on an isentrope, ie $\Delta s = 0$, then $(\varepsilon_{ph1} - \varepsilon_{ph2}) = (\varepsilon_{ph1} - \varepsilon_{ph2})'$, ie no correction is required.

Since, as can be seen from (3.58), the correction is a linear function of s and T_0 it is a simple matter to produce a supplementary correction diagram in which $\Delta\varepsilon$ can be given[2.2] as a function of s for different constant values of T_0.

As follows from the Gouy–Stodola relation, the irreversibility of a process is a function of T_0. Consequently, the value of the process irreversibility determined from the chart can be multiplied by T_0'/T_0 to give the value corrected for the actual temperature, ie:

$$I' = \frac{T_0'}{T_0} I \quad (3.60)$$

Thus, if the sole purpose of the analysis is the determination of process irreversibilities, the method of making the correction $\Delta\varepsilon$, described above, is not necessary.

τ–$\dot{H}$ diagram

Consider an element of a duct (Fig 3.24(a)) carrying a fluid subjected to a heat flux $\dot{Q}_A$ over an area dA:

$$đ\dot{Q} = \dot{Q}_A \, dA$$

Applying the steady flow energy equation for the case when changes in kinetic and potential energy are negligible:

$$đ\dot{Q} = \dot{m} \, dh$$
$$= d\dot{H}$$

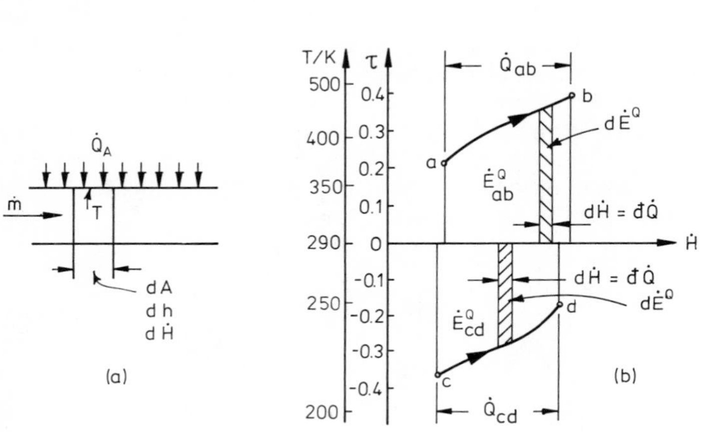

Fig 3.24 Representation of thermal exergy on a τ–$\dot{H}$ diagram.

The expression for dimensionless exergetic temperature has the form:

$$\tau = \frac{T - T_0}{T}$$

For a fixed temperature T_0, τ is a function of T only. If T is the temperature at which the thermal energy is received by the element, the rate of gain of exergy by the fluid can be written as follows:

$$\mathrm{d}\dot{E}^Q = \tau \mathrm{d}\dot{H} \tag{3.61}$$

Using τ–$\dot{H}$ co-ordinates, we can represent $\mathrm{d}\dot{E}^Q$ as an elemental area (Fig 3.24(b)). This is illustrated both for the case when $T > T_0$ and when $T < T_0$. In the first case the variation of τ with enthalpy in the process is given by the curve a–b and in the second case by c–d; the corresponding heat transfer rates for the stream are given by Q_{ab} and Q_{cd} respectively. The total gain of exergy for a process will be obtained by integrating (3.61). Hence, for the two processes:

$$\dot{E}^Q_{ab} = \int_a^b \tau \mathrm{d}\dot{H} \tag{3.62}$$

$$\dot{E}^Q_{cd} = \int_c^d \tau \mathrm{d}\dot{H} \tag{3.63}$$

Clearly in each case $\dot{E}^Q$ is given on the τ–$\dot{H}$ diagram by the area lying between the process curve, the abscissa axis (corresponding to $\tau = 0$) and the limiting ordinates. For $T > T_0$, both τ and $\mathrm{d}\dot{H}$ are positive and hence $\dot{E}^Q_{ab}$ represents a *gain* of exergy by the fluid. For $T < T_0$, τ is now negative whereas $\mathrm{d}\dot{H}$ is still taken positive and hence $\dot{E}^Q_{cd}$ is negative, ie it represents a transfer of exergy *from* the fluid.

Plotting the process line in τ–$\dot{H}$ co-ordinates can be facilitated by drawing, parallel with the τ co-ordinate, a T co-ordinate which gives the absolute temperature corresponding to any value of τ (Fig 3.24(b)). Of course, the temperature scale is non-linear.

An alternative method of constructing a τ–$\dot{H}$ diagram for a counterflow heat exchanger with fluids of the same heat capacities (ie $\dot{m}_a c_{Pa} = \dot{m}_b c_{Pb}$) as an example, is shown in Fig 3.25. In such cases, the variation of temperature with enthalpy rate is given by two straight parallel lines. The construction of the τ–$\dot{H}$ curves is shown for point X. As shown above, the area under curve a1–a2 represents the rate of decrease in exergy, $\dot{E}_a$, of stream *a* and the area under curve b1–n2 represents $\dot{E}_b$.

For most liquid streams in heat transfer processes, the irreversibility rate due to pressure losses, $\dot{I}^{\Delta P}$, is negligible compared with that due to heat transfer over a finite temperature difference, $\dot{I}^{\Delta T}$. Whenever $\dot{I}^{\Delta P} \ll \dot{I}^{\Delta T}$, the τ–$\dot{H}$ diagram can be used to determine the irreversibility rate of the process. Applying the exergy balance to the heat exchanger, which will be assumed to be adiabatic ($\dot{Q}_0 = 0$):

$$(\dot{E}_{a1} - \dot{E}_{a2}) - (\dot{E}_{b2} - \dot{E}_{b1}) = \dot{I}^{\Delta T} \tag{3.64}$$

But:

$$\dot{E}_{a1} - \dot{E}_{a2} = \dot{E}^Q_a \tag{3.65}$$

and:

$$\dot{E}_{b2} - \dot{E}_{b1} = \dot{E}^Q_b \tag{3.66}$$

Substituting (3.65) and (3.66) in (3.64):

$$\dot{E}^Q_a - \dot{E}^Q_b = \dot{I}^{\Delta T} \tag{3.67}$$

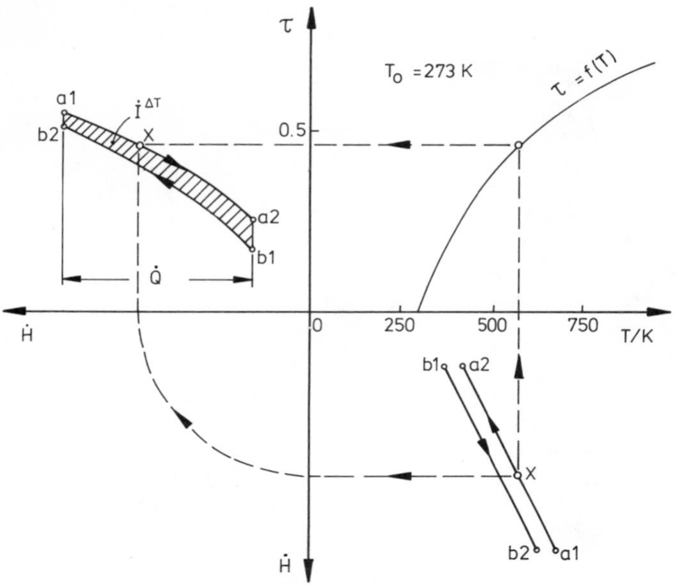

Fig 3.25 Construction of a τ–$\dot{H}$ diagram for a counter flow heat exchanger.

Now, from (3.67), the magnitude of $\dot{I}^{\Delta T}$ is given on the τ–$\dot{H}$ diagram by the difference in the two areas representing $\dot{E}^Q_a$ and $\dot{E}^Q_b$ on the diagram and is shown cross-hatched.

This type of graphical method of determination of $\dot{I}^{\Delta T}$ can be particularly valuable when the T–$\dot{H}$ relationship is complex, eg in a multistream heat exchanger network, but can be easily represented graphically. The τ–$\dot{H}$ diagram can be also useful in comparing different heat exchanger arrangements to minimise irreversibility rate.

$(-1/T)$–$\dot{H}$ diagram (The Thring diagram)

Figure 3.26(a) shows a representation in the $(-1/T)$–$\dot{H}$ co-ordinates[3.8] of a process in which two streams exchange heat at constant pressures. The negative sign attached to the $1/T$ co-ordinate ensures that the curve for the fluid being heated is lower. The corresponding absolute temperatures are shown next to $(-1/T)$ to facilitate plotting the process curves. In these co-ordinates an elementary strip corresponding to an increment in $\dot{H}$, $d\dot{H} = đ\dot{Q}$, and limited in the vertical direction by the two process curves represents an increment in the entropy production rate. This can be proved by applying (1.60) to the elementary section of the heat exchanger shown in Fig 3.26(b):

$$đ\dot{\Pi}^{\Delta T} = d\dot{S}_a - d\dot{S}_b \tag{3.68}$$

The superscript ΔT indicates that the irreversibility is due solely to the heat transfer over a finite temperature difference.

Considering the elementary section of each channel separately (Fig 3.26(c)):

$$d\dot{S}_a = -\frac{đ\dot{Q}}{T_a}$$
$$d\dot{S}_b = \frac{đ\dot{Q}}{T_b} \tag{3.69}$$

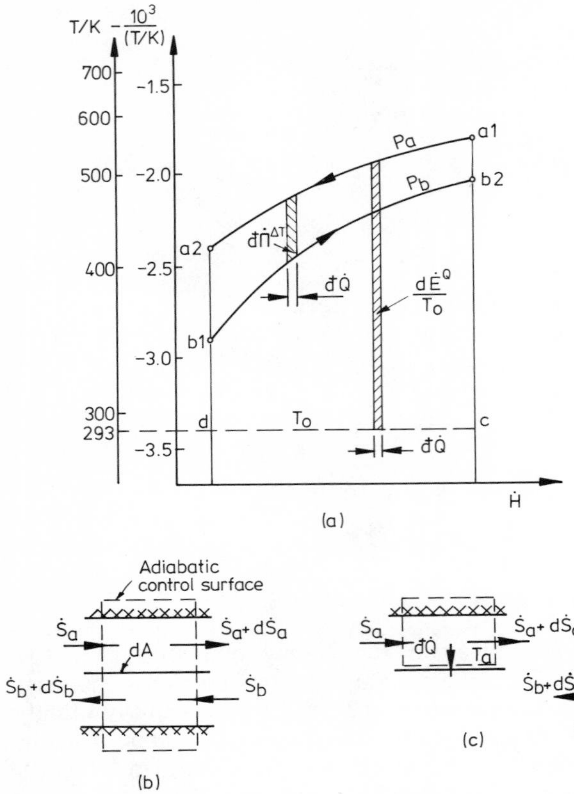

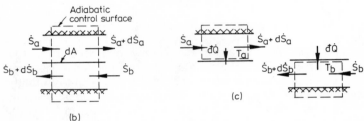

Fig 3.26 Representation of entropy production due to irreversible heat transfer on the $(-1/T)$–$\dot{H}$ diagram. (The Thring diagram.)

Substituting (3.69) in (3.68):

$$đ\dot{\Pi}^{\Delta T} = đ\dot{Q}\left(\frac{1}{T_b} - \frac{1}{T_a}\right) \tag{3.70}$$

which corresponds to the representation shown in Fig 3.26(a). Hence the total entropy production rate for the process $\dot{\Pi}^{\Delta T}$ is represented by the area between the two process curves and the limiting co-ordinates. Once $\dot{\Pi}^{\Delta T}$ is evaluated, the corresponding value of the irreversibility rate follows from the Gouy–Stodola relation:

$$\dot{I}^{\Delta T} = T_0 \dot{\Pi}^{\Delta T} \tag{3.71}$$

Clearly, the entropy production rate is independent of the environmental temperature. Hence the applicability of this graphical construction is not confined to any particular value of T_0.

Using similar arguments, it can be shown that an elementary strip on the $(-1/T)$–$\dot{H}$ diagram corresponding to an increment in $\dot{H}$ ($d\dot{H} = đ\dot{Q}$) and limited in the vertical direction by the process curve of a stream and a selected value of the environmental temperature T_0 represents an infinitesimal change in the thermal exergy of the fluid divided by T_0, ie:

$$\frac{\mathrm{d}\dot{E}^{Q}}{T_0} = \mathrm{d}\dot{H}\left(\frac{1}{T_0} - \frac{1}{T}\right) \tag{3.72}$$

Hence, the thermal exergy transferred by a stream is proportional to the area lying between the process line of the stream and a horizontal line representing T_0. The coefficient of proportionality depends on T_0 but the shapes of the curves do not. Thus, in this construction, the $(-1/T)$–$\dot{H}$ diagram can be drawn without reference to the value of T_0 which will be used subsequently in the calculation of either $\dot{I}^{\Delta T}$ or $\dot{E}^{Q}$.

The limitation of the $(-1/T)$–$\dot{H}$ and the τ–$\dot{H}$ diagrams is that they cannot be used to represent the effect of pressure losses due to fluid friction on the process irreversibility. However, in applications where irreversibility due to heat transfer over a finite temperature difference plays a dominant role, eg in heat exchangers, and industrial furnaces, these types of diagrams are most useful[3.8].

Exergy-composition (ε–x) diagrams for binary mixtures

Processes involving separation of components from mixtures occur in a number of industrial applications. Binary (two-component) mixtures, eg absorption refrigeration cycles, separation of oxygen from air, and separation of deuterium from hydrogen, are both common in such applications and relatively simple to study.

The ε–h diagram is not suitable for the study of separation processes since the mole fraction, x, defining composition of the mixture is needed as an additional co-ordinate. This leads to a three-dimensional representation, in ε–h–x co-ordinates (Fig 3.27). Because of the difficulty of representing the various constant-property surfaces on a two-dimensional diagram, Fig 3.27 shows only the traces which these surfaces make on the ε–h, ε–x and h–x co-ordinate planes. The co-ordinates of a point in the three-

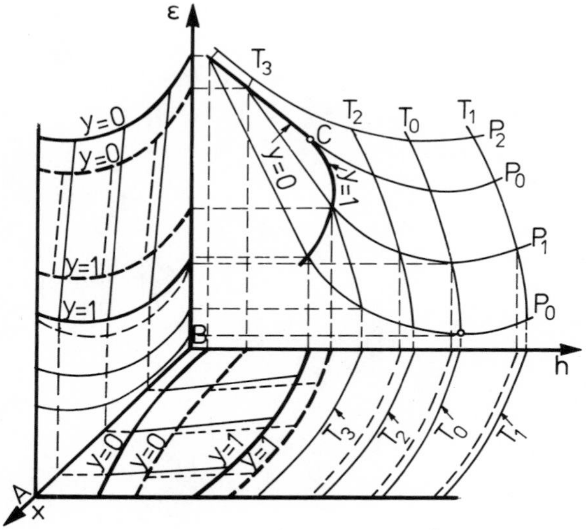

Fig 3.27 ε_{ph}–h–x co-ordinate system for a binary mixture. (Reproduced from Ref [2.2], by permission.)

dimensional space defines the state of a mixture of composition defined by the value of the mole fraction, x, of one of the components. Usually the more volatile of the two components is used for this purpose. When $x=0$ or $x=1$, the ε–h property planes correspond to the pure components. As will be seen from Fig 3.27, the property plane corresponding to $x=0$ in this representation is similar to the one shown for a pure component in Fig 3.23(b).

Since a three-dimensional system of co-ordinates is not practicable, this information is usually presented in two-dimensional co-ordinates for fixed values of pressure. Such a diagram may be considered to have been generated from the three-dimensional representation from projections, on either h–x or ε–x plane, of the lines of intersection of a selected $P=\text{const}$ surface with the $T=\text{const}$ surfaces and the saturated liquid ($y=0$) and saturated vapour ($y=1$) surfaces. The resulting, well-known property diagrams in the h–x co-ordinates can be found in books on chemical thermodynamics. Such a diagram is shown in Fig 3.27 on the h–x co-ordinate plane (horizontal). The other diagram[3.9], with ε and x co-ordinates is also shown in Fig 3.27. The dotted lines in these diagrams correspond to pressure P_1, whilst the unbroken lines give the states for pressure P_0. Figure 3.28 is a more detailed version of the ε–x diagram for $P_0=\text{const}$. Note that the isotherm $T_0=\text{const}$ occupies the lowest position on the diagram. The isotherms corresponding to other temperatures for both $T>T_0$ and $T<T_0$ lie above the $T_0=\text{const}$ line; the greater the difference $|T-T_0|$ for a given isotherm, the higher its position on the diagram. The isotherms are concave upwards because the exergy of pure components is higher than that of any mixture made up from them at the same pressure and temperature. Since in the case illustrated the diagram is constructed for P_0, the isotherm $T_0=\text{const}$ gives the value of chemical exergy for a mixture of given composition. In the special case of a mixture consisting of O_2 and N_2 only, the minimum point marked by 0, on the isotherm $T_0=\text{const}$ gives the chemical exergy of

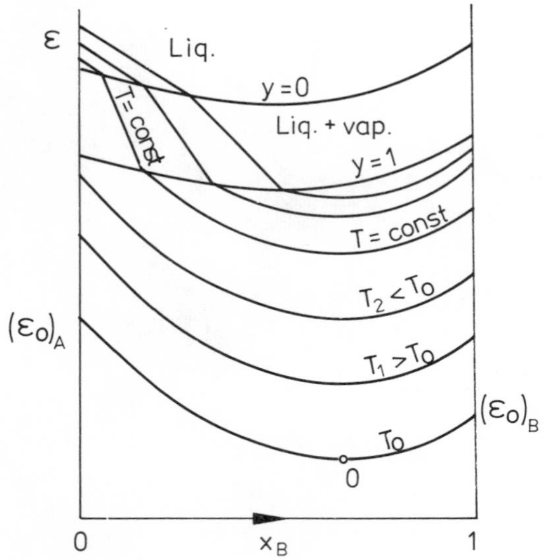

Fig 3.28 Exergy-composition diagram for a binary mixture, for $P=P_0$. (Reproduced from Ref [2.2], by permission.)

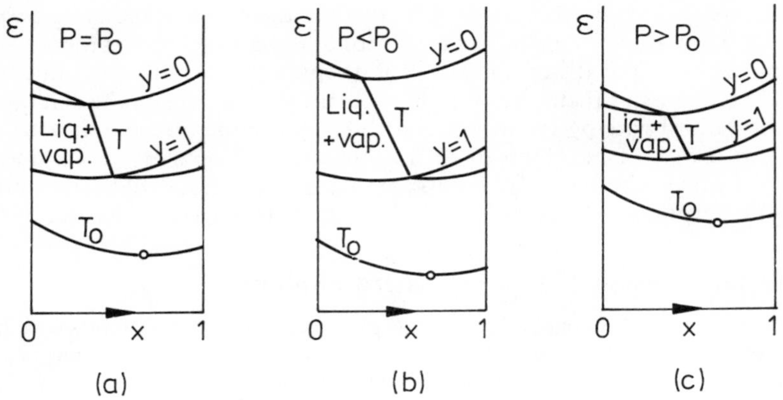

Fig 3.29 Effect of pressure on the distribution of constant-property lines on a ε_{ph}–x diagram. (Reproduced from Ref [2.2], by permission.)

air, zero. Thus in the case of air the abscissa axis should pass through this minimum point on the T_0=const isotherm. This, of course, only applies if we accept the approximation that air consists solely of O_2 and N_2.

Above the gaseous region (superheated vapour) is the wet vapour region, limited by the two saturation lines. The saturated liquid line ($y=0$) lies above the saturated vapour line ($y=1$). The region extending immediately above the line $y=0$ corresponds to the liquid phase.

Figure 3.29 shows the effect of pressure on the distribution of lines on an ε–x diagram. Reducing pressure lowers all isotherms. The two saturation lines (for $y=0$ and $y=1$) are brought close together with increasing pressures.

It is usual to show isotherms and saturation lines for a number of different pressures on the same ε–x diagram. Using such a diagram one can determine the minimum work of separation of components of a mixture for any initial state of the mixture and of the products of the separation process.

The method of construction of an isotherm on an ε–x diagram is shown in Fig 3.30. Assuming the mixture to consist of ideal gases, we can use (2.21a). The points 1 and 2 mark the exergy of the pure components and the line joining them represents variation in the exergy of a hypothetical mechanical mixture (one that can be separated without a

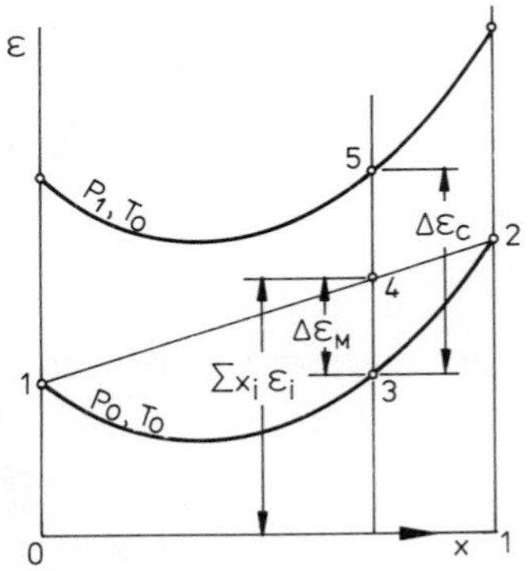

Fig 3.30 Construction of an isotherm on an exergy–composition diagram. (Reproduced from Ref [2.2], by permission.)

work input). The exergy of such a mixture is given by the simple sum of the exergies of the components of the mixture, $\sum x_i \varepsilon_i$, which corresponds to the first term on the RHS of (2.25). The second term on the RHS of this expression represents loss of exergy due to mixing, $\Delta\varepsilon_M$, and has to be subtracted from the first term. This is shown on the diagram as a reduction in the ordinate by the distance 4–3. An isotherm corresponding to a higher pressure $P_1 > P_0$ can be obtained by raising the first point by a distance corresponding to the reversible work $\Delta\varepsilon_C$ necessary to raise the pressure of the mixture of given composition from P_0 to P_1, which in the diagram is shown as the distance 3–5.

3.7 Thermodynamic feasibility of new thermal plants

The archives of patent offices are packed with draft proposals for thermal plants which, in one way or another, violate the laws of thermodynamics. Such proposals also appear from time to time in technical literature. As the implications of the Second Law are generally less well understood than those of the First Law, it is usually the former which is disregarded by such inventors. Devices which violate the Second Law are known as perpetual motion machines of the second kind, or PMM2 (see also Section 1.11). Whereas some of these proposed plants will appear wrong at a glance to the trained thermodynamicist, other devices may not appear to be so obviously wrong. In the latter case some thermodynamic analysis may be required before their thermodynamic feasibility is disproved.

As the exergy balance is based on both the First as well as the Second Law of Thermodynamics, it is ideal for analysing the thermodynamic feasibility of thermal plants. Two thermal plant proposals will now be examined as examples of such analysis.

Locomotive power plant

This power plant* (Fig 3.31) consists of a steam plant operating on the Rankine cycle with superheat and a vapour compression heat pump. Heat rejected by the steam condenser upgraded by the heat pump is used as the main heat input, $\dot{Q}_H$, in the steam

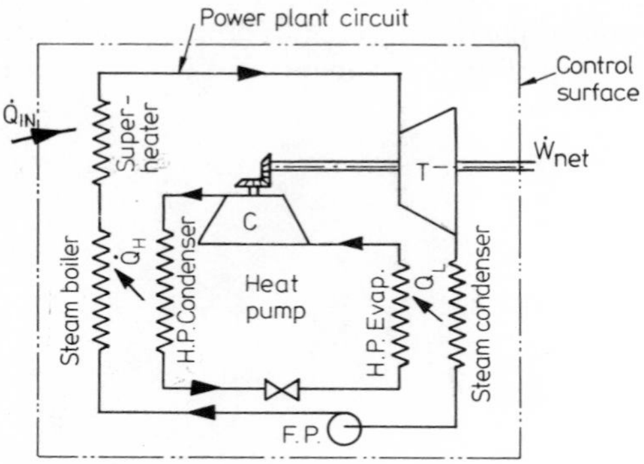

Fig 3.31 Proposal for a locomotive power plant.

* This thermal plant proposal was sent by a reader to a reputable engineering journal. Several other readers criticised the plant in subsequent issues of the journal, offering *extensive calculations* to prove their point.

boiler. The only heat input from an outside source which is envisaged in this proposal is $\dot{Q}_{IN}$, which is supplied to the superheater. The power required to drive the heat pump compressor is taken off from the turbine of the steam plant. The obvious fault, to a trained thermodynamicist, is that power is generated while the plant interacts with only one TER, ie no heat is rejected. This, clearly, directly violates the Planck statement of the Second Law and hence the plant must be rejected on thermodynamic grounds. In the following proof the Second Law test is applied through the exergy balance.

Energy balance:

$$\dot{Q}_{IN} = \dot{W}_{NET} \tag{3.73}$$

Exergy balance:

$$\dot{E}^{Q}_{IN} = \dot{W}_{NET} + \dot{I} \tag{3.74}$$

Eliminating $\dot{W}_{NET}$ from (3.73) and (3.74):

$$\dot{E}^{Q}_{IN} - \dot{Q}_{IN} = \dot{I} \tag{3.75}$$

But, when $T > T_0$, the magnitude of thermal exergy is always smaller than that of the associated heat transfer (since $\tau < 1$), hence:

$$\dot{E}^{Q}_{IN} - \dot{Q}_{IN} < 0 \tag{3.76}$$

This leads to:

$$\dot{I} < 0$$

which would violate the Second Law. Hence, the proposed plant is thermodynamically not feasible.

'Atmospheric heat' power plant

The inventor of this plant, originally discussed by Brodyanskii[2.2], conceived the idea of using the plentiful and freely available energy of the atmospheric air as the sole energy source of his Rankine cycle vapour power plant (Fig 3.32). 'Atmospheric heat' is upgraded in the open circuit heat pump to some temperature $T > T_0$ and then transferred as $\dot{Q}_H$ to the working fluid of the power plant. According to the proposal, a fraction of the power generated by the power plant turbine would be utilised to drive the heat pump. An electric motor (not shown on the diagram), would be used to drive the heat pump during a start up period.

Since the total enthalpy rate, $\dot{H}_1$, of the air entering the heat pump would be greater than that, $\dot{H}_2$, of the stream discharged into the environment, the energy balance:

$$\dot{H}_1 - \dot{H}_2 = \dot{Q}_0 - \dot{W}_{NET} \tag{3.77}$$

does not reveal any difficulties about the operation of the proposed power plant. Applying now the exergy balance to the control region:

$$\dot{E}_1 + \dot{E}^{Q}_{0} = \dot{W}_{NET} + \dot{E}_2 + \dot{I} \tag{3.78}$$

But the exergy $\dot{E}_1$ of the atmospheric air entering the heat pump is zero and so is $\dot{E}^{Q}_{0}$ associated with the heat rejected by the power plant condenser. Hence:

$$0 = \dot{W}_{NET} + \dot{E}_2 + \dot{I} \tag{3.79}$$

Now, since $\dot{E}_2 > 0$ and $\dot{I} > 0$ we must conclude that:

$$\dot{W}_{NET} < 0 \tag{3.80}$$

which measn that the 'atmospheric heat' power plant cannot deliver any net power output.

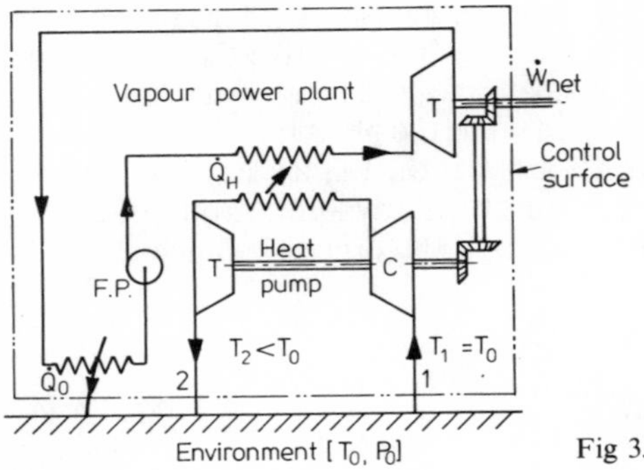

Fig 3.32 Proposal for 'atmospheric heat' power plant[2.2].

The alternative to the above type of analysis is to carry out detailed calculations both for the heat pump and for the power plant for some given set of operating parameters. The results of such calculations should lead to the conclusion, that *at the given operating parameters*, the plant would not produce a net power output. Hence, this method of analysis, while lacking in generality, would at the same time require much more time and effort.

It should be stressed that this type of analysis can only give an answer with regard to thermodynamic feasibility of plant, not about technical or economic feasibility.

Chapter 4 Exergy analysis of simple processes

Most large power, chemical and process plants can be considered to comprise a number of simple processes. The most common are: expansion, compression, heat exchange, mixing, separation of constituents of mixtures, and chemical reactions including combustion of fuels. This chapter studies these basic processes and analyses them on their own, using the techniques of the exergy method, to prepare ground for the analysis of complex systems.

4.1 Expansion processes

The purpose and characteristics of expansion processes in power plant are essentially different to those found in refrigeration or cryogenic plant. Therefore these two application areas will be considered separately.

Expansion processes in power plants

In power plants expansion generally occurs at temperatures above the environmental temperature. Except for throttling, the purpose of an expansion process is to deliver power at the expense of a reduction in the exergy of the stream of the working fluid. Most common expanders are rotodynamic and are usually treated as adiabatic. Expansion can occur as a single or a multi-stage expansion process.

Single stage expansion process

A simplified diagram of a turbine is shown in Fig 4.1(a). The turbine may or may not exchange heat with the environment. In either case we have $\varepsilon_0^Q = 0$. Hence the exergy balance for the control surface in terms of specific quantities is:

$$\varepsilon_1 - \varepsilon_2 = w_{ex} + i \tag{4.1}$$

As the desired output from the device is w_{ex}, it follows that the necessary input is the *reduction* in the exergy of the stream, $\varepsilon_1 - \varepsilon_2$. Since the process always involves some degree of irreversibility, a part of the input is dissipated. This relationship is shown on a Grassmann diagram in Fig 4.1(b). If the process involves heat exchange with the environment ('heat loss'), $q = -q_L$. Except in the special case of a reversible isothermal expansion occurring at $T = T_0$, this 'heat loss' will contribute to the overall process irreversibility. This can be seen from the Gouy–Stodola relation (3.15), which in the

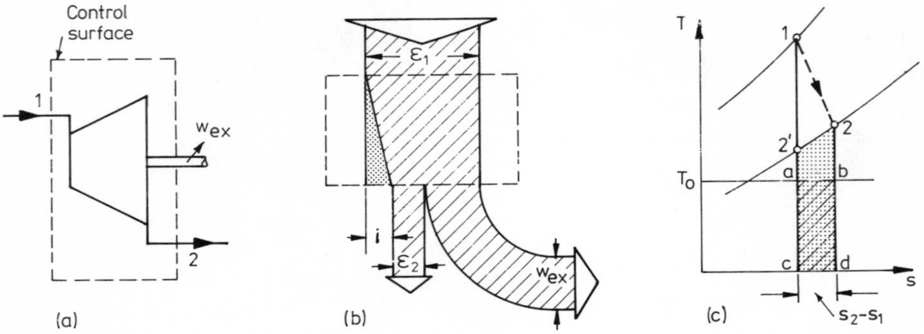

Fig 4.1 Single stage expansion in an adiabatic turbine.

present case, when $T_r = T_0$, becomes:

$$i = T_0(s_2 - s_1) + q_L \tag{4.2}$$

If the turbine is considered to be adiabatic, ie $q_L = 0$, the Gouy–Stodola relation for the process takes the form:

$$i = T_0(s_2 - s_1)_{adia} \tag{4.3}$$

Having identified above the input $(\varepsilon_1 - \varepsilon_2)$ and output (w_{ex}) of the expander, the rational efficiency for this process (Section 3.4), takes the alternative forms:

$$\psi = \frac{w_{ex}}{\varepsilon_1 - \varepsilon_2} \tag{4.4}$$

$$\psi = 1 - \frac{i}{\varepsilon_1 - \varepsilon_2} \tag{4.5}$$

Considering only the process and excluding the effect of mechanical friction in the bearings and take $q_L = 0$, the rational efficiency in its form given by (4.4) can be written:

$$\psi = \frac{h_1 - h_2}{\varepsilon_1 - \varepsilon_2} \tag{4.6}$$

For comparison, the well-established criterion of performance, the isentropic efficiency η_s, can, with reference to Fig 4.1(c), be put in the well-known form:

$$\eta_s = \frac{h_1 - h_2}{h_1 - h_{2'}} \tag{4.7}$$

Now, (4.6) and (4.7) are two criteria of performance which assess the perfection of the process on different bases. ψ compares the actual process with a reversible process with the same inlet and exit states. η_s compares the actual process with an isentropic process starting from the same inlet state but ending in a different exit state, though at the same exit pressure as the actual process. To analyse further these criteria of performance, (4.6) and (4.7) are rearranged:

$$\psi = \frac{h_1 - h_2}{(h_1 - h_2) + T_0(s_2 - s_1)} \tag{4.8}$$

$$\eta_s = \frac{h_1 - h_2}{(h_1 - h_2) + (h_2 - h_{2'})} \tag{4.9}$$

Note that the quantity which makes ψ smaller than 1 is $T_0(s_2 - s_1)$ which is the specific irreversibility of this process, shown as a cross-hatched area on the T–s diagram in Fig 4.1(c). The quantity which makes the value of η_s less than 1 is $(h_2 - h_{2'})$, shown as a dotted area in Fig 4.1(c). This enthalpy difference can be regarded as *frictional reheat.* Because of frictional reheat, the enthalpy and the exergy of the working fluid in the final state of the actual process are greater than they would have been under isentropic conditions. When the final state of an expansion process corresponds to the initial state of another process, eg in multi-stage turbines, this difference in enthalpy or exergy can be utilised. Consequently we must not regard the whole of the frictional reheat as a loss. Figure 4.1(c) shows that the difference in the areas representing frictional reheat and process irreversibility (area 2′2ba2′) corresponds to the exergy of frictional reheat. Denoting $h_1 - h_{2'} = \Delta h_s$ and:

$$h_2 - h_{2'} = r \tag{4.10}$$

then:

$$w_{ex} = h_1 - h_2 = \Delta h_s - r \tag{4.11}$$

Substituting (4.11) in (4.1) and further denoting:

$$\varepsilon_1 - \varepsilon_2 = \Delta\varepsilon_{1-2}$$

gives:

$$\Delta h_s - \Delta\varepsilon_{1-2} = r - i$$

Noting, further, that for an isentropic process $\Delta h_s = \Delta\varepsilon_s$, then:

$$\Delta\varepsilon_s - \Delta\varepsilon_{1-2} = r - i$$

This relationship, shown on an exergy–enthalpy diagram in Fig 4.2, clearly demonstrates that the $(r - i)$ part of frictional reheat is recoverable; this recoverable part increases with the temperature at which the turbine exhausts (Fig 4.1(c)) since the area 2′2ba2′ increases with the mean temperature $\bar{T}_2 = (T_2 + T_{2'})/2$. The same conclusion will be reached by considering Fig 4.2. As the process is moved towards higher temperatures, ie to the right of the ε–h diagram, the slope of the $P = \text{const}$ lines increases and hence, for a given value of η_s and pressure ratio, $(r - i)$ becomes larger.

Making use of (4.8) and (4.9):

$$\psi = \frac{\eta_s}{i/r + \eta_s(1 - i/r)} \tag{4.12}$$

which is a relationship between ψ and η_s as a function of the ratio i/r. From Fig 4.1(c):

$$r \simeq \frac{T_2 + T_{2'}}{2}(s_2 - s_1)$$

and:

$$i = T_0(s_2 - s_1)$$

Hence:

$$r/i \simeq \bar{T}_2/T_0 \tag{4.13}$$

Thus r/i can be considered a dimensionless, mean exhaust temperature.

The relationship given by (4.12) is plotted in Fig 4.3 for three values of η_s and a range of values of r/i. For any fixed value of η_s, the rational efficiency increases with r/i (or $\bar{T}_2/T_0$), the two criteria of performance being equal when $r/i = 1$. This leads to the

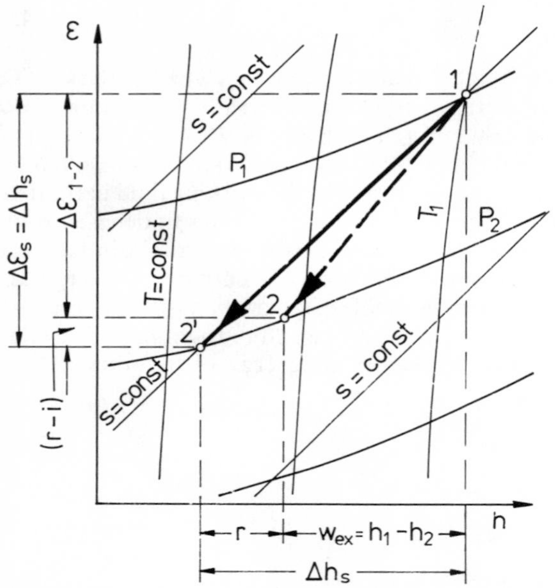

Fig 4.2 Single stage adiabatic expansion process in ε–h co-ordinates.

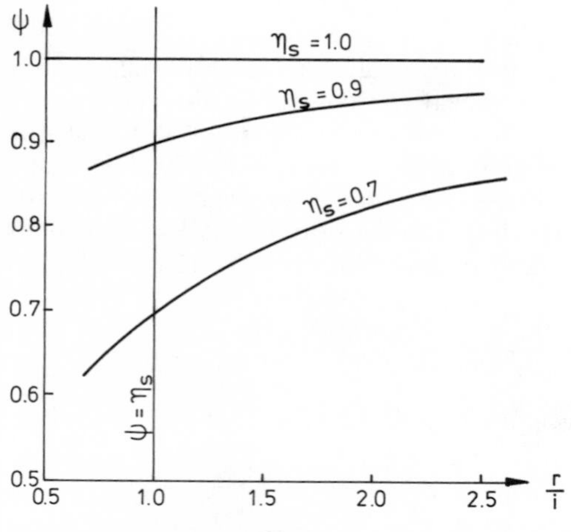

Fig 4.3 Relationship between isentropic efficiency and rational efficiency for an expansion process.

conclusion that a turbine stage with a low isentropic efficiency is more tolerable at a high temperature than at a low temperature.

Although kinetic energy of the working fluid was not included in this discussion, its effect should not be neglected in turbine calculations, particularly if high accuracy is required.

Multi-stage expansion process

The relationship between stage efficiency and overall efficiency is often required when analysing the performance of a multi-stage turbine. When using the isentropic efficiency as a criterion of performance, this relationship of a turbine with equal stage efficiencies η_s is given[1.4] by:

$$\eta_{ov} = \eta_s R_F \tag{4.14}$$

where η_{ov} is the overall isentropic efficiency and R_F is known as the reheat factor. R_F depends both on the properties of the working fluid and the magnitude of the stage efficiencies. Its value is always greater than unity. The effect of frictional reheat on the different stages is shown for a three stage expansion in Fig 4.4. Note that the higher the

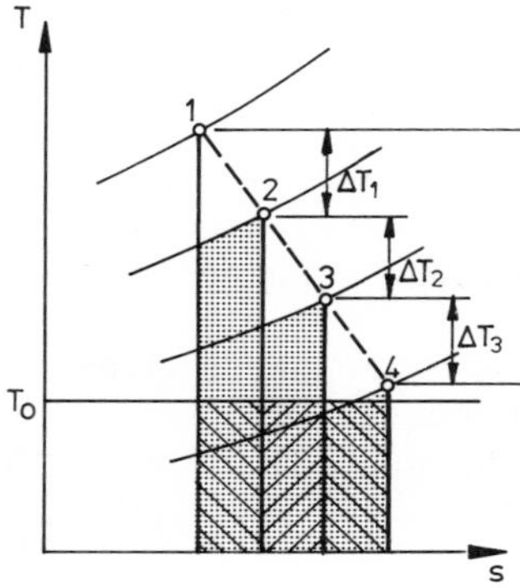

Fig 4.4 Multi-stage expansion process in an adiabatic turbine.

stage exhaust temperature, the greater is the thermal exergy of frictional reheat given by $(r-i)$ and shown in the diagram as the difference between the dotted and cross-hatched areas. In the last stage of expansion here, the exhaust temperature T_4 is only slightly greater than T_0 and $(r-i)\simeq 0$. Note that it is the variability of the effect of frictional reheat on stage performance which makes the relationship between stage efficiencies and the overall isentropic efficiency so involved.

The relationship between the overall rational efficiency and stage rational efficiencies is rather more direct. The overall rational efficiency can be expressed as[4.1]:

$$\psi_{ov} = \frac{\Delta h_{ov}}{\Delta \varepsilon_{ov}} \tag{4.15}$$

The numerator in this expression, the overall enthalpy drop, can be expressed as the sum of the stage enthalpy drops, ie:

$$\Delta h_{ov} = \Delta h_1 + \Delta h_2 + \Delta h_3 + \cdots \tag{4.16}$$

Similarly, the overall exergy drop in the denominator of (4.15) can be written as a sum of the stage exergy drops:

$$\begin{aligned}\Delta \varepsilon_{ov} &= \Delta \varepsilon_1 + \Delta \varepsilon_2 + \Delta \varepsilon_3 + \cdots \\ &= \sum \Delta \varepsilon_i \end{aligned} \tag{4.17}$$

Substituting (4.16) and (4.17) in (4.15) and rearranging:

$$\psi_{ov} = \frac{\Delta h_1}{\Delta \varepsilon_1} \frac{\Delta \varepsilon_1}{\sum \Delta \varepsilon_i} + \frac{\Delta h_2}{\Delta \varepsilon_2} \frac{\Delta \varepsilon_2}{\sum \Delta \varepsilon_i} + \frac{\Delta h_3}{\Delta \varepsilon_3} \frac{\Delta \varepsilon_3}{\sum \Delta \varepsilon_i} + \cdots$$

$$= \psi_1 \frac{\Delta \varepsilon_1}{\sum \Delta \varepsilon_i} + \psi_2 \frac{\Delta \varepsilon_2}{\sum \Delta \varepsilon_i} + \psi_3 \frac{\Delta \varepsilon_3}{\sum \Delta \varepsilon_i} + \cdots \tag{4.18}$$

Thus, from expression (4.18), the overall rational efficiency is equal to the weighted average of stage rational efficiencies. In the special case of a turbine with equal stage rational efficiencies:

$$\psi_{ov} = \psi_i \tag{4.19}$$

It can be shown that the line of condition for a turbine with equal stage rational efficiencies is a straight line when plotted on the ε–h, ε–s, or h–s diagrams and, if the working fluid is a perfect gas, on the T–s diagram.

External (exhaust) irreversibility in an expansion process

When an expander exhausts to atmosphere, eg in open cycle gas turbines, pneumatic motors and non-condensing steam plant, the resulting mixing of the working fluid with the atmosphere leads to external irreversibility. The mixing region may be treated as one of the sub-regions into which the plant is divided.

Figure 4.5(a) shows two sub-regions A and B containing the expander and mixing process respectively. The exergy ε_2 of the air leaving the turbine is all lost in sub-region B and so the exergy ε_0 of the air which diffuses across the boundary is zero. Hence, with $\varepsilon^Q = 0$ the exergy balance for the combined control region is:

$$\varepsilon_1 = w_{ex} + i_A + i_B \tag{4.20}$$

The two components, i_A and i_B, of total irreversibility are shown as cross-hatched areas in Fig 4.5(b). Since exhaust air in state 2 is at pressure P_0, exergy ε_2 which is dissipated through mixing consists principally of the thermal component of exergy, $\varepsilon^{\Delta T}$ (see (2.13)).

Since the function of a pneumatic motor is to deliver work, the exergy output is w_{ex}. It should, therefore, be clear from (4.20) that the only term which can be considered to be the exergy input to the expander is ε_1. Hence, the two forms of the rational efficiency for this expander are:

$$\psi = \frac{w_{ex}}{\varepsilon_1} \tag{4.21}$$

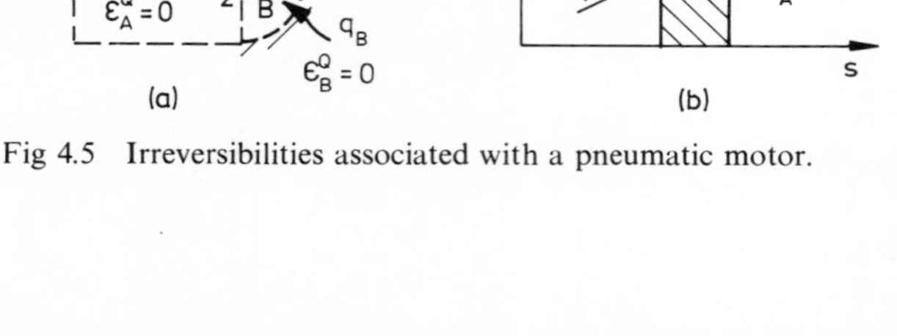

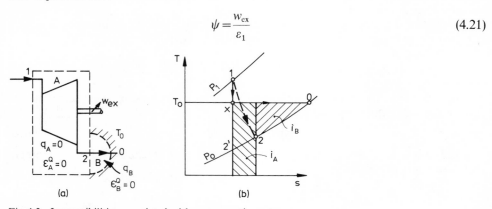

Fig 4.5 Irreversibilities associated with a pneumatic motor.

$$\psi = 1 - \frac{i_A + i_B}{\varepsilon_1} \tag{4.22}$$

In general, in a mixing process, the kinetic energy of the stream may have to be taken into account. With working fluids other than air, the chemical exergy of the fluid must be included in the exergy terms to take account of the differences in composition between the working fluid and the environment.

Example 4.1

An adiabatic turbine (Fig 4.5(a)) is supplied with air at the rate of 0.033 kg/s at 3.5 bar and 303 K. The air expands in the turbine to a pressure of 1 bar and 238 K and is then discharged into the ambient air. The environmental temperature and pressure are 283 K and 1 bar respectively.

(i) Specify on a T–s diagram ideal processes which, if used in the expansion of air, would produce maximum power output at given supply conditions. Hence, calculate the maximum power output.

(ii) Calculate, as a fraction of the exergy of the air at the inlet to the turbine: shaft power; irreversibility rate within the turbine; external irreversibility rate due to mixing of the air discharged from the turbine with the ambient air.

(iii) Calculate the isentropic efficiency and the rational efficiency of the expansion process inside the turbine. Take air to be perfect gas with $c_P = 1.00$ kJ/kgK and $\gamma = 1.4$.

Solution: (i) One set of fully (internally and externally) reversible processes is shown in Fig 4.5(b). These are: $1 \to x$—frictionless adiabatic expansion; $x \to 0$—reversible isothermal expansion at T_0. Another possible set of processes which would result in the production of maximum power output is shown in Fig 2.7(b). In either case:

$$\dot{W}_{MAX} = \dot{E}_1$$

$$= \dot{m} c_P \left[(T_1 - T_0) - T_0 \ln \frac{T_1}{T_0} + \frac{\gamma - 1}{\gamma} T_0 \ln \frac{P_1}{P_0} \right] \tag{a}$$

Substituting the given numerical values:

$$\dot{W}_{MAX} = \underline{3.3646 \text{ kW}}$$

(ii) Referring to Fig 4.5(a), since $\dot{E}_0 = 0$ and $\dot{E}^Q = 0$, the exergy balance for the control region consisting of sub-regions A and B is:

$$\dot{E}_1 = \dot{W}_{ex} + \dot{I}_A + \dot{I}_B \tag{b}$$

Dividing (b) throughout by $\dot{E}_1$:

$$1 = \psi_{AB} + \delta_A + \delta_B \tag{c}$$

where:

$$\psi_{AB} = \dot{W}_{ex} / \dot{E}_1 \tag{d}$$

$$\delta_A = \dot{I}_A / \dot{E}_1 \tag{e}$$

$$\delta_B = \dot{I}_B / \dot{E}_1 \tag{f}$$

Clearly, ψ_{AB} is the rational efficiency based on $\dot{E}_1$ as the exergy input and δ_A and δ_B are the efficiency defects for the two sub-regions.

Now, since, $\dot{E}_B^Q = 0$ and $\dot{E}_0 = 0$:

$$\dot{E}_2 = \dot{I}_B \tag{g}$$

$\dot{E}_2$ can be calculated from an expression similar to (a). Hence:

$$\dot{E}_2 = \dot{I}_B = 0.1325 \text{ kW}$$

For sub-region A, with $\dot{Q}_A = 0$,

$$\dot{I}_A = T_0(\dot{S}_2 - \dot{S}_1)$$

$$= \dot{m}c_P T_0 \left(\ln \frac{T_2}{T_1} - \frac{\gamma - 1}{\gamma} \ln \frac{P_2}{P_1} \right)$$

$$= 1.0871 \text{ kW}$$

Also, for an adiabatic turbine we can write:

$$\dot{W}_{ex} = \dot{m}c_P(T_1 - T_2)$$

$$= 2.145 \text{ kW}$$

Hence:

$$\underline{\psi_{AB} = 0.6375}$$

$$\underline{\delta_A = 0.3231}$$

$$\underline{\delta_B = 0.0394}$$

(iii) The isentropic efficiency of the turbine is:

$$\eta_s = \frac{T_1 - T_2}{T_1 - T_{2'}}$$

where:

$$T_{2'} = T_1 (P_0/P_1)^{\frac{\gamma - 1}{\gamma}}$$

$$= 211.8 \text{ K}$$

Hence, from (h):

$$\underline{\eta_s = 0.7127}$$

The internal rational efficiency of the turbine (excluding the effect of $\dot{I}_B$) is:

$$\psi_A = \frac{\dot{W}_{ex}}{\dot{E}_1 - \dot{E}_2}$$

$$\underline{= 0.6637}$$

The difference between η_s and ψ_A has been explained above with reference to expressions (4.8) and (4.9).

Throttling process in power plants

Throttling is generally used in power plants to reduce the pressure of a working medium to reduce its exergy and hence the power output of the plant. This is a wasteful way of controlling power output but it is used frequently because of its simplicity.

The exergy balance for a throttling valve (Fig 4.6(a)) can be written:

$$\varepsilon_1 = \varepsilon_2 + i \tag{4.23}$$

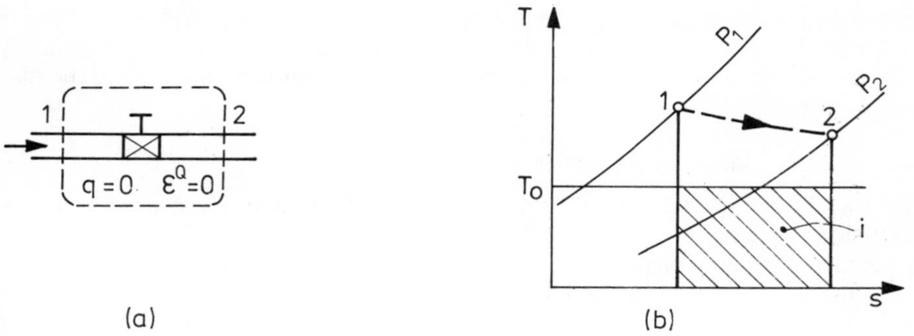

Fig 4.6 Adiabatic throttling process.

The exergy drop between states 1 and 2 is dissipated through viscous friction. The irreversibility of the process is represented graphically in Fig 4.6(b). As a throttling process has no useful output expressible in terms of exergy, its rational efficiency is zero.

Expansion processes in low temperature systems

In thermal plant such as refrigerators, heat pumps and gas liquefaction plants, expansion processes occur mostly at temperatures below ambient. The primary purpose of such expansion processes is the production of a cooling effect. Any shaft power which may be produced during the process can be regarded as a useful by-product. In exergy terms *the purpose of an expansion process in a low temperature system is to obtain the maximum possible increase in the thermal component of exergy* $\varepsilon^{\Delta T}$ *of the stream at the expense of a given reduction in the pressure component of exergy,* $\varepsilon^{\Delta P}$.

Figure 4.7 shows the two components of exergy on a ε–h diagram. According to the definition in (2.13), $\varepsilon_1^{\Delta T}$ corresponds to the work obtainable from the stream in a reversible isobaric process between the initial temperature T_1 and T_0. The process is shown in Fig 4.7 between states 1 and x. The other component, $\varepsilon^{\Delta P}$, from (2.14), corresponds to the work obtainable from the stream in a reversible isothermal process at T_0 between P_1 and P_0 (process x–0 on the diagram). The sum of the two components gives the physical exergy of the stream:

$$\varepsilon_1 = \varepsilon_1^{\Delta T} + \varepsilon_1^{\Delta P} \tag{4.24}$$

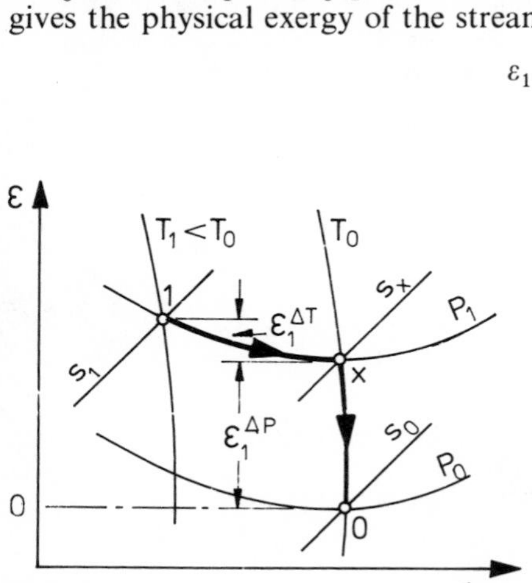

Fig 4.7 Representation of the thermal and the pressure components of physical exergy on a ε–h diagram.

Note from Fig 4.7 that the slopes of the isobars for $T < T_0$ are negative, while at temperatures above T_0 (Fig 4.2) the slopes of the isobars are positive. This difference is of fundamental importance in the consideration of the expansion processes in the two temperature ranges.

Rational efficiency for a cryoexpander

Figure 4.8(a) shows a control region for a general case of an expansion process in a sub-environmental temperature range. Because of the special function of an expander in a low temperature system, it may be appropriate to call it a *cryoexpander*. Expressing the exergy of the streams entering and leaving the control region in terms of their two components as shown in (4.24), the exergy balance for this case can be written, using specific quantities, as:

$$\varepsilon_1^{\Delta T} + \varepsilon_1^{\Delta P} + \varepsilon^{Q} = \varepsilon_2^{\Delta T} + \varepsilon_2^{\Delta P} + w_{ex} + i \tag{4.25}$$

Here, thermal exergy transfer, ε^{Q}, would take place only if there is heat transfer from the cooled object to the expanding fluid. In practice this occurs in, say, a reciprocating expander, where some cooling is achieved by heat transfer across the cylinder wall from

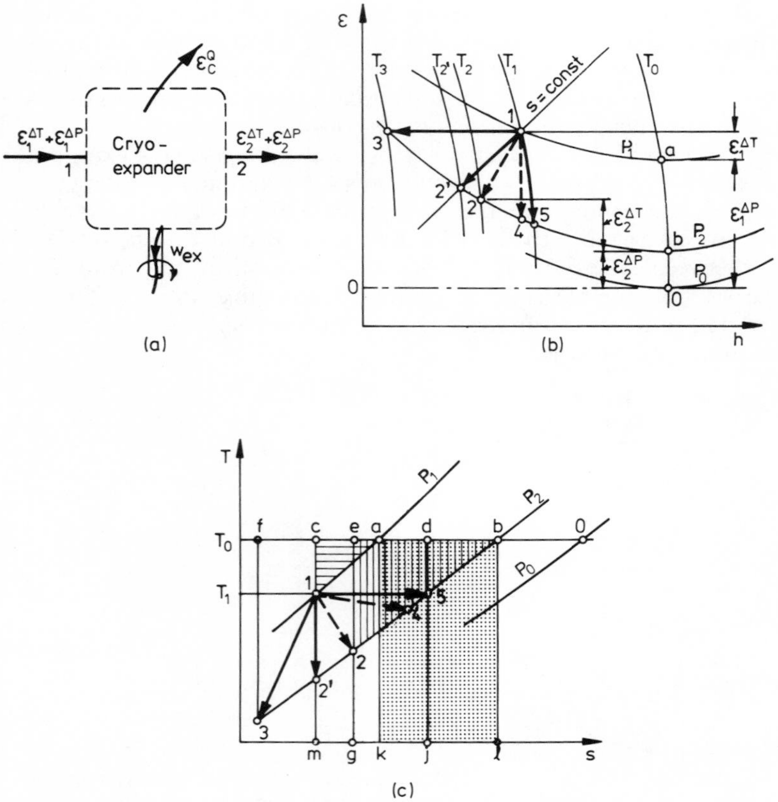

Fig 4.8 Expansion process at sub-environmental temperatures.

the working medium to a heat transfer medium. When this occurs at $T < T_0$, then for a positive heat transfer thermal exergy flux is negative, ie *from* the control region. Hence, using the notation $\varepsilon^Q = -\varepsilon_c^Q$ in (4.25) and rearranging:

$$(\varepsilon_2^{\Delta T} - \varepsilon_1^{\Delta T}) + \varepsilon_c^Q + w_{ex} = (\varepsilon_1^{\Delta P} - \varepsilon_2^{\Delta P}) - i \tag{4.26}$$

The output terms collected on the LHS of (4.26) are: increase in the thermal component of exergy of the stream; thermal exergy transfer to the cooled object; shaft work.

The input, the first term on the RHS of (4.26) is the drop in the pressure component of exergy of the stream. Knowing the input and output terms, the general expression for the rational efficiency is:

$$\psi = \frac{(\varepsilon_2^{\Delta T} - \varepsilon_1^{\Delta T}) + \varepsilon_c^Q + w_{ex}}{\varepsilon_1^{\Delta P} - \varepsilon_2^{\Delta P}} \tag{4.27}$$

From (4.26) and (4.27), an alternative expression is:

$$\psi = 1 - \frac{i}{\varepsilon_1^{\Delta P} - \varepsilon_2^{\Delta P}} \tag{4.28}$$

These expressions for ψ may be further rearranged into forms which may be more easily interpreted in some applications[2.2]. Before doing so, however, consider graphical representation of the expansion processes.

A number of possible expansion processes are plotted, in ε–h and T–s co-ordinates in Fig 4.8(b) and (c) respectively. Among these, 1–2 represents the general case of an irreversible expansion process. The two components of exergy, $\varepsilon^{\Delta T}$ and $\varepsilon^{\Delta P}$, have been marked for both states, 1 and 2, in Fig 4.8(b). In the T–s diagram (Fig 4.8(c)), the exergy components $\varepsilon_1^{\Delta T}$ and $\varepsilon_2^{\Delta T}$ are represented by the triangular areas 1ca1 and 2eb2 respectively. The exergy input, $(\varepsilon_1^{\Delta P} - \varepsilon_2^{\Delta P})$ is given for an ideal gas (satisfying the condition $h = f(T)$) by the dotted rectangular area ablka. For any, real gas*:

$$\varepsilon_1^{\Delta P} - \varepsilon_2^{\Delta P} = \varepsilon_a - \varepsilon_b$$

or using a briefer form of notation:

$$\Delta\varepsilon_{1-2}^{\Delta P} = \Delta\varepsilon_{a-b} \tag{4.29}$$

Also:

$$\begin{aligned} \varepsilon_2^{\Delta T} - \varepsilon_1^{\Delta T} &= \Delta\varepsilon_{2-b} - \Delta\varepsilon_{1-a} \\ &= \Delta\varepsilon_{a-b} - \Delta\varepsilon_{1-2} \end{aligned} \tag{4.30}$$

Substituting (4.29) and (4.30) in (4.27):

$$\psi = \frac{\Delta\varepsilon_{a-b} - \Delta\varepsilon_{1-2} + \varepsilon_c^Q + w_{ex}}{\Delta\varepsilon_{a-b}} \tag{4.31}$$

The expression for ψ given by (4.28) can be similarly modified to:

$$\psi = 1 - \frac{i}{\Delta\varepsilon_{a-b}} \tag{4.32}$$

---

* It is helpful at this stage to identify the quantities $\varepsilon_1^{\Delta P}$, $\varepsilon_2^{\Delta P}$, ε_a and ε_b on Fig 4.8(b). Note that the exergy component $\varepsilon^{\Delta T}$ is equal to zero at all the points of the isotherm $T_0 = \text{const}$ (hence $\varepsilon_a = \varepsilon_a^{\Delta P}$ and $\varepsilon_b = \varepsilon_b^{\Delta P}$) and that the value of $\varepsilon^{\Delta P}$ is constant for all points of a given isobar. Hence, $\varepsilon_1^{\Delta P} = \varepsilon_a$ and $\varepsilon_2^{\Delta P} = \varepsilon_b$.

The expressions for ψ given above are quite general and can be applied to different types of expansion processes occurring at sub-environmental temperatures.

For a working fluid which can be taken to be an ideal gas, the increase in the thermal component of exergy, $\varepsilon_2^{\Delta T}-\varepsilon_1^{\Delta T}$, of the stream corresponds to the area 25de2*. This also corresponds to the drop in the exergy of a stream during the isobaric process 2–5. Hence:

$$\varepsilon_2^{\Delta T}-\varepsilon_1^{\Delta T}=\varepsilon_2-\varepsilon_5=\Delta\varepsilon_{2-5} \tag{4.33}$$

From (4.30) and (4.33):

$$\Delta\varepsilon_{a-b}-\Delta\varepsilon_{1-2}=\Delta\varepsilon_{2-5}$$

or:

$$\Delta\varepsilon_{a-b}=\Delta\varepsilon_{1-5} \tag{4.34}$$

Hence, for an expansion process occurring in the ideal gas region:

$$\psi_{\mathrm{id.gas}}=\frac{\Delta\varepsilon_{2-5}+\varepsilon_c^Q+w_{ex}}{\Delta\varepsilon_{1-5}} \tag{4.35}$$

Here the function played by the isotherm $T_0=\mathrm{const}$ is replaced in (4.35) by the isotherm $T_1=\mathrm{const}$.

Adiabatic cryoexpanders

Rotodynamic expanders can usually be regarded as adiabatic so $q=0$ and hence $\varepsilon^Q=0$. In the ideal case of an isentropic expansion, characterised by the absence of any irreversible phenomena such as friction or heat transfer over finite temperature difference, the process irreversibility $i=0$ and hence $\psi=1$. This type of process is represented in Figs 4.8(b) and (c) by the process 1–2′. For this process expression (4.26) takes the form:

$$\varepsilon_1^{\Delta P}-\varepsilon_2^{\Delta P}=\varepsilon_2^{\Delta T}-\varepsilon_1^{\Delta T}+w_{ex}$$

which shows that in the ideal case of a reversible adiabatic process, all the exergy input is converted into an increase in the thermal component of exergy of the stream and shaft work.

A real, irreversible adiabatic expansion process may be represented by process 1–2 in Figs 4.8(b) and (c) and rational efficiency calculated from any of the expressions for ψ given above with $\varepsilon_c^Q=0$. The irreversibility of the process can be written from the Gouy–Stodola relation in the form:

$$i=T_0(s_2-s_1) \tag{4.36}$$

which is represented in Fig 4.8(c) by the rectangular area cegmc.

Expansion with heat transfer

In a cryoexpander, such as a reciprocating expander, with a relatively large area of contact between the working fluid and the enclosing walls for the mass throughput, significant heat transfer can take place between the working fluid and a heat transfer medium circulated through the enclosing wall. The heat exchange from the heat transfer medium to the working fluid accounts for that part of the useful cooling effect

* This follows from the fact that for an ideal gas the areas d5bd and c1ac are equal.

denoted by ε_c^Q in the exergy balance (4.25) and in the expressions for the rational efficiency (4.27). In the limiting case of an isothermal process (process 1–5 in Figs 4.8(b) and (c)) the increase in the thermal component, $(\varepsilon_2^{\Delta T} - \varepsilon_1^{\Delta T})$, of exergy of the stream becomes either insignificant or zero for an ideal gas. The cooling effect is thus accomplished principally through heat transfer during expansion and is expressed in terms of exergy by ε_c^Q. Isothermal expansion can be approximated by multi-stage expansion with inter-stage heating. This arrangement is similar to one used in power plants except that, since the heating occurs at $T < T_0$, the exergy flow has the opposite sign.

Constant-exergy expansion process

Process 1–3 on the ε–h diagram in Fig 4.8(b) reveals an interesting possibility of a process during which the exergy of the working fluid remains constant[2.2]. The attraction of such a process is that it achieves the greatest possible increase in the thermal component of exergy of the stream for a given pressure drop. The exit temperature, T_3, of the working fluid is, in this case, the lowest attainable. This process is also shown in Fig 4.8(c). The thermal component of exergy corresponding to the final state, 3, is represented by the triangle 3bf3. Among important characteristics of this process are reduction in the entropy of the working fluid and full reversibility (internal and external).

Clearly, under these conditions a reduction in the entropy of the fluid can only occur through a reversible heat transfer from the working fluid to the environment. (Note that the thermal exergy flux, ε^Q, associated with the heat transfer is zero.) A model of an ideal device by which such a process could be approximated is shown in Fig 4.9(a). The process could be realised in a series of repeated modules each consisting of a reversible, adiabatic expander and a heat pump which utilises the work output of the expander to cool its exhaust, rejecting the heat to the environment at T_0. The succession of

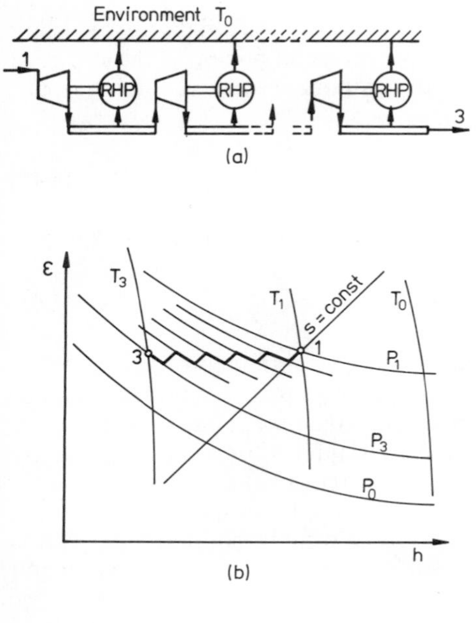

Fig 4.9 A model of an ideal constant-exergy expansion process.

alternating isentropic expansion and isobaric cooling which approximates to the constant-exergy process is shown in the ε–h co-ordinates in Fig 4.9(b). The rational efficiency of the ideal constant-exergy expansion process is one. This process is mainly of theoretical interest because of the mechanical complexity of the plant and, as far as is known, has not been put as yet to a practical use.

Adiabatic throttling process

An adiabatic throttling process, characterised by $h = \text{const}$, is shown in Figs 4.8(b) and (c) by the process 1–4. If a cooling effect is to be obtained in the throttling process, the isenthalpic Joule–Thomson coefficients $\mu_h \equiv (\partial T/\partial P)_h$ must be positive in the range of properties under consideration.

When used for pressure reduction or power output control, adiabatic throttling is purely dissipative. However, when applied at sub-environmental temperatures to produce a cooling effect, the throttling process can have a very high rational efficiency, particularly when the process takes place in the wet vapour region[2.2].

The irreversibility of the throttling process, 1–4, can be obtained either from the exergy balance or the Gouy–Stodola relation. In the first instance, with $\varepsilon_c^Q = 0$ and $w_{ex} = 0$:

$$i = \varepsilon_1 - \varepsilon_4 \tag{4.37}$$

In the second instance, with $q = 0$:

$$i = T_0(s_4 - s_1) \tag{4.38}$$

The rational efficiency in the form given by (4.27), with $\varepsilon_c^Q = 0$ and $w_{ex} = 0$, yields:

$$\psi = \frac{\varepsilon_4^{\Delta T} - \varepsilon_1^{\Delta T}}{\varepsilon_1^{\Delta P} - \varepsilon_4^{\Delta P}} \tag{4.39}$$

As this form of the rational efficiency reveals, the function of the adiabatic throttling process is to convert the available difference in the pressure component of the exergy of the stream (the input) to an increase in its thermal component of exergy (the output). Using (4.29) we can rewrite (4.39) as:

$$\psi = \frac{\varepsilon_4^{\Delta T} - \varepsilon_1^{\Delta T}}{\Delta\varepsilon_{a-b}} \tag{4.40}$$

The alternative expression for ψ can be written for the process 1–4 from (4.32) as:

$$\psi = 1 - \frac{i_{1-4}}{\Delta\varepsilon_{a-b}} \tag{4.41}$$

A throttling process in the wet vapour region is shown for air in T–s co-ordinates in Fig 4.10. Since the diagram is drawn to scale, the cross-hatched areas are proportional to the magnitudes of the exergy terms and irreversibility of the process, as shown. Hence, the expressions for the rational efficiencies given by (1) and (2) on the diagram are graphical interpretations of the corresponding algebraical forms, as expressed by (4.40) and (4.41) respectively. Determination of areas is, in general, laborious, but the second version involves rectangles of the same height. Therefore, what is required is the widths of the rectangles or entropy differences $(s_4 - s_1)$ and $(s_b - s_a)$. Consequently, this version

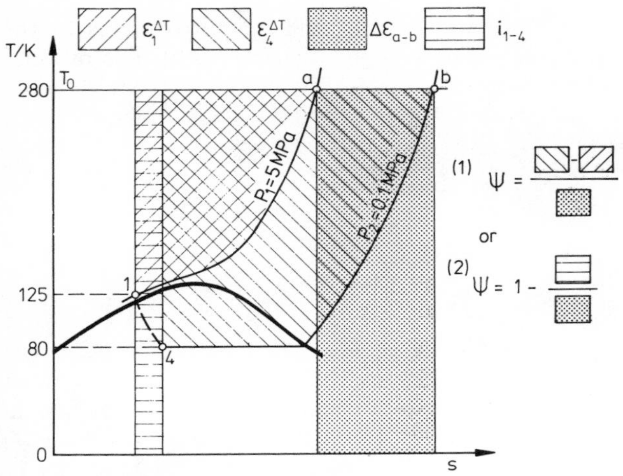

Fig 4.10 Graphical technique for evaluating the rational efficiency of a throttling process for air.

of the rational efficiency is:

$$\psi = 1 - \frac{s_4 - s_1}{s_b - s_a} \tag{4.42}$$

These entropy differences can be obtained directly from a chart or evaluated from tables of properties of the fluid. In the case illustrated in Fig 4.10, the rational efficiency of the adiabatic throttling process is 74%. As will be seen from the diagram, ψ could be further improved by lowering the initial temperature T_1 giving a lower value of i_{1-4}. Brodyanskii[2.2] calculated the rational efficiency of the throttling process for some selected inlet and outlet states for air, ammonia, Freon-12 and helium. These, under some conditions, exceeded 90% reaching 98% in some cases. Much lower rational efficiencies, 1%, were obtained for processes taking place fully in the gaseous phase, above the critical temperature.

This apparently paradoxical result, ie high values of rational efficiency for a process which is generally regarded as inherently irreversible, can be explained by reference to the physical nature of the process. As the temperature of the gas at the inlet to the valve decreases, the volume change involved in the throttling process between two particular pressures also decreases. Now, the irreversibility of this process may be said to be due to failure to utilise the expansion of the gas to produce (maximum) work. As this work, for given pressure limits, decreases with the decreasing volume change, so does the irreversibility associated with process.

4.2 Compression processes

In applications such as power plants, compressed air installations, gas pipelines and air liquefaction plants compression usually starts at approximately environmental temperature. In refrigeration plants and heat pumps compression processes start at temperatures lower than T_0 but usually end above it. In multiple cascade refrigeration plant, the compression processes of some of the cycles corresponding to lower temperatures may occur fully at $T < T_0$.

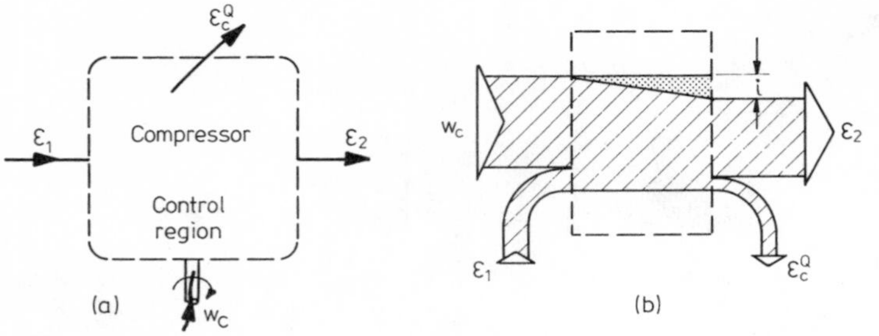

Fig 4.11 Compression process (a) control region (b) Grassmann diagram.

The exergy balance for a single stage compressor can be written, with reference to the control region shown in Fig 4.11(a) using specific quantities as:

$$\varepsilon_1 + w_c = \varepsilon_c^Q + \varepsilon_2 + i \qquad (4.43)$$

A pictorial representation of the exergy balance is shown on a Grassmann diagram in Fig 4.11(b).

w_c and ε_c^Q have been taken, as should be clear from Fig 4.11(a), as negative quantities. A negative value of ε_c^Q would correspond to 'heat loss' by the control region when $T > T_0$ and to 'heat gain' when $T < T_0$.

Compression processes taking place above the environmental temperature

Various simple compression processes are shown in Fig 4.12(a) in T–s co-ordinates for $T_1 > T_0$ and in Fig 4.12(b) in ε–h co-ordinates for $T_1 = T_0$. These processes can be put into two categories, those which occur in adiabatic, usually rotodynamic, compressors and those which occur with some degree of cooling, say in reciprocating, water-cooled compressors.

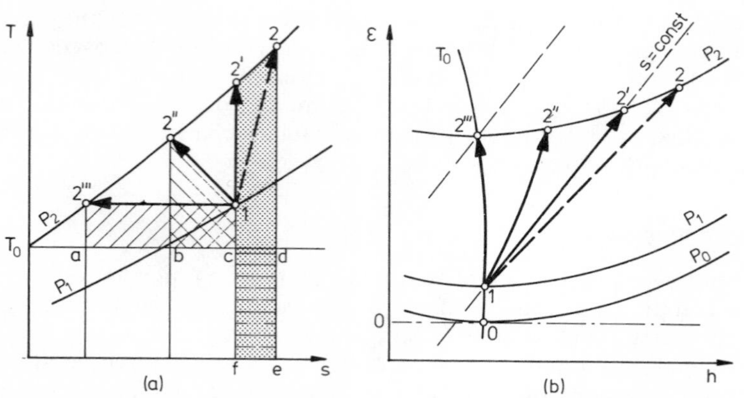

Fig 4.12 Compression process (a) in T–s co-ordinates for $T_1 > T_0$ (b) in ε–h co-ordinates for $T_1 = T_0$.

Adiabatic compressors

In the absence of heat transfer $\varepsilon_c^Q = 0$, and the exergy balance (4.43) may be put in the form:

$$w_c = \varepsilon_2 - \varepsilon_1 + i$$

In this expression, the increase in the exergy of the stream, $\varepsilon_2 - \varepsilon_1$, can be identified as the desired output and the shaft work, w_c, as the necessary input. Hence the two forms of the rational efficiency for the case of an adiabatic compressor can be expressed as:

$$\psi = \frac{\varepsilon_2 - \varepsilon_1}{w_c} \tag{4.44}$$

$$\psi = 1 - \frac{i}{w_c} \tag{4.45}$$

The first version of ψ may be compared with the expression for the isentropic efficiency of this process. With reference to Fig 4.12(a) and using the same method as for the adiabatic expansion process (see Section 4.1):

$$\psi = 1 - \frac{T_0(s_2 - s_1)}{h_2 - h_1} \tag{4.46}$$

$$\eta_s = 1 - \frac{h_2 - h_{2'}}{h_2 - h_1} \tag{4.47}$$

From (4.46), any reduction in ψ below unity is directly proportional to the numerator of the second term on the RHS, which from the Gouy–Stodola relation is identifiable as the process irreversibility, ie:

$$i = T_0(s_2 - s_1) \tag{4.48}$$

The irreversibility, i, of the adiabatic compression process is represented by the area cdefc in Fig 4.12(a).

A similar role to that played by i in the expression for ψ is played in the expression for η_s by $(h_2 - h_{2'})$, the frictional reheat of the compression process, r, ie:

$$r = h_2 - h_{2'} \tag{4.49}$$

This is represented by the dotted area 2′2ef2′ in Fig 4.12(a). From (4.46) and (4.47):

$$\psi = 1 - \frac{i}{r}(1 - \eta_s) \tag{4.50}$$

This relationship is plotted in Fig 4.13 for a range of values of r/i, which (Section 4.1) may be looked upon as the dimensionless mean exit temperature (since $r/i \simeq (T_2 + T_2)/(2T_0)$). From the curves, note that for any value of η_s, ψ increases with r/i. Thus very efficient compressors (in terms of η_s) are needed at sub-environmental temperatures, if the plant overall efficiency is not to be unduly affected by low compressor ψ.

Non-adiabatic compressors

A general frictionless compression process is called a polytropic process and can be described by the relation $Pv^n = \text{const}$, where n is a constant. In the special case when

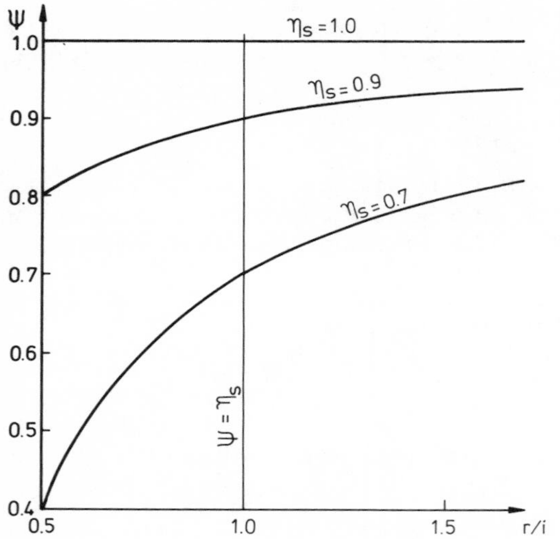

Fig 4.13 Relationship between isentropic efficiency and rational efficiency for a compression process.

$n=\gamma$ and the working fluid is a perfect gas, the general case corresponds to an isentropic process, and when $n=1$ it defines the isothermal process. Processes 1–2′, 1–2″ and 1–2‴ in Fig 4.12 represent an isentropic process, a general polytropic process and an isothermal process respectively. When $T>T_0$, the heat transferred from the working fluid corresponds to some value of ε_c^Q. This thermal exergy may be used for, say, water heating or simply dissipated in the environment. For processes 1–2″ and 1–2‴ their values of ε_c^Q are represented by the cross-hatched areas 12″bc1 and 12‴ac1 respectively. If ε_c^Q in these processes is dissipated, the process suffers an additional irreversibility of external type equal to ε_c^Q, which generally should be included in the value of i, effectively including the region where this dissipation occurs in the control region of the compressor. The special case for which there is no irreversibility associated with 'heat loss', is one when the process is isothermal and takes place at $T=T_0$, as shown by process 1–2‴ in Fig 4.12(b). This process is fully reversible, so from (4.43), $\varepsilon_{2'''}-\varepsilon_1=(w_c)_{T_0}$. Since the isentropic process is also fully reversible, $\varepsilon_{2'}-\varepsilon_1=(w_c)_S$. Thus the work input in these two reversible processes can be read from the $\varepsilon-h$ diagram. Comparing these two processes in Fig 4.12(b) shows that *when $T_1 \geqslant T_0$, the work done between given pressure limits in the frictionless, isentropic process is greater than that in the frictionless isothermal process, ie* $(w_c)_S>(w_c)_T$. The difference between $(w_c)_S$ and $(w_c)_T$ is equal to the loss of exergy $(\varepsilon_{2'}-\varepsilon_{2'''})$ of the gas during an isobaric process at P_2 between states 2′ and 2‴, which is equal to the change in the *thermal* component of exergy during the process $1-2'$. In most applications, when $T_1 \geqslant T_0$ it is the pressure component of exergy of the stream which is being used. The thermal component of exergy of the stream in pneumatic installations and gas pipelines is either dissipated in an aftercooler following the compression process or is lost subsequently by natural 'heat loss' to the environment. In gas turbine plant with heat exchange, the plant performance is improved by arranging the compression process to take place in two or three stages with intercooling. The ideal version of this arrangement is the Ericsson cycle in which the multi-staging with intercooling is taken to its logical conclusion and the compression process becomes isothermal. Thus the isothermal compression process in most areas of application when $T_1 \geqslant T_0$ can be regarded as the ideal model and *isothermal efficiency* is commonly used as a criterion of performance for non-adiabatic compressors. The usual definition is:

$$\eta_{iso} = \frac{\text{isothermal work}}{\text{actual work}}$$

where the 'actual work' may be interpreted either as 'actual indicated work' or 'actual shaft work'. Taking the latter interpretation, η_{iso} can be written for an ideal gas as:

$$\eta_{iso} = \frac{RT_1 \ln(P_2/P_1)}{w_c} \tag{4.51}$$

This traditional criterion of performance will be now compared with the rational efficiency for a compression process in which the gas temperature, although not necessarily constant throughout the process, is the same at inlet and outlet. Accordingly, the overall change in exergy can be expressed as:

$$\varepsilon_2 - \varepsilon_1 = \varepsilon_2^{\Delta P} - \varepsilon_1^{\Delta P} \tag{4.52}$$

Further, assume that the thermal energy rejected by the working fluid is not put to any useful end and is dissipated in the environment, ie $\varepsilon_c^Q = 0$. With these assumptions, the exergy balance (4.43) takes the form:

$$\underset{\text{output}}{\varepsilon_2^{\Delta P} - \varepsilon_1^{\Delta P}} = \underset{\text{input}}{w_c} - i$$

This gives the rational efficiency for this process as:

$$\psi = \frac{\varepsilon_2^{\Delta P} - \varepsilon_1^{\Delta P}}{w_c} \tag{4.53}$$

In the special case when the working fluid is an ideal gas, the change in the pressure component of exergy takes the form:

$$\varepsilon_2^{\Delta P} - \varepsilon_1^{\Delta P} = RT_0 \ln(P_2/P_1) \tag{4.54}$$

Substituting (4.54) in (4.53):

$$\psi = \frac{RT_0 \ln(P_2/P_1)}{w_c} \tag{4.55}$$

Once the actual work of compression, w_c, is determined experimentally or by calculation, the rational efficiency for a non-adiabatic compressor can be calculated from (4.55). From (4.51) and (4.55) the relationship between η_{iso} and ψ is:

$$\psi = \frac{T_0}{T_1} \eta_{iso} \tag{4.56}$$

It is clear from this that for η_{iso} an internally reversible, ie frictionless, isothermal process is the ideal model. For ψ, the ideal model is a frictionless isothermal process *with external reversibility*. Hence, an internally reversible isothermal process which takes place at $T > T_0$ will, if ε_c^Q is not utilised, have a rational efficiency of less than unity.

ψ given by (4.55) is applicable both to single stage non-adiabatic compression processes and to multi-stage compression processes with intercooling and aftercooling. The latter type of process is shown for two compression stages in Fig 4.14, in which

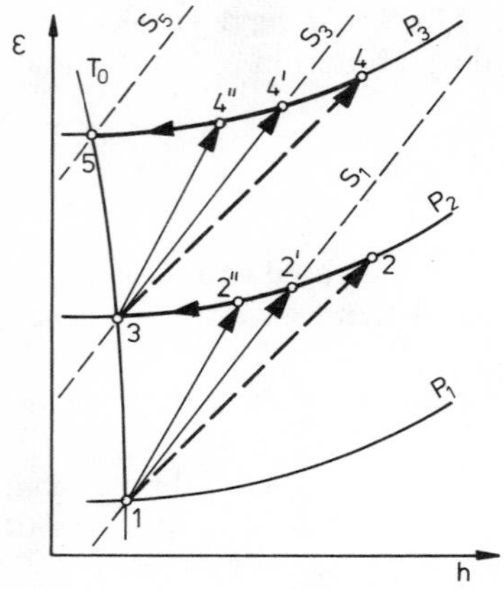

Fig 4.14 Two-stage compression with intercooling and aftercooling.

three possible types of processes have been indicated:

(i) Irreversible adiabatic process 12345.
(ii) Frictionless adiabatic (isentropic) process 12′34′5.
(iii) Frictionless polytropic, $1<n<\gamma$, process 12″34″5.

In cases (i) and (ii), the adiabatic compression processes, the work done in each stage can be read from the increments of enthalpy on the abscissa axis. Also, the heat transferred in the intercooler and the aftercooler can be taken from the h-axis, for all three types of processes, and the corresponding thermal exergy changes read from the ε co-ordinate. These exergy changes are in fact equal to the irreversibilities taking place in these heat exchangers. If the viscous pressure losses are marked on the diagram, the irreversibility due to this effect can be included in the value of irreversibilities read from the ε co-ordinate. Note that intercooling and aftercooling at temperatures close to T_0, where the slope of the P=const curves is almost horizontal, involves relatively low irreversibilities.

Generally, more relevant and useful information is obtained from the analysis of compression processes by the exergy method than by applying the First Law only. As Fig 4.14 shows, the change in enthalpy of the gas between state 1 and 5 is about zero and gives no indication about the change in its work potential.

Minimum work of compression

As noted above, the minimum work of compression between given pressure limits corresponds to a frictionless isothermal process at temperature T_0. The case of a more general non-isothermal frictionless process, but one for which $T_1=T_2$, is illustrated in T–s co-ordinates in Fig 4.15(a). Since in this case $T \neq T_0$ the question of external irreversibility arises. This is eliminated by using the thermal energy rejected during the process in the RHE (Fig 4.15(b)) to produce work, $w=\varepsilon_c^Q$, which is used to reduce the necessary work input to the compressor. Under conditions of full reversibility, internal and external, the work input to the plant is equal to the ideal minimum work. Applying the exergy balance (4.43) to the control region comprising the compressor and the RHE, with $i=0$ and $\varepsilon_c^Q=\varepsilon_0^Q=0$:

$$\varepsilon_2-\varepsilon_1=w_c=[w_c]_{MIN}$$

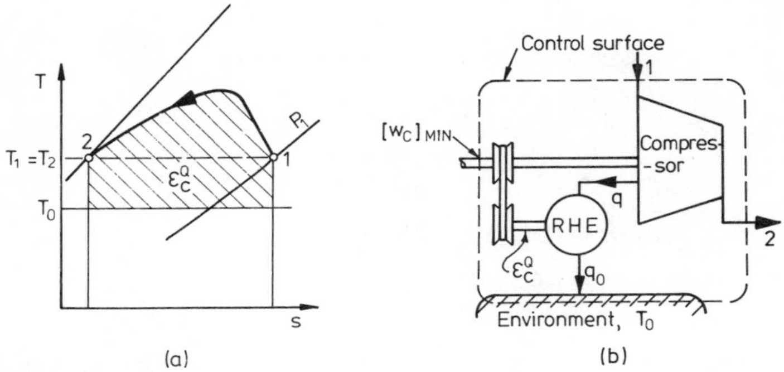

Fig 4.15 Fully reversible non-adiabatic compression process, $T_1 = T_2$.

Now, since, $T_2 = T_1$, (4.52) and, if the gas is ideal, expression (4.54) can be used, giving:

$$[w_c]_{MIN} = \varepsilon_2^{\Delta P} - \varepsilon_1^{\Delta P}$$
$$= RT_0 \ln(P_2/P_1) \quad (4.57)$$

Multi-stage adiabatic compression process without intercooling or aftercooling

This type of process, which involves many stages in adiabatic axial-flow compressors, is shown in *T–s* co-ordinates in Fig 4.16 for a 3-stage compression process. The frictional reheat, *r*, for each stage is represented by a dotted area in the Δs interval corresponding to the particular stage. Also, the irreversibility, *i*, of each stage, is shown as a cross-hatched area. Note that the higher the stage exit temperature, the greater the value of the frictional reheat for that stage, for a given isentropic efficiency. Because of the variability of the effect of frictional reheat on stage performance, the relationship between isentropic efficiency of the stage and the overall efficiency of the compressor is rather complex.

To determine the relationship between the rational efficiencies of the compressor stages and the rational efficiency of the whole compressor, define the latter as:

$$\psi_{ov} = \frac{\Delta\varepsilon_{ov}}{w_c}$$

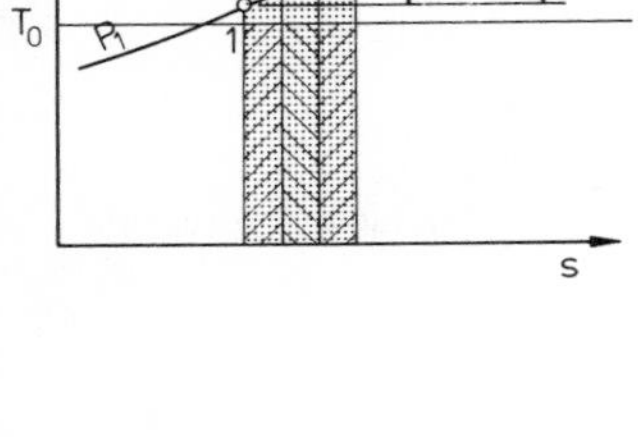

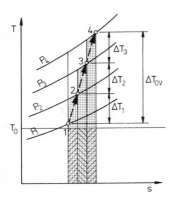

Fig 4.16 Multi-stage, adiabatic compression process.

where:

$$\Delta\varepsilon_{ov} = \sum \Delta\varepsilon_i$$

and:

$$w_c = \sum \Delta h_i$$

Hence ψ_{ov} is:

$$\psi_{ov} = \frac{\Delta\varepsilon_1}{\Delta h_1}\frac{\Delta h_1}{\sum \Delta h_i} + \frac{\Delta\varepsilon_2}{\Delta h_2}\frac{\Delta h_2}{\sum \Delta h_i} + \frac{\Delta\varepsilon_3}{\Delta h_3}\frac{\Delta h_3}{\sum \Delta h_i} + \cdots$$

Defining the rational efficiency of the *i*-th stage as:

$$\psi_i = \frac{\Delta\varepsilon_i}{\Delta h_i}$$

the overall rational efficiency takes the form:

$$\psi_{ov} = \psi_1 \frac{\Delta h_1}{\sum \Delta h_i} + \psi_2 \frac{\Delta h_2}{\sum \Delta h_i} + \psi_3 \frac{\Delta h_3}{\sum \Delta h_i} + \cdots \tag{4.58}$$

From (4.58), the overall rational efficiency of the compressor can be expressed as a function of the rational efficiencies of its different stages. The greater the enthalpy rise in a given stage, the greater the effect of the rational efficiency of that stage on the overall rational efficiency.

Compression processes at sub-environmental temperatures

At sub-environmental temperatures an isothermal process, shown as 1–2″ in Fig 4.17, is difficult to arrange since the thermal energy rejected at T_1 would have to be delivered to the environment at T_0 by a refrigerator. However, in most cases of compression at sub-environmental temperatures, it is necessary to increase not only the pressure but also the temperature of the working fluid. Consequently, it is the adiabatic rather than the isothermal compression processes which are in general use. A reversible adiabatic process, 1–2′, and an irreversible one, 1–2, both taking place entirely in the sub-environmental temperature range, are shown in Fig 4.17. Since, under these conditions, the value of frictional reheat (shown as a dotted area) is less than that of the process irreversibility (shown shaded), the rational efficiency has a lower value than the

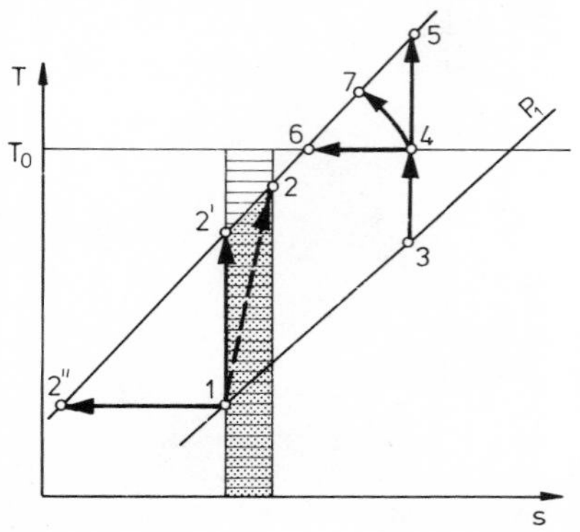

Fig 4.17 Compression process at sub-environmental temperatures.

isentropic efficiency for the process. This is illustrated in Fig 4.13 for $(r/i)<1$. The expressions for the rational efficiency of adiabatic compression processes (4.44) and (4.45) formulated for the temperature range above the environmental temperature also apply at lower temperatures.

In most simple refrigeration plants and heat pumps, compression starts at $T<T_0$ and ends at $T>T_0$ (isentropic process 3–5 in Fig 4.17). As was indicated earlier, in compression processes which take place at $T>T_0$ the work expended on increasing the thermal component of exergy is usually wasted. Therefore, the ideal process, under these conditions consists of an isentropic process between states 3 and 4, followed by an isothermal process at $T=T_0$ between states 4 and 6. In practice, the use of a water-cooled reciprocating compressor may result in the second stage of the compression process conforming to the polytropic law as shown by process 4–7. This may be regarded as a practical approximation to the ideal isothermal compression process.

4.3 Heat transfer processes

Heat transfer processes may be usefully divided into two groups.

Group 1: In this group of heat transfer processes the thermal component of exergy of one stream increases at the expense of a reduction in the thermal component of exergy of another stream. Thus in this case a heat exchanger transfers thermal exergy between two streams through conductive, convective or radiant heat transfer. Since there is a useful output expressible in terms of exergy, a rational efficiency can be formulated.

Group 2: This group of heat transfer processes is characterised by transfer of thermal energy to or from the environment. In some heat transfer processes such as those found in cooling towers and refrigerator condensers, the stream, at $T>T_0$ transfers heat *to* the environment. In heat pump evaporators, however, the stream is at a temperature lower than T_0 and heat transfer is *from* the environment. In either case the exergy of the stream decreases as a result of the heat transfer, and since the exergy of the environment can never increase, heat transfer processes in this group lead only to dissipation of exergy. As there is no output expressible in terms of exergy, there is no rational efficiency for a heat transfer process in this group. However, the magnitude of the irreversibility rate calculated for the process can be used to assess the process, particularly when it is compared with the value of intrinsic irreversibility (Section 3.3) calculated for the heat exchanger.

Forms of irreversibility in heat transfer processes

Loss of efficiency of heat exchangers is caused by different forms of irreversibility due to:

(i) Heat transfer over a finite temperature difference.
(ii) Pressure losses.
(iii) Thermal interaction with the environment.
(iv) Streamwise conduction in the walls of the heat exchanger.

Irreversibility due to heat transfer over a finite temperature difference

This is the principal form of irreversibility in heat exchangers and is due to the reduction in the quality of thermal energy as it is transferred from a higher to a lower temperature. In all real heat exchangers a finite temperature difference between the heat transfer media is required for a finite heat transfer rate so this form of irreversibility is unavoidable. In some processes, eg heat transfer between liquid streams and between evaporating and condensing fluids, this form of irreversibility may account for most of the irreversibility rate of the process.

From the Gouy–Stodola relation the irreversibility rate due to heat transfer over a finite temperature difference, $\dot{I}^{\Delta T}$, can be written as:

$$\begin{aligned}\dot{I}^{\Delta T} &= T_0 \dot{\Pi}^{\Delta T} \\ &= T_0[(\dot{S}_{2b} - \dot{S}_{1b}) + (\dot{S}_{2a} - \dot{S}_{1a})]\end{aligned} \tag{4.59}$$

Figure 4.18 shows in T–$\dot{S}$ co-ordinates the heat transfer processes for a counter-flow and a parallel-flow heat exchanger when the only form of irreversibility is $\dot{I}^{\Delta T}$. Under

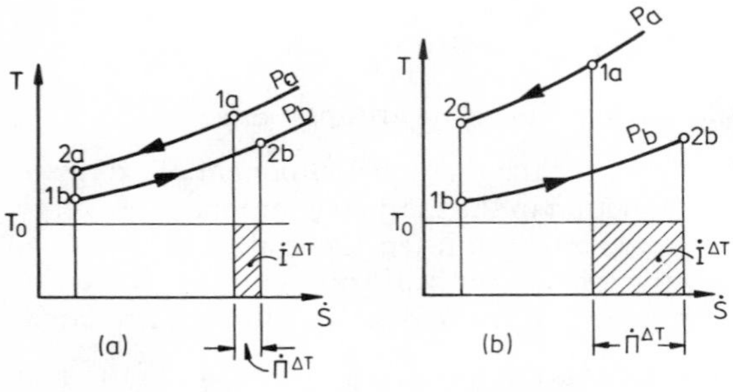

Fig 4.18 Isobaric heat transfer (a) counter-flow and (b) parallel-flow heat exchanger.

these circumstances the heat transfer for either stream is represented by the area under the process line, and hence the entropy production rate for the two streams can be constructed for the two cases as shown in the diagrams; as follows from (4.59), the cross-hatched areas represent $\dot{I}^{\Delta T}$. In the parallel-flow heat exchanger $T_{2a} > T_{2b}$, so for the same minimum temperature difference between the streams, the counter-flow heat exchanger incurs a smaller irreversibility. Further, in the latter case, when the two streams have the same heat capacities ($\dot{m}_a c_{Pa} = \dot{m}_b c_{Pb}$), the temperature difference, ΔT, between them is the same throughout the heat exchanger and, as $\Delta T \rightarrow 0$, the irreversibility rate tends to zero, ie $\dot{I}^{\Delta T} \rightarrow 0$.

Under conditions of isobaric heating and cooling, the changes in the exergy of the streams are equal to the changes in the thermal components of exergy, ie:

$$\dot{E}_{1a} - \dot{E}_{2a} = \Delta\dot{E}_a^{\Delta T}$$

and:

$$\dot{E}_{2b} - \dot{E}_{1b} = \Delta\dot{E}_b^{\Delta T}$$

Hence, with $\dot{W}_x = 0$ and $\dot{E}^Q = 0$, the exergy balance for a heat exchanger which suffers only irreversibility due to heat transfer over a finite temperature difference, can be written as:

$$\Delta\dot{E}_a^{\Delta T} - \Delta\dot{E}_b^{\Delta T} = \dot{I}^{\Delta T} \tag{4.60}$$

(4.60) provides an alternative expression for $\dot{I}^{\Delta T}$ in terms of the changes in the thermal component of exergy of the streams.

The above analysis is equally applicable to heat exchangers operating below T_0, eg

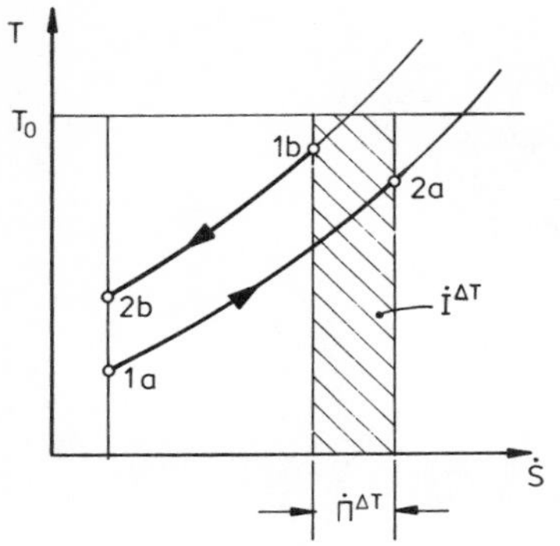

Fig 4.19 Isobaric heat transfer at sub-environmental temperatures.

the counter-flow heat exchanger in Fig 4.19. As heat transfer takes place here from stream b to stream a, the transfer of exergy takes place from stream a to stream b. It follows, therefore, that $\Delta\dot{E}_a^{\Delta T} > \Delta\dot{E}_b^{\Delta T}$ and both the expressions for $\dot{I}^{\Delta T}$, (4.59) and (4.60), are also applicable to heat transfer at $T < T_0$.

Heat transfer with pressure losses

In heat exchangers involving gaseous streams, pressure losses can contribute significantly to the overall irreversibility rate and cannot be nglected in an analysis of the process. For heat transfer between streams a and b without stray heat transfer losses, ie $\dot{Q}_0 = 0$ and, with $\dot{W}_x = 0$, the exergy balance can be written as:

$$(\dot{E}_{1a} - \dot{E}_{2a}) - (\dot{E}_{2b} - \dot{E}_{1b}) = \dot{I} \tag{4.61}$$

This type of heat transfer process is represented in the $\dot{E}$–$\dot{H}$ co-ordinates in Fig 4.20. As shown, the exergy change for either stream can be resolved into the thermal component $\Delta\dot{E}^{\Delta T}$, and the pressure component, $\Delta\dot{E}^{\Delta P}$.

Rewriting the exergy changes in (4.61) in terms of the two components:

$$(\Delta\dot{E}_a^{\Delta T} + \Delta\dot{E}_a^{\Delta P}) - (\Delta\dot{E}_b^{\Delta T} - \Delta\dot{E}_b^{\Delta P}) = \dot{I} \tag{4.62}$$

Note, both from the diagram and from (4.62), that the change in exergy of the high temperature stream, stream a, is increased as a result of the frictional pressure drop but

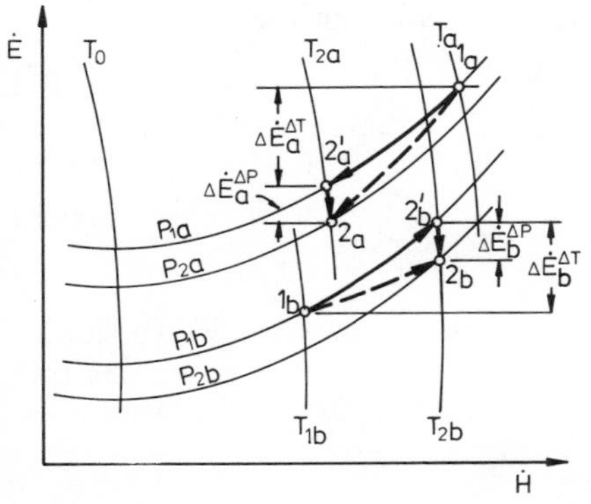

Fig 4.20 Heat transfer process in $\dot{E}$–$\dot{H}$ co-ordinates.

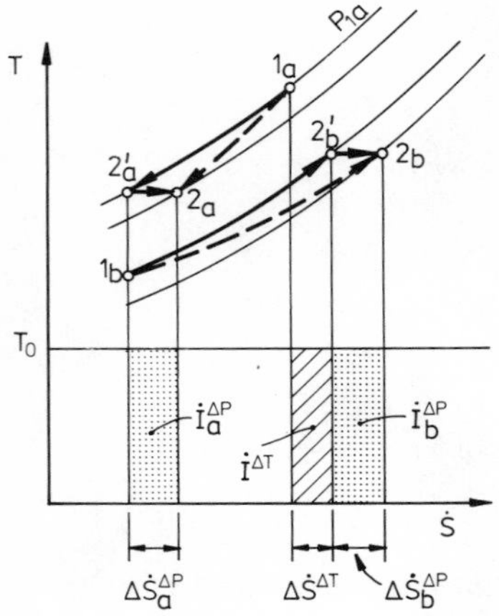

Fig 4.21 Heat transfer process in T–$\dot{S}$ co-ordinates.

in the lower temperature stream, stream b, the pressure drop results in a lower change in exergy than under isobaric conditions. Rearranging the terms in (4.62):

$$(\Delta\dot{E}_a^{\Delta T} - \Delta\dot{E}_b^{\Delta T}) + (\Delta\dot{E}_a^{\Delta P} + \Delta\dot{E}_b^{\Delta P}) = \dot{I} \tag{4.63}$$

The first half of the LHS of (4.63) may be said* to correspond to that part of the total irreversibility rate which arises from the transfer of thermal exergy, and will be denoted as $\dot{I}^{\Delta T}$, as shown in (4.60). The second half of (4.63) is that part due to pressure drops arising from viscous friction and will be denoted by $\dot{I}^{\Delta P}$, ie:

$$(\Delta\dot{E}_a^{\Delta P} + \Delta\dot{E}_b^{\Delta P}) = \dot{I}^{\Delta P} \tag{4.64}$$

The apportioning of the total irreversibility of a heat transfer process to the two components $\dot{I}^{\Delta T}$ and $\dot{I}^{\Delta P}$ can also be carried out using the Gouy–Stodola relation. Referring to Fig 4.21 and assuming $\dot{Q}_0 = 0$:

$$\dot{I} = T_0[(\dot{S}_{2b} - \dot{S}_{1b}) - (\dot{S}_{1a} - \dot{S}_{2a})] \tag{4.65}$$

By replacing the actual processes with reversible isothermal and isobaric processes the entropy changes in (4.65) can be rewritten:

$$\dot{S}_{1a} - \dot{S}_{2a} = (\dot{S}_{1a} - \dot{S}_{2'a}) - (\dot{S}_{2a} - \dot{S}_{2'a}) \tag{4.66a}$$

$$\dot{S}_{2b} - \dot{S}_{1b} = (\dot{S}_{2'b} - \dot{S}_{1b}) + (\dot{S}_{2b} - \dot{S}_{2'b}) \tag{4.66b}$$

Substituting (4.66a) and (4.66b) in (4.65) gives the last relation in the form:

$$\dot{I} = T_0[\Delta\dot{S}^{\Delta T} + \Delta\dot{S}_a^{\Delta P} + \Delta\dot{S}_b^{\Delta P}] \tag{4.67}$$

where:

$$\Delta\dot{S}^{\Delta T} = (\dot{S}_{2'b} - \dot{S}_{1b}) - (\dot{S}_{1a} - \dot{S}_{2'a}) \tag{4.68a}$$

$$\Delta\dot{S}_a^{\Delta P} = \dot{S}_{2a} - \dot{S}_{2'a} \tag{4.68b}$$

$$\Delta\dot{S}_b^{\Delta P} = \dot{S}_{2b} - \dot{S}_{2'b} \tag{4.68c}$$

In accordance with (4.59), $\Delta\dot{S}^{\Delta T}$, as given by (4.68a) corresponds to the irreversibility associated with heat transfer over a finite temperature difference. The remaining two entropy changes, $\Delta\dot{S}_a^{\Delta P}$, and $\Delta\dot{S}_b^{\Delta P}$, are associated with irreversibility due to pressure

* See below for a discussion of the limitations of this assumption.

losses in streams a and b respectively. Both the entropy changes and the related components of the irreversibility rate of the process are represented graphically in Fig 4.21. Hence:

$$\dot{I} = \dot{I}^{\Delta T} + \dot{I}_a^{\Delta P} + \dot{I}_b^{\Delta P} \tag{4.69}$$

The division of the total irreversibility rate into the two components $\dot{I}^{\Delta T}$ and $\dot{I}^{\Delta P}$ has been carried out above on the assumption that the effects of the two forms of irreversibilities can be separated. In a real gas, changes in pressure due to viscous dissipation result in temperature changes which in turn affect the magnitude of the thermal component of exergy. However, this effect is of only secondary importance in gaseous streams at moderate densities. Since, for an ideal gas, enthalpy is a function of temperature only, the two irreversibilities can be separated completely; the definitions of $\dot{I}^{\Delta T}$ and $\dot{I}^{\Delta P}$ derived from the exergy balance, (4.60) and (4.64), and from the Gouy–Stodola, (4.67) and (4.69), are in complete agreement. Assuming the gases to be ideal, either definition will give the following expressions for the two types of components of irreversibility rate:

$$\dot{I}^{\Delta T} = T_0 \left[\dot{m}_b \int_{T_{1b}}^{T_{2b}} c_{Pb} \frac{dT}{T} - \dot{m}_a \int_{T_{2a}}^{T_{1a}} c_{Pa} \frac{dT}{T} \right] \tag{4.70a}$$

$$\dot{I}_a^{\Delta P} = \dot{m}_a \, R_a \, T_0 \ln(P_{1a}/P_{2a}) \tag{4.70b}$$

$$\dot{I}_b^{\Delta P} = \dot{m}_b \, R_b \, T_0 \ln(P_{1b}/P_{2b}) \tag{4.70c}$$

From a comparison of (4.70b) and (4.70c) with (4.57), the terms $\dot{I}_a^{\Delta P}$ and $\dot{I}_b^{\Delta P}$ can be considered the minimum power necessary to reverse the effect of the pressure losses.

The two components of irreversibility $\dot{I}^{\Delta T}$ and $\dot{I}^{\Delta P}$ are not independent, since an increase in stream velocities will lead to an increase in pressure losses and heat transfer coefficients. Therefore, increasing stream velocities increases $\dot{I}^{\Delta P}$ and reduces $\dot{I}^{\Delta T}$, although in general not by the same amount. The two increments are the same in magnitude but of opposite sign at the point of minimum irreversibility. A technique for optimising the geometrical parameters of a heat exchanger in which the irreversibility rate is minimised, will be described in Chapter 6.

The above analysis can be easily extended to cover heat transfer processes occurring at sub-environmental temperatures.

Irreversibility due to thermal interaction with the environment

The control region containing a heat exchanger shown in Fig 4.22(a) is divided into two parts. Part A contains the heat exchanger itself while part B corresponds to a region of temperature gradients between the heat exchanger surface at an average temperature $\bar{T}_s$ and the environment at T_0. The irreversibility rate taking place in part B is equal to the thermal exergy flux due to the 'heat loss' at the surface temperature $\bar{T}_s$. Denoting this component of irreversibility rate as $\dot{I}_s^Q$:

$$\dot{I}_s^Q = \dot{E}_s^Q = \dot{Q}_s \frac{\bar{T}_s - T_0}{\bar{T}_s} \tag{4.71}$$

Clearly, the accuracy of the calculated value of the component $\dot{I}_s^Q$ will depend on the accuracy of the estimate of $\dot{Q}_s$ and $\bar{T}_s$. Should detailed information regarding the distribution of the heat exchanger surface temperature T_s and that of the surface heat

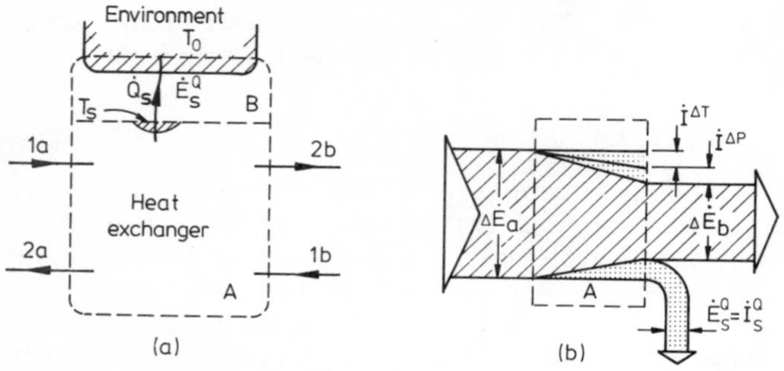

Fig 4.22 Heat exchange with 'heat losses', (a) control region, (b) Grassmann diagram.

flux $\dot{Q}_A$ be available the following expression based on Eq (2.3) can be used:

$$\dot{I}^Q_s = \dot{E}^Q_s = \int_A \left(\frac{T_s - T_0}{T_s}\right)\dot{Q}_A \mathrm{d}A \tag{4.72}$$

The exergy balance and the three components of the irreversibility rate are shown graphically in the form of a Grassmann diagram in Fig 4.22(b).

For a heat exchanger operating in the sub-environmental temperature range, the irreversibility rate $\dot{I}^Q_s$ is due to heat transfer from the environment to the heat exchanger. The direction of $\dot{E}^Q_s$ will be the same as for a heat exchanger operating at $T > T_0$, since as can be seen from (4.71) a change in the sign of $\dot{Q}_s$ is accompanied by a change in the sign in the difference $(T_s - T_0)$. Thus the above analysis is equally applicable to heat exchangers operating at sub-environmental temperatures.

Irreversibility due to streamwise conduction in the walls of the heat exchanger

In all heat exchangers in which the fluid temperature changes in the direction of flow, a temperature gradient is established in the heat exchanger walls parallel to the flow. This leads to heat conduction in the wall, at right angles to the main heat transfer direction, which, as in any process involving heat conduction over a finite temperature gradient, results in an additional irreversibility. The evaluation of this type of irreversibility may be quite complicated. However, an approximate assessment of the effect of streamwise conduction on the effective temperature difference between the fluid streams can be obtained by considering[2.2] a streamwise section of the wall separating the two fluids shown in Fig 4.23. As will be seen, the transverse temperature gradient, $\partial T/\partial y$ has been combined vectorially with the streamwise temperature gradient, $\partial T/\partial x$, to give the resultant, $\partial T/\partial r$, which is inclined at an angle α to the first vector. Assuming linear variation in temperature in both x and y directions, then the isothermal lines will be straight and at a right angle to the vector $\partial T/\partial r$ as shown in the figure. The resultant heat conduction will take place in a direction normal to the isothermal lines. The effect of streamwise conduction is to increase the effective temperature gradient and the length of the path of heat conduction (from a–b to c–b). As a result of this heat will be transferred, in effect, over a larger temperature difference than would have been the case in the absence of streamwise conduction. Clearly, the above discussion concerns only that part of the overall temperature difference between the streams which is associated with heat conduction through the solid wall. The temperature drops across the thermal boundary layers and any contribution due to fouling of the heat transfer surfaces must be dealt with separately.

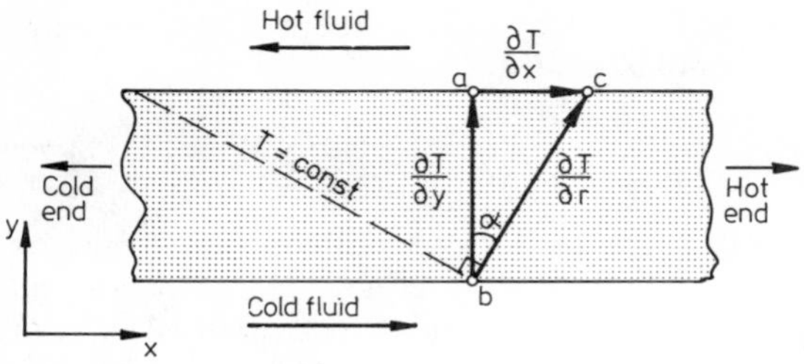

Fig 4.23 Temperature gradients in a streamwise cross-section of the wall between heat transfer media.

The contribution to heat exchanger irreversibility due to streamwise conduction can be obtained by using the *effective temperature drop* based on $\partial T/\partial r$ and the increased path of heat conduction c–b in the calculation of $\dot{I}^{\Delta T}$. This contribution is generally more significant in short heat exchangers than in long ones, and at lower temperatures, eg in cryogenic applications.

Rational efficiency of a simple surface heat exchanger

The rational efficiency of a surface heat exchanger consisting of two streams can be formulated using the exergy balance (4.61) which is applicable to the control region shown in Fig 4.24. Assume that *exergy is being transferred from stream* a *to stream* b. Since the function of a non-dissipative heat exchanger is to transfer exergy between the streams (with a minimum of irreversibility), the heat exchanger input can be identified in (4.61) as the reduction in the exergy of stream a and the output as the increase in the exergy of stream b. Thus:

$$\text{Input} = \dot{E}_{1a} - \dot{E}_{2a} \tag{4.73}$$

$$\text{Output} = \dot{E}_{2b} - \dot{E}_{1b} \tag{4.74}$$

This leads to the following alternative forms of rational efficiency:

$$\psi = \frac{\dot{E}_{2b} - \dot{E}_{1b}}{\dot{E}_{1a} - \dot{E}_{2a}} \tag{4.75}$$

$$\psi = 1 - \frac{\dot{I}}{\dot{E}_{1a} - \dot{E}_{2a}} \tag{4.76}$$

The form of rational efficiency given by (4.75) was originally proposed by Bruges[4.2] who called it 'energy transfer ratio'.

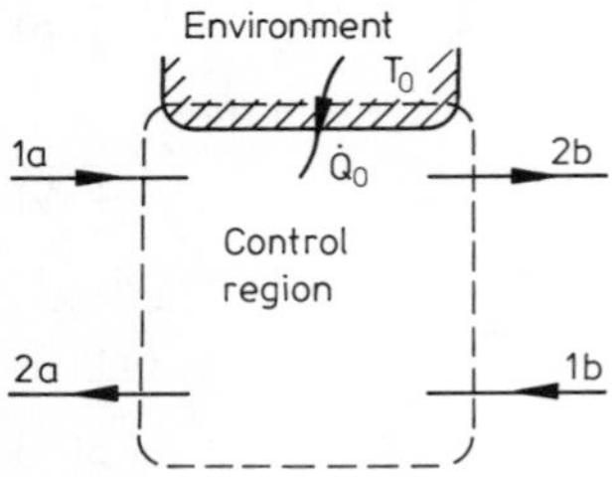

Fig 4.24 Control region for a simple two-stream heat exchanger.

Expressing the 'input' and 'output' in (4.75) in terms of the thermal and pressure components of the changes in exergy as in (4.62):

$$\psi = \frac{\Delta \dot{E}_b^{\Delta T} - \Delta \dot{E}_b^{\Delta P}}{\Delta \dot{E}_a^{\Delta T} + \Delta \dot{E}_a^{\Delta P}} \tag{4.77}$$

Note the effect of $\Delta \dot{E}_b^{\Delta P}$ and $\Delta \dot{E}_a^{\Delta P}$, the exergy changes associated with pressure losses. As will be seen from (4.77) pressure losses of stream b tend to reduce the numerator and those of streams a tend to increase the denominator, so that the overall effect is to reduce ψ. It is in heat exchangers with gaseous streams that the $\Delta \dot{E}^{\Delta P}$ terms become important in the exergy balance; occasionally $\Delta \dot{E}_b^{\Delta T} < \Delta E_b^{\Delta P}$ when ψ becomes negative. Under such circumstances, the increase in the thermal component of exergy of the output stream is less than the minimum power necessary to compensate for the pressure drop in that stream. This is most likely in heat exchangers operating at near-environmental temperatures when, for a given heat transfer rate, the thermal exergy flux has a relatively low value. Although heat exchangers operating with $\psi < 0$ may not make much thermodynamic sense, they sometimes make economic sense. For example, consider an energy recovery scheme in which fresh air, which is to be used for air-conditioning is heated by waste gases from an industrial process plant. If the air stream is heated to 35°C when the outside air temperature is 10°C, the thermal exergy associated with a heat transfer rate of, say, 100 kW is only (100 kW)25/283 = 8.8 kW. Now, if the fan power necessary to operate the heat exchanger was as much as 5 kW per gas stream, ie a total of 10 kW, it would still give a ratio of heating rate to power input of 10 which is much better than the coefficient of performance of currently available domestic heat pumps operating under such conditions.

Rational efficiency of a heat exchanger with compressors

Except where fluid flow is by natural convection, all heat exchangers must be provided with pumps or compressors to give the fluids the pressure rise necessary for given operating conditions. Although the pump or compressor is usually situated at some remote position in the plant and often serves a number of heat exchangers and plant components, it is convenient and instructive to consider a heat exchanger with separate compressors provided for either stream (Fig 4.25(a)) in this analysis. Each compressor is sited at the exit of the stream from the heat exchanger and operates on a pressure ratio just sufficient to compensate the stream for the loss of pressure in the heat exchanger. Consequently each stream leaves the control region, defined by control surface B, at the same pressure as its entry pressure, ie: $P_{1a} = P_{2'b}$ and $P_{1b} = P_{2'b}$. Consequently, $\Delta \dot{E}_a^{\Delta P} = 0$, $\Delta \dot{E}_b^{\Delta P} = 0$ and:

$$\dot{E}_{1a} - \dot{E}_{2'a} = \Delta \dot{E}_a^{\Delta T} \tag{4.78a}$$

$$\dot{E}_{2'b} - \dot{E}_{1b} = \Delta \dot{E}_b^{\Delta T} \tag{4.78b}$$

The exergy balance with respect to the control surface B can be written:

$$\Delta \dot{E}_a^{\Delta T} + \dot{W}_a + \dot{W}_b - \dot{I} = \Delta \dot{E}_b^{\Delta T} \tag{4.79}$$

In this plant configuration, the input and output are:

$$\text{Input} = \Delta \dot{E}_a^{\Delta T} + \dot{W}_a + \dot{W}_b \tag{4.80a}$$

$$\text{Output} = \Delta \dot{E}_b^{\Delta T} \tag{4.80b}$$

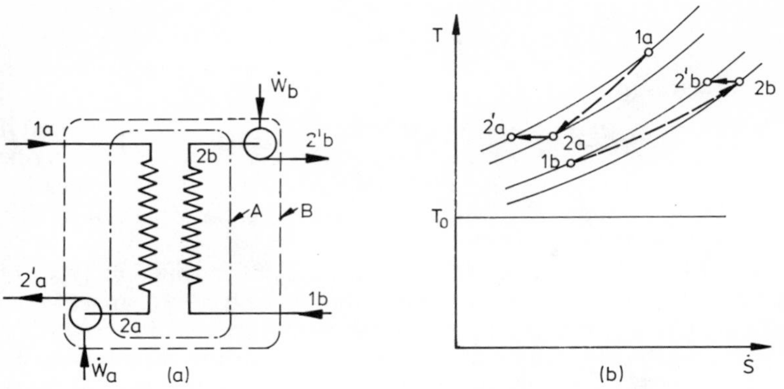

Fig 4.25 Heat exchanger with compressors, (a) control region, (b) processes in T–$\dot{S}$ co-ordinates.

and hence the two alternative forms of rational efficiency can be expressed as:

$$\psi = \frac{\Delta\dot{E}_b^{\Delta T}}{\Delta\dot{E}_a^{\Delta T} + \dot{W}_a + \dot{W}_b} \tag{4.81}$$

$$\psi = 1 - \frac{\dot{I}}{\Delta\dot{E}_a^{\Delta T} + \dot{W}_a + \dot{W}_b} \tag{4.82}$$

Using the expression for the rational efficiency of a non-adiabatic compressor given by (4.53):

$$\dot{W}_a = \Delta\dot{E}_a^{\Delta P}/\psi_a \tag{4.83a}$$

$$\dot{W}_b = \Delta\dot{E}_b^{\Delta P}/\psi_b \tag{4.83b}$$

The rational efficiencies given by (4.81) and (4.82) can, therefore, include the effect of the irreversibility incurred in the compression processes associated with the operation of the heat exchanger, provided that suitable values can be assigned to the compressor rational efficiencies ψ_a and ψ_b.

Now consider the special case when the compressors used are fully reversible, internally and externally. Under these circumstances (see (4.57)), $\psi_a = 1$ and $\psi_b = 1$ and hence:

$$\dot{W}_a = \Delta\dot{E}_a^{\Delta P} \tag{4.84a}$$

$$\dot{W}_b = \Delta\dot{E}_b^{\Delta P} \tag{4.84b}$$

Substituting (4.84a) and (4.84b) in (4.81) and (4.82):

$$\psi = \frac{\Delta\dot{E}_b^{\Delta T}}{\Delta\dot{E}_a^{\Delta T} + \Delta\dot{E}_a^{\Delta P} + \Delta\dot{E}_b^{\Delta P}} \tag{4.85}$$

$$\psi = 1 - \frac{\dot{I}}{\Delta\dot{E}_a^{\Delta T} + \Delta\dot{E}_a^{\Delta P} + \Delta\dot{E}_b^{\Delta P}} \tag{4.86}$$

Since the compressors included in control region B have been assumed to operate fully reversibly, the rational efficiencies given by (4.85) and (4.86) may be regarded as forms of rational efficiencies of a heat exchanger, alternative to those given by (4.75) and (4.76).

Note that the difference between these two forms of rational efficiencies lies in the different assumptions which are used in them as to what constitutes the exergy outputs of a heat exchanger. Whereas in (4.85) this is taken to be the change in the thermal component of exergy, $\Delta\dot{E}_b^{\Delta T}$, of the output stream, in (4.75) both $\Delta\dot{E}_b^{\Delta T}$ and $\Delta\dot{E}_b^{\Delta P}$ are included. The use of $\Delta\dot{E}_b^{\Delta T}$ as the output may be justified logically on the grounds that it is the function of a heat exchanger to transfer *thermal exergy*. A drop in $\Delta\dot{E}_b^{\Delta P}$ may be regarded as being of no relevance to the function of the heat exchanger. Consequently, the terms $\Delta\dot{E}_a^{\Delta P}$ and $\Delta\dot{E}_b^{\Delta P}$, which represent the pressure losses essential for the operation of the heat exchanger, as in (4.85) and (4.86).

Both these forms of rational efficiencies have their uses, but in different types of analyses. (4.75) and (4.76) are more appropriate when analysing a heat exchanger in the context of a thermal plant where the definition of input and output is dictated by the *actual* exergy flows crossing a control surface. When considering a heat exchanger on its own, however, forms of rational efficiency given by (4.85) and (4.86) may have advantages. The numerator in (4.85), $\Delta\dot{E}_b^{\Delta T}$, can never become negative so the value of ψ varies within the limits $0 \leqslant \psi \leqslant 1$ according to the degree of thermodynamic perfection of the heat exchanger.

Finally, it should be pointed out at this stage that the concept of efficiency defect may be useful in representing different types of losses occurring in a heat exchanger. Substituting (4.72) in (4.86):

$$1 = \psi + \delta^{\Delta T} + \delta^{\Delta P} + \delta^{Q} \tag{4.87}$$

where the different contributions to the efficiency defect of the heat exchanger correspond to the components of the irreversibility rate divided by the exergy input. (4.87) can be represented on a Grassmann diagram or in the form of a pie diagram.

Effect of operating temperature range on heat exchanger performance

Chapter 2 showed that the amount of thermal exergy associated with a given heat transfer varies with temperature (Fig 2.3(d)). To determine how this affects heat exchanger performance, consider two gaseous streams at T_1 and T_2. Thermal exergy is transferred from stream 1 to stream 2. If the streams undergo only small temperature changes, T_1 and T_2 may be taken as the mean temperatures of the streams, with good accuracy, during the heat transfer process*. The heat transfer process will involve two forms of irreversibility $\dot{I}^{\Delta T}$ due to heat transfer over a finite temperature difference and $\dot{I}_1^{\Delta P}$ and $\dot{I}_2^{\Delta P}$ due to pressure losses in the two streams. Thus the total irreversibility rate is:

$$\dot{I} = \dot{I}^{\Delta T} + \dot{I}_1^{\Delta P} + \dot{I}_2^{\Delta P} \tag{a}$$

As shown by Eqs (3.70) and (3.71), $\dot{I}^{\Delta T}$ can be expressed in the following form:

$$\dot{I}^{\Delta T} = \dot{Q}T_0\left(\frac{1}{T_2} - \frac{1}{T_1}\right) \tag{b}$$

Denoting:

$$T_1 - T_2 = \Delta T \tag{c}$$

* Alternatively, consider a short section of a heat exchanger in which the average temperatures of the two streams are T_1 and T_2.

and by assuming that $\Delta T \ll T_1$:

$$\dot{I}^{\Delta T} = \dot{Q} T_0 \frac{\Delta T}{T_1^2} \tag{d}$$

Assuming that the two forms of irreversibility $\dot{I}_1^{\Delta P}$ and $\dot{I}_2^{\Delta P}$ have equal shares in the overall irreversibility:

$$\dot{I}_1^{\Delta P} + \dot{I}_2^{\Delta P} = \dot{I}^{\Delta T} \tag{e}$$

In general, the relationship between $\dot{I}^{\Delta T}$ and $\dot{I}^{\Delta P}$ for a heat exchanger of optimum geometry depends on the size of the heat transfer surface. The relative magnitudes of the two forms of irreversibility given above are typical of some compact heat exchangers.

From (a), (d) and (e):

$$\dot{I}/\dot{Q} = 2T_0 \frac{\Delta T}{T_1^2} \tag{f}$$

showing how strongly T_1 affects the irreversibility rate for given value of ΔT, T_0 and $\dot{Q}$.

From the expression for the rational efficiency of the heat transfer process, (4.86), and using (4.64):

$$\psi = 1 - \frac{\dot{I}^{\Delta T} + \dot{I}_1^{\Delta P} + \dot{I}_2^{\Delta P}}{\dot{Q}\dfrac{T_1 - T_0}{T_1} + \dot{I}_1^{\Delta P} + \dot{I}_2^{\Delta P}} \tag{g}$$

Hence, from (a) and (e) and (g):

$$\psi = 1 - \frac{2T_0 \dfrac{\Delta T}{T_1^2}}{\dfrac{T_1 - T_0}{T_1} + T_0 \dfrac{\Delta T}{T_1^2}} \tag{h}$$

This has been used to plot (Fig 4.26) the temperature difference, ΔT, between the streams necessary to maintain a rational efficiency of 0.9 over a wide range of

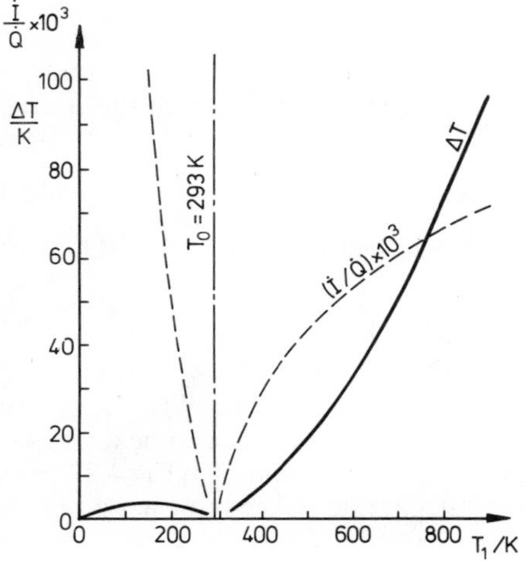

Fig 4.26 Temperature difference and the corresponding dimensionless irreversibility rate as a function of temperature for $\psi = 0.9$.

temperature T_1. Also shown is the ratio $\dot{I}/\dot{Q}$ corresponding to the different values of ΔT. Note that high heat exchanger rational efficiencies can be obtained at high values of T_1 even when the temperature differences and pressure losses are quite high. The situation is quite different at sub-environmental temperatures. At cryogenic temperatures ΔT_1 must be limited to a few degrees K if the same high rational efficiency is to be maintained. The curve showing the variation of $\dot{I}/\dot{Q}$ is of the same type as $\dot{E}^Q/\dot{Q}$ (Fig 2.3(d)) but to a different vertical scale. This follows because $\dot{I}/\dot{Q}$ represents here a constant proportion of the thermal exergy input which is lost in the process. Both the curves have been omitted in the immediate vicinity of $T_1 = T_0$ since in this region the thermal exergy tends to zero, and the corresponding rational efficiency is of little practical significance.

Figure 4.26 is based on a number of simplifying assumptions so the information conveyed can only be regarded as qualitative. In particular it must be emphasised that, since it has been assumed that the stream undergoes only small temperature changes during the heat transfer process, the ΔT and $\dot{I}/\dot{Q}$ are definable in terms of one temperature only, namely T_1. In practice, each gas stream undergoes substantial temperature changes, covering a range of values of T_1. Consequently the information provided in Fig 4.26 may be looked upon as average values typical of various temperature ranges, for which $T_{\text{average}} = T_1$. Also, only two of the four different forms of irreversibility have been taken into account.

Dissipative heat transfer processes

Some aspects of the analysis applied to Group 1 heat exchange processes can also be used for the analysis of dissipative heat transfer processes (Group 2); the analysis of irreversibility due to pressure losses is equally applicable. However, since the main function of such heat exchangers is to exchange heat with the environment, the two remaining components of irreversibility rate $\dot{I}^{\Delta T}$ and $\dot{I}^{Q}$ become identical. Further, as mentioned earlier, since these heat exchangers have no output expressible in terms of exergy we cannot formulate for them rational efficiencies. However, these processes should take place with a minimum of irreversibility so their performance can be assessed using the irreversibility rate and the dimensionless criteria derived from it. Some of the more common dissipative heat transfer processes are considered here.

Condenser of a vapour-compression refrigerator

Refrigerator condensers reject thermal energy to the environment with a minimum of irreversibility. Using the Gouy–Stodola relation, the irreversibility rate for the control region shown in Fig 4.27(a) is:

$$\dot{I} = \dot{Q}_c - T_0(\dot{S}_1 - \dot{S}_2) \tag{4.88}$$

Since the thermal exergy associated with $\dot{Q}_c$ is zero, the exergy balance for the control region reduces to:

$$\dot{E}_1 - \dot{E}_2 = \dot{I} \tag{4.89}$$

illustrating the dissipative nature of the process, in which the total change in the exergy of the refrigerant, as it passes through the condenser, is lost in irreversibility. In general, some of the irreversibility will be due to pressure losses and some due to heat transfer over a finite temperature difference. Using (4.88), the latter is indicated in Fig 4.27(b) for the case when pressure losses may be regarded as negligible. $\dot{I}^{\Delta T}$ is represented in the diagram by the dotted area.

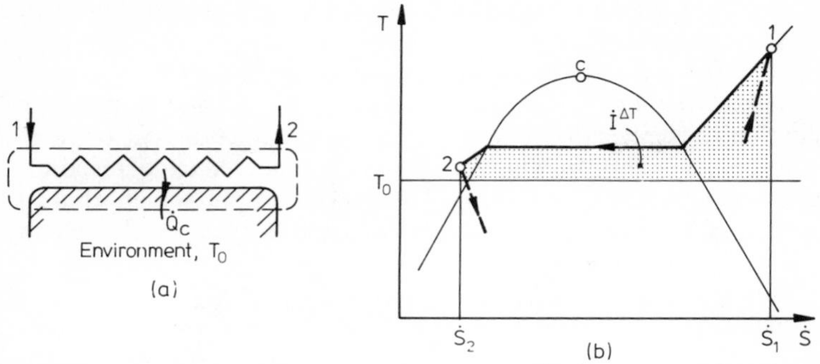

Fig 4.27 Heat transfer process in a condenser of a vapour-compression refrigerator.

Evaporator of a vapour-compression heat-pump

The evaporator of a heat-pump draws the thermal energy which is freely available in the environment into the heat pump so that it can be up-graded to the required temperature. The irreversibility rate associated with this plant component (Fig 4.28(a)) can be expressed from the Gouy–Stodola relation as:

$$\dot{I} = T_0(\dot{S}_2 - \dot{S}_1) - \dot{Q}_E \tag{4.90}$$

The exergy balance for the heat pump evaporator is given by (4.89). The total irreversibility rate, as above comprises $\dot{I}^{\Delta T}$ and $\dot{I}^{\Delta P}$. (4.90) is shown in Fig 4.28(b) for the frictionless heat transfer process, when $\dot{I} = \dot{I}^{\Delta T}$. For this purpose (4.90) has been put into the form:

$$\dot{I}^{\Delta T} = T_0\left[(\dot{S}_2 - \dot{S}_1) - \frac{\dot{Q}_E}{T_0}\right]$$
$$= T_0\dot{\Pi}^{\Delta T} \tag{4.90a}$$

In Fig 4.28(b), the cross-hatched rectangle eadfe, lying under T_0=const, has been drawn equal in area to 12cf1 which represents the heat transfer rate $\dot{Q}_E$ in the

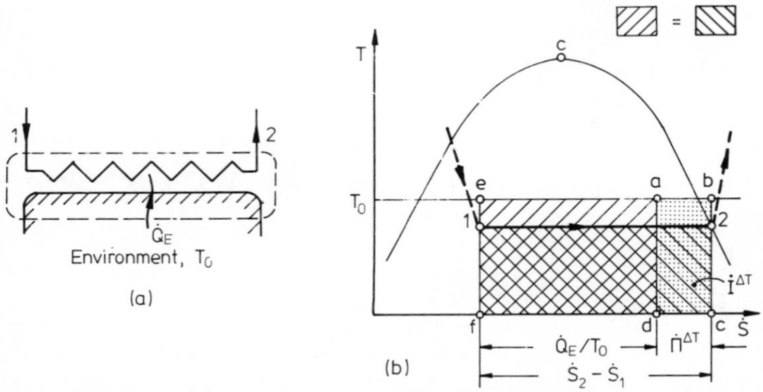

Fig 4.28 Heat transfer process in an evaporator of a vapour-compression heat pump.

evaporator. The rest of the construction should be clear from the diagram. The dotted rectangular area abcda represents $\dot{I}^{\Delta T}$.

Alternatively, using the original form of the expression for $\dot{I}^{\Delta T}$ (with $\dot{I}^{\Delta P}=0$) as given by (4.90), a form of representation of $\dot{I}^{\Delta T}$ of the type used in Fig 4.27(b) may be produced. It can be shown easily that the irreversibility rate $\dot{I}^{\Delta T}$ for the evaporator is also given in Fig 4.28(b) by the rectangular area eb21e. This area may be considered to correspond to a Carnot cycle operating between T_0 and $T_{1,2}$. The power output which could be developed by a hypothetical reversible engine operating on this cycle would be equal to $\dot{I}^{\Delta T}$.

Steam power plant condenser with a cooling tower

As the function of rejection of thermal energy from a steam power plant is performed jointly by the steam condenser and the cooling tower it is convenient to put them together in one control region (Fig 4.29(a)). The cooling tower is taken to be of 'dry' type so that evaporative losses need not be considered. Power input $\dot{W}_F$ is provided to operate air fans and circulating pumps. The control region includes the region where

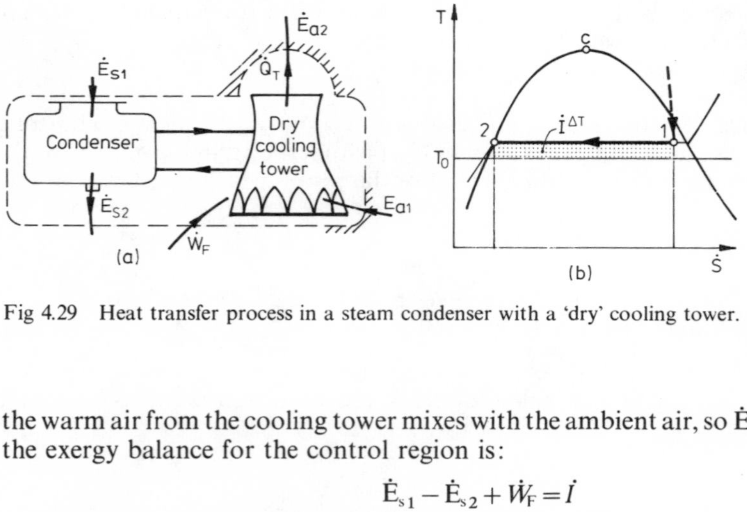

Fig 4.29 Heat transfer process in a steam condenser with a 'dry' cooling tower.

the warm air from the cooling tower mixes with the ambient air, so $\dot{E}_{a2}=\dot{E}_{a1}=0$. Hence the exergy balance for the control region is:

$$\dot{E}_{s1}-\dot{E}_{s2}+\dot{W}_F=\dot{I} \tag{4.91}$$

which demonstrates the dissipative nature of the processes. With $\dot{S}_{a1}=\dot{S}_{a2}$, the Gouy–Stodola relation yields:

$$\dot{I}=\dot{Q}_T-T_0(\dot{S}_{s1}-\dot{S}_{s2}) \tag{4.92}$$

Figure 4.29(b) shows (4.92) for the case when frictional dissipation in the process is negligible.

Evaporative cooling tower

In this type of cooling tower the mass transfer takes place between the hot water and the stream of air. The control surface enveloping the cooling tower (Fig 4.30) includes the region of mixing between the hot, humid air and the ambient air, so $\dot{E}_{a2}=\dot{E}_{a1}=0$.

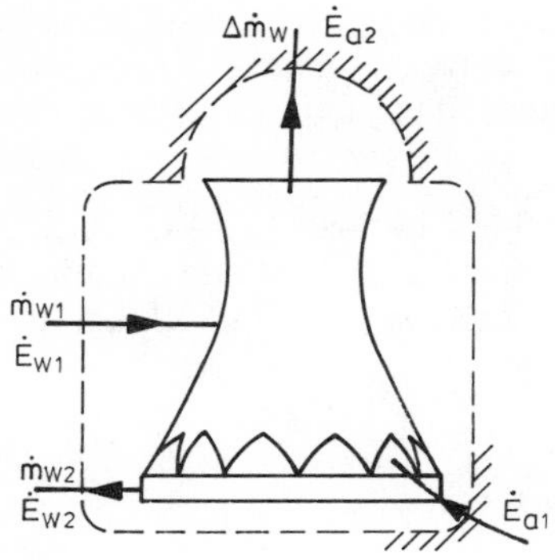

Fig 4.30 Natural-draught, evaporative cooling tower.

Consequently, the exergy balance for the control region takes the form:

$$\dot{I} = \dot{E}_{w1} - \dot{E}_{w2}$$

$$= \dot{m}_{w1}\varepsilon_{w1} - \dot{m}_{w2}\varepsilon_{w2} \quad (4.93)$$

The rate of loss of water through evaporation is:

$$\Delta\dot{m}_w = \dot{m}_{w1} - \dot{m}_{w2} \quad (4.94)$$

Hence (4.93) can be rewritten:

$$\dot{I} = \dot{m}_{w2}(\varepsilon_{w1} - \varepsilon_{w2}) + \Delta\dot{m}_w\,\varepsilon_{w1} \quad (4.95)$$

$\Delta\dot{m}_w\,\varepsilon_{w1}$ represents irreversibility due to evaporation and subsequent mixing of the water vapour with ambient air. Since ε_{w1} stands on its own, ie it is not subtracted from another exergy term, the chemical exergy does not cancel out. Hence in this case ε_{w1} must include chemical exergy of water substance (see Appendix A for a discussion and a method of calculation).

Example 4.2

In a vapour compression heat pump using R12 (CF_2Cl_2) as the working fluid, the condenser is an air-cooled counter-flow heat exchanger. In a test run the operating parameters obtained were:

Fluid	Mass flow rate (kg/s)	Inlet Temperature	Inlet Pressure (bar)	Outlet Temperature	Outlet Pressure (bar)
R12	0.125	45°C	9.607	35°C	9.607
Air	1.0	18°C	1.045	35°C	1.035

Calculate:

(i) Heat transfer rate to the environment.
(ii) Irreversibility rate of the heat exchange process.
(iii) Component of the irreversibility rate due to pressure losses.
(iv) Component of the irreversibility rate due to heat transfer to the environment.
(v) Rational efficiency of the heat exchanger.

Assume air to be a perfect gas with $c_P = 1.0$ kJ/kgK, $\gamma = 1.4$. The enthalpy and entropy of R12 in the compressed liquid state may be taken to be equal to the respective saturated liquid properties at the same temperature. The temperature of the surroundings is 278 K.

Solution: As the saturation temperature of R12 at 9.607 bar is[2.4] 40°C, R12 is superheated by 5 K at the inlet to the condenser and sub-cooled by the same amount at the outlet. Thus: $s_{R\,1} = 0.6945$ kJ/kgK, $s_{R\,2} = 0.2559$ kJ/kgK, $h_{R\,1} = 207.05$ kJ/kg, and $h_{R\,2} = 69.55$ kJ/kg.

(i) From the SFEE:

$$\dot{Q}_0 = \dot{m}_a c_{Pa}(T_{a2} - T_{a1}) + \dot{m}_R(h_{R2} - h_{R1})$$
$$= \underline{-0.1875 \text{ kW}}$$

(ii) From the Gouy–Stodola relation:

$$\dot{I} = T_0[\dot{m}_a(s_{a\,2} - s_{a\,1}) + \dot{m}_R(s_{R\,2} - s_{R\,1}) - \dot{Q}_0$$

where for air:

$$s_{a\,2} - s_{a\,1} = c_{Pa}\left[\ln\frac{T_{a\,2}}{T_{a\,1}} - \frac{\gamma - 1}{\gamma}\ln\frac{P_{a\,2}}{P_{a\,1}}\right]$$
$$= 0.05953 \text{ kJ/kgK}$$

Substituting numerical values:

$$\dot{I} = [1.0 \times 0.05953 - 0.125 \times 0.4386]278 + 0.1875$$
$$= \underline{1.4955 \text{ kW}}$$

(iii) Considering air as an ideal gas, from (4.70a):

$$\dot{I}_a^{\Delta P} = \dot{m}_a \frac{\gamma - 1}{\gamma} c_{Pa} T_0 \ln\frac{P_{1a}}{P_{2a}}$$
$$= 1.0\frac{0.4}{1.4}1.0 \times 278 \ln\frac{1.045}{1.035}$$
$$= \underline{0.7637 \text{ kW}}$$

Hence, $\dot{I}_a^{\Delta P}$ accounts for more than half the total irreversibility rate.

(iv) Taking the mean surface temperature of the heat exchanger as 310 K, the loss of exergy due to 'heat loss' is (from (4.71)):

$$\dot{I}_s^{Q} = 0.1875\frac{310 - 278}{310}$$
$$= 0.0194 \text{ kW}$$

Because the heat exchanger operates over a range of temperatures near environmental, this component of irreversibility is relatively small. Although the calculated value is only approximate, it is a useful indication of the relative importance of this contribution to the total irreversibility rate. The component $\dot{I}^{\Delta T}$ is, by difference:

$$\dot{I}^{\Delta T} = \dot{I} - \dot{I}^{\Delta P} - \dot{I}_s^{Q}$$

$$= 1.4955 - 0.7637 - 0.0194$$
$$= \underline{0.7124\ \text{kW}}$$

(v) Taking (4.86) as the basis, the rational efficiency of the condenser, with $\Delta \dot{E}_R^{\Delta P} = 0$, is:

$$\psi = 1 - \frac{\dot{I}}{\Delta \dot{E}_R^{\Delta T} + \Delta \dot{E}_a^{\Delta P}}$$

where $\Delta \dot{E}_a^{\Delta P} = \dot{I}_a^{\Delta P}$ (see (4.64)) and has been calculated above.
With $\Delta \dot{E}_R^{\Delta P} = 0$:

$$\Delta \dot{E}_R = \Delta \dot{E}_R^{\Delta T} = \dot{m}_R[(h_{R1} - h_{R1}) - T_0(s_{R1} - s_{R2})]$$
$$= 1.94662\ \text{kW}$$

Substituting:

$$\psi = 1 - \frac{1.4955}{1.9462 + 0.7637}$$
$$= \underline{0.448}$$

4.4 Mixing and separation processes

Mixing

Mixing occurs spontaneously when substances* are put into physical contact with each other†. The mixing process has two distinct aspects, the intermingling of the molecules of the substances and the exchange of energy between the streams involved in the process. Molecular aspects are irrelevant when the streams are of the same chemical composition while no energy will exchange if the streams are initially at the same pressure and temperature.

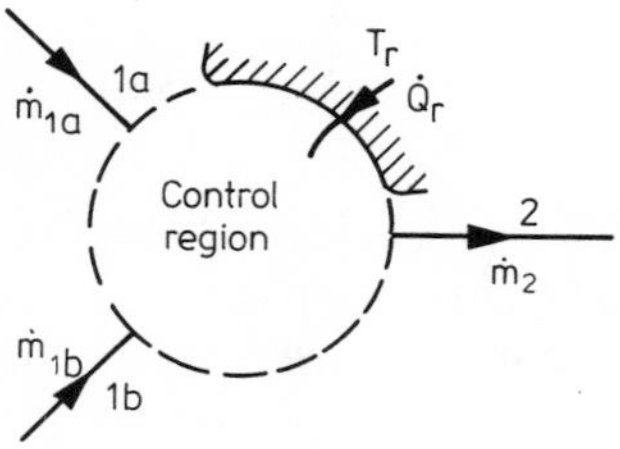

Fig 4.31 Control region for a steady-flow mixing process.

Consider a steady-flow mixing process with heat exchange involving two streams of non-reacting substances (Fig 4.31). Using the notation shown in the figure, the Gouy–Stodola relation for the mixing process with heat transfer is:

$$\dot{I} = T_0\left[\dot{m}_{1a}(s_2 - s_{1a}) + \dot{m}_{1b}(s_2 - s_{1b}) - \frac{\dot{Q}_r}{T_r}\right] \tag{4.96}$$

* Solids and immiscible liquids are excluded from these considerations.

† When interfacial phenomena are present or the molecules of these substances are subjected to intermolecular forces or differential body forces due to gravitational, centrifugal or electric fields, the mixing process may be slowed down, stopped or indeed reversed.

The exergy balance for the mixing process corresponding to the control region shown in Fig 4.31 is:

$$\dot{E}_{1a} + \dot{E}_{1b} - \dot{E}_2 + \dot{E}^Q = \dot{I} \tag{4.97}$$

In a general mixing process of this type, the irreversibility of the process is due to:

(i) The process of intermingling molecules of different species through molecular diffusion. A measure of this contribution to the process irreversibility is the work necessary in a reversible process to undo the mixing process, ie to separate the resultant components.
(ii) The process of energy transfer between streams which initially are not in thermal or mechanical equilibrium.
(iii) Heat transfer with finite temperature gradients.
(iv) Viscous dissipation during mixing. This results in a pressure drop between the inlets to the mising space and the outlet.

In theory, reversible mixing and separation can be arranged through the use of reversible compressors or expanders and semi-permeable membranes. Such an arrangement, used to derive the chemical exergy of a mixture of ideal gases, is shown in Fig 2.13. The main difficulty in using such techniques is the lack of suitable semi-permeable membranes.

For streams comprising molecules of one species, mixing can take place reversibly provided both streams are initially at the same pressure and temperature. For example, mixing dry saturated steam and wet steam at the same pressure (or temperature) is a reversible process since, in principle, such a mixture can be separated into the original components with no work input. The same is true of streams of mixtures of the same components which are initially of the same pressure and temperature but in different phases (eg binary mixtures in liquid and vapour phases). Such mixing occurs under conditions of chemical equilibrium characterised by the equality of the chemical potential of all the substances in the phases involved in the process[1.7].

Mixing processes common in thermal and chemical plants include open-type feed heaters, steam ejectors and distillation columns.

Rational efficiency of a separation process

The term 'separation process' will be used here in the most general sense to cover such applications as sea water desalination, petroleum fractionation or air separation. Although multi-stage distillation is the most common technique used, vapour-compression distillation, reverse osmosis or multi-flash distillation are used in certain areas. Thermodynamically, there are some common features which characterise these plants, although not all are found in every type of plant. Figure 4.32 shows the control region for such a generalised separation plant. The mixture stream is shown separated into two product streams, although in particular plants the number of useful product streams may be only one or several. There is also a stream of waste material which is dumped into the environment—brine returned to the sea from a desalination plant or nitrogen rejected to atmosphere from an air separation plant. The exergy of each of these streams is greater than zero, and in different circumstances they might be regarded as useful product streams; where there is no demand for this particular product, however, they must be regarded as waste streams. The region where the waste stream mixes with the environment (the semi-circular area on the diagram) is included in the control region so that the associated irreversibility is included in the total

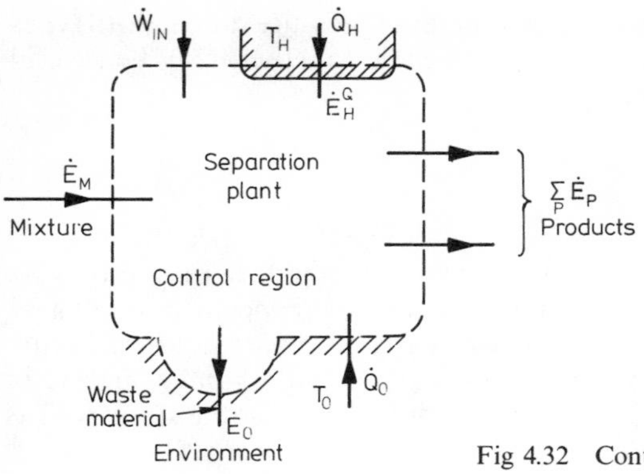

Fig 4.32 Control region for the generalised separation plant.

irreversibility of the plant*. Another, related form of irreversibility is that associated with heat transfer with the environment, $\dot{Q}_0$. The latter is positive for cryogenic processes ($T < T_0$) but negative in, say, petroleum fractionation ($T > T_0$). In general the input to the plant is in two forms, shaft power, $\dot{W}_{IN}$, and heating, $\dot{Q}_H$, provided at temperature T_H. From this description of a generalised separation plant, its exergy balance can be expressed as:

$$\sum_P \dot{E}_P - \dot{E}_M = \dot{E}_H^Q + \dot{W}_{IN} - \dot{I} \tag{4.98}$$

Considering the function of the plant, and from the exergy balance (4.98), the exergy input and output are:

$$\text{Input} = \dot{E}_H^Q + \dot{W}_{IN}$$

$$\text{Output} = \sum_P \dot{E}_P - \dot{E}_M$$

Hence, the two forms of the rational efficiency are:

$$\psi = \frac{\sum_P \dot{E}_P - \dot{E}_M}{\dot{E}_H^Q + \dot{W}_{IN}} \tag{4.99a}$$

$$\psi = 1 - \frac{\dot{I}}{\dot{E}_H^Q + \dot{W}_{IN}} \tag{4.99b}$$

In this case output in the form $(\sum_P \dot{E}_P - \dot{E}_M)$ rather than as $\sum_P \dot{E}_P$ was chosen since this eliminates the effect on ψ of chemical exergy. The chemical exergy of some mixtures, eg crude oil, is many times larger than the exergy used in the operation of the separation plant and thus, if it were not eliminated, it might mask the effect of irreversibilities on the rational efficiency of the process.

The output, which forms the numerator of (4.99a), is equal to the minimum power input for separating the mixture into the required products. This is usually quite small in relation to the actual exergy input and so the rational efficiency of most separation processes is low, usually less than 10%, because large heat transfer rates are involved

* When analysing the plant it will generally be of some interest to calculate this form of irreversibility as a separate item in order to know its contribution to the overall plant irreversibility.

and, in some types of plants, compression of gases over large pressure ratios. Both types of processes are associated with irreversibility rates which, together, may be several times the minimum power of separation.

Exergy of separation

Exergy of separation of a mixture into its components can be determined directly and conveniently from the exergy–concentration diagram for the mixture in question or, indirectly, from separate enthalpy–concentration and entropy–concentration charts. For simplicity, the method of analysis will be demonstrated on a binary mixture. Figure 4.33 shows a simplified molar exergy–concentration chart[4.3] for a binary mixture of the type discussed in Section 3.6 (see Fig 3.28). Such charts show the isotherms as well as

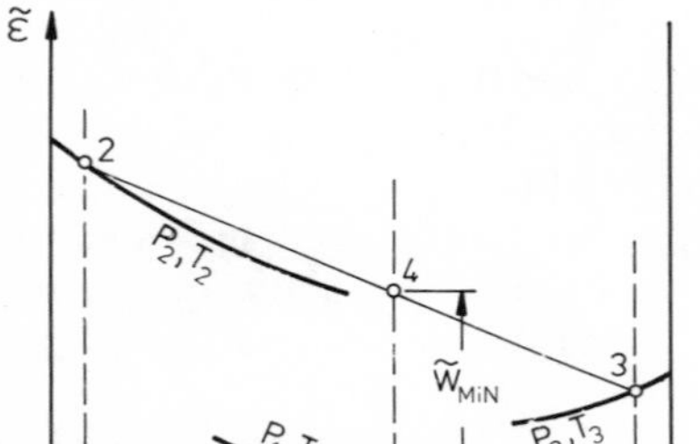

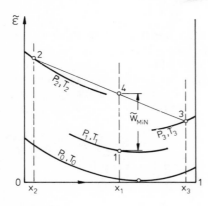

Fig 4.33 Determination of the minimum work of separation[2.2].

the saturation lines (for $y=0$ and $y=1$) for a number of different pressures. Thus, it is possible to consider a process in which the state of the mixture and the products correspond to different pressures, temperatures, and phases (vapour or liquid). Let 1 in Fig 4.33 represent the state of the mixture while 2 and 3 represent the states of the two products, the mole fraction x being that of the more volatile of the components. To show that point 4 on the diagram gives the state which corresponds to the equivalent molar exergy of the two products, note the overall molar balance:

$$\dot{n}_1 = \dot{n}_2 + \dot{n}_3 \tag{4.100}$$

and the molar balance for one of the components, say the more volatile:

$$\dot{n}_1 x_1 = \dot{n}_2 x_2 + \dot{n}_3 x_3 \tag{4.101}$$

The equivalent molar exergy $\tilde{\varepsilon}_4$ of the product streams is defined as:

$$(\dot{n}_2 + \dot{n}_3)\tilde{\varepsilon}_4 = \dot{n}_2 \tilde{\varepsilon}_2 + \dot{n}_3 \tilde{\varepsilon}_3 \tag{4.102}$$

From (4.100), (4.101) and (4.102):

$$\frac{\tilde{\varepsilon}_4 - \tilde{\varepsilon}_3}{x_3 - x_1} = \frac{\tilde{\varepsilon}_2 - \tilde{\varepsilon}_4}{x_1 - x_2} \tag{4.103}$$

(4.103) proves that state 4 corresponds to the point of intersection between a straight line, known as a tie line, joining 2 and 3 and the ordinate passing through point 1. The difference in molar exergy between the products and the mixture, equal to the minimum power of separation, can be written:

$$\dot{W}_{MIN} = (\dot{n}_2\tilde{\varepsilon}_2 + \dot{n}_3\tilde{\varepsilon}_3) - \dot{n}_1\tilde{\varepsilon}_1$$

Using (4.100) and (4.102) this expression can be rewritten in the following form:

$$\tilde{w}_{MIN} = \frac{\dot{W}_{MIN}}{\dot{n}_1} = \tilde{\varepsilon}_4 - \tilde{\varepsilon}_1 \tag{4.104}$$

Thus, the vertical distance between points 4 and 1 on the diagram gives the minimum (reversible) molar work of separation. Note that this quantity may be positive, negative or zero depending on whether point 1 is below point 4, above it or the two coincide, respectively. If $\tilde{\varepsilon}_1 = \tilde{\varepsilon}_4$ ($\tilde{w}_{MIN} = 0$), the exergy of the mixture is high enough for the separation to take place under reversible conditions without any further input of exergy. Under actual conditions we must have $\tilde{\varepsilon}_1 > \tilde{\varepsilon}_4$ and the difference in molar exergy $(\tilde{\varepsilon}_1 - \tilde{\varepsilon}_4)$ must be high enough to compensate for losses due to process irreversibilities. This arrangement is used in air separation plants where the air enters the distillation column at a very low temperature and at a high pressure so that both thermal and pressure components of exergy are high. Thus, when considering the rational efficiency of the distillation column on its own, it would be appropriate to consider the exergy associated with the air stream entering the column as the exergy input, and the exergy of the separated products as the output. This is further justified because the chemical exergy of air is zero. Thus, for an air distillation column the rational efficiency has the forms:

$$\psi_{AIR} = \frac{\sum_{P} \dot{E}_P}{\dot{E}_M} \tag{4.105a}$$

$$\psi_{AIR} = 1 - \frac{\dot{I}}{\dot{E}_M} \tag{4.105b}$$

Adiabatic distillation column

The most common separation process is multi-stage distillation. The principal plant component is the distillation column, known also as a rectifier or a fractionating column. Different types of distillation columns and separation techniques in general use are described in textbooks on chemical engineering practice[4.4]. Here analysis using the exergy method will be illustrated on the type of multi-stage distillation column shown in Fig 4.34. A reboiler and a partial condenser are used so that product

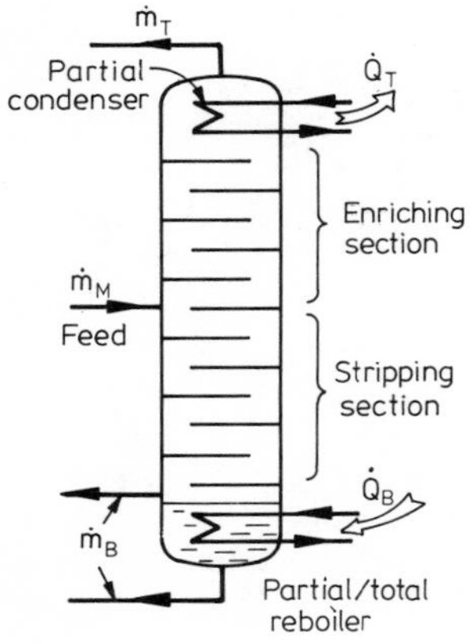

Fig 4.34 Schematic diagram of a distillation column.

leaving the top of the column is a saturated vapour and product leaving the bottom is a saturated liquid or saturated vapour. The distillation process takes place under adiabatic conditions through the exchange of mass and heat between the downward stream of liquid and the upward stream of vapour. All heating necessary for the process takes place in the reboiler where a proportion of the less volatile product, liquefied in the column, is evaporated. This vapour rises through the stages providing the necessary 'heating', while the rest is drawn off as product. At the top of the column, part of the vapour of the more volatile product is condensed and allowed to fall through the stages providing the necessary 'cooling'. The returned liquid is known as reflux and the ratio of the mass flow rate of the reflux to that of the vapour drawn off is called reflux ratio. In an adiabatic distillation process all the heating, $\dot{Q}_B$, is provided at temperature T_B, which corresponds to the maximum temperature of the distillation process; all cooling, $\dot{Q}_T$, is provided at T_T which is the minimum temperature.

The column can operate under two sets of conditions, ie when the whole process takes place at a temperature $T > T_0$ and when $T < T_0$. The heat and exergy transfers for the two cases are represented[4.3] in T–$\dot{S}$ co-ordinates in Fig 4.35. The columns are shown upside-down next to the T–$\dot{S}$ diagram so that the temperatures in the columns correspond to those shown on the diagram. All quantities marked with ′ refer to the operation of the column at $T > T_0$ and those marked with ″ to $T < T_0$. The subscripts B, T and M refer to the bottom product, top product and mixture respectively.

The steady flow energy equation for a control region containing the distillation column, neglecting changes in kinetic and potential energy, can be written as:

$$\dot{Q}_B - \dot{Q}_T = \dot{H}_T + \dot{H}_B - \dot{H}_M \tag{4.106}$$

where (Fig 4.35) $\dot{Q}_B$ represents heat transfer to the column and $\dot{Q}_T$ from the column*.

The exergy balance for the control region of either column is:

$$\dot{E}_B^Q + \dot{E}_T^Q = \dot{E}_T + \dot{E}_B - \dot{E}_M + \dot{I} \tag{4.107}$$

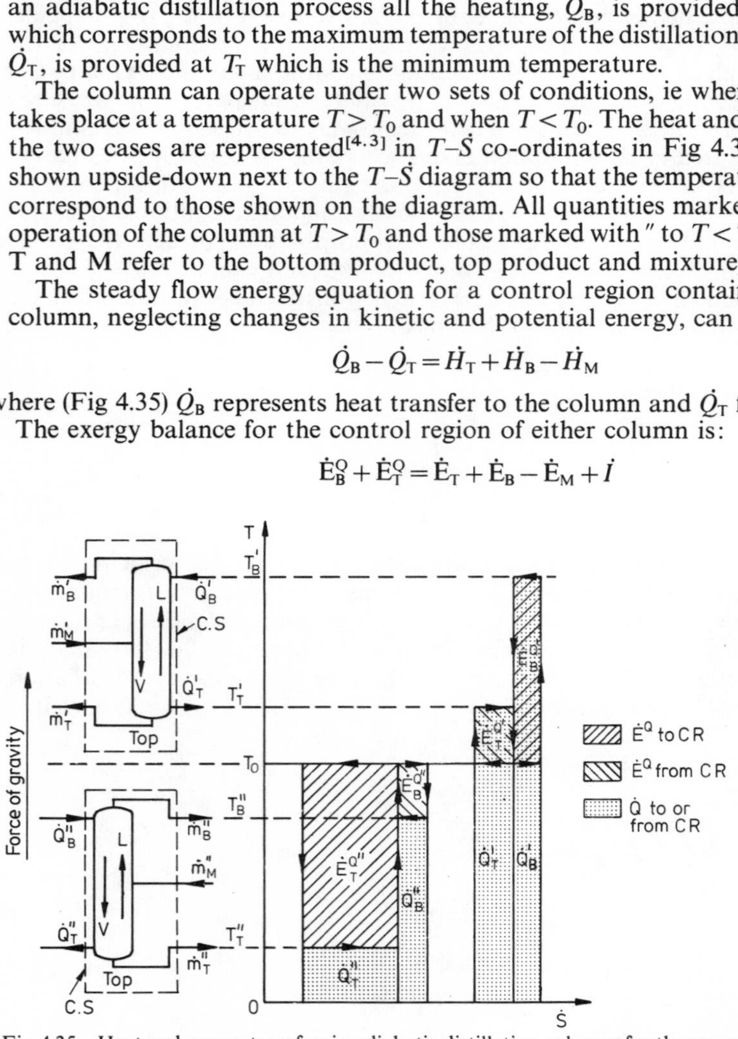

Fig 4.35 Heat and exergy transfers in adiabatic distillation columns for the cases when $T > T_0$ and $T < T_0$.

* The relative magnitudes of $\dot{Q}_B$ and $\dot{Q}_T$ depend on the actual distillation process. For illustrative purposes the areas corresponding to these two heat transfers have been drawn equal in Fig 4.35.

where $\dot{E}_T^Q$ and $\dot{E}_B^Q$ may be positive or negative depending on which case is being considered.

The quantity $(\dot{E}_T + \dot{E}_B - \dot{E}_M)$ on the RHS of (4.107) is the difference between the combined exergy of the products and that of the mixture. This quantity, and its determination (per mass of the mixture) from an exergy–concentration diagram, is discussed above. The LHS of (4.107) represents the net thermal exergy input to the column. For operation at $T > T_0$, the exergy input to the reboiler is greater than the exergy rejected in the condenser, so that the net thermal exergy input is $(\dot{E}_B^{Q'} - \dot{E}_T^{Q'})$. For operation at sub-environmental temperatures, this relationship is reversed and the net thermal exergy input is given by $(\dot{E}_T^{Q''} - \dot{E}_B^{Q''})$. These relationships are shown on a T–$\dot{S}$ diagram (Fig 4.35) where each quantity of thermal exergy, $\dot{E}^Q$, is represented as an area of a Carnot cycle, direct or reversed, which would be necessary to accomplish a given thermal exergy transfer at the required temperature, in conjunction with the environment as a source or sink of zero grade energy. Note that the net exergy input at $T < T_0$ is much greater than that at $T > T_0$ for given heat input* $\dot{Q}_B$ to the reboiler (or as the two are equal, heat output $\dot{Q}_T$ from the condenser). Since all the auxiliary processes, such as heat transfer and compression, associated with supply of the exergy to the column are inherently more irreversible for $T < T_0$ than $T > T_0$, it should be clear that the former is the more exergy intensive type of process of the two.

A model for an ideal non-adiabatic distillation process

To get a better understanding of the form of irreversibility inherent in adiabatic distillation columns, consider a model in which a multi-stage distillation process takes place reversibly. For full reversibility, the heat transfer at each stage must take place reversibly and the mass exchange between the vapour stream and liquid stream must occur under conditions of thermodynamic equilibrium. As with the adiabatic column, temperature increases in the downward direction since this is the direction of flow (under gravity) of the liquid stream which gets gradually richer in the component with the higher boiling point. Figure 4.36 shows the N-th stage of the column where mixing takes place at temperature T_N. For reversible mixing, the liquid stream in state b descending from the $(N+1)$-th stage must be first preheated to T_N and similarly the vapour stream in state a rising from the $(N-1)$-th stage must be precooled to T_N before entering the mixing region. These processes are illustrated in temperature–concentration co-ordinates in Fig 4.37.

On entering the mixing region, mass transfer takes place under equilibrium conditions. The liquid stream leaves in state d enriched in the less volatile component and the vapour stream state c enriched in the more volatile component.

The heat transfers, $\dot{Q}_N^L$ and $\dot{Q}_N^V$, occurring at this distillation stage must take place reversibly. Thus suitable thermal energy reservoirs at the required temperatures are needed and the heat transfers must occur over infinitesimal temperature differences. Considering a column with an infinite number of stages, and hence infinitesimal temperature differences between the stages, the two separate heat transfers may be replaced by $\dot{Q}_{NET}$ at temperature T_N†.

The net heat transfer will be positive (to the control region (CR)) in the stripping stages

* For convenience, the areas used in Fig 4.35 to represent heat transfers at $T > T_0$ are nearly twice as large as those used for $T < T_0$. Represented to the same scale, the difference in the net values of exergy input would appear even greater.

† In a real column with a large, although finite number of stages this should lead to a relatively small increase in process irreversibility.

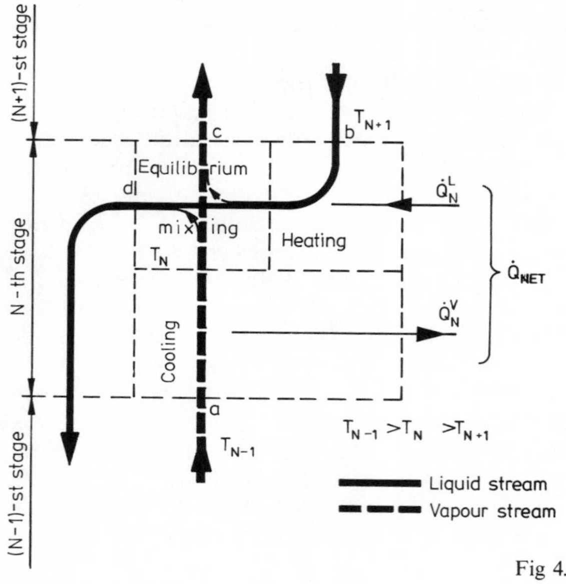

Fig 4.36 Ideal, non-adiabatic distillation stage.

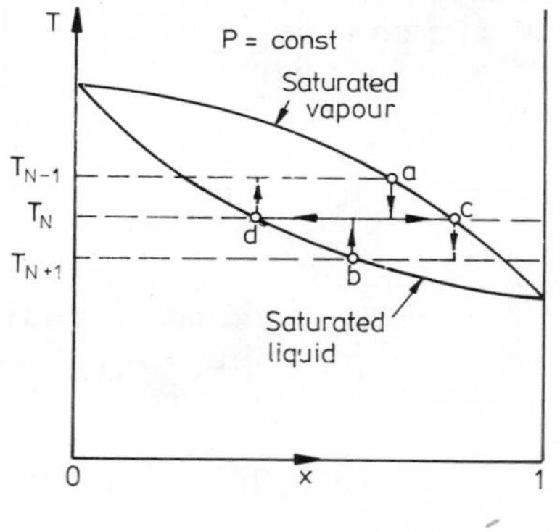

Fig 4.37 Equilibrium mixing process.

and negative (from the CR) in the enriching stages. Figure 4.38 is a qualitative representation of these heat and exergy transfers for the operating temperatures above and below T_0. As in Fig 4.35, the distillation columns are shown upside-down so that the temperatures in the columns correspond to those on the adjoining T–$\dot{S}$ diagram.

Because a large number of stages was assumed, the curve TMB representing the heat transfer process is smooth rather than stepped. Comparing the T–$\dot{S}$ diagrams (Figs 4.35 and 4.38) shows that, for given heat transfer rates $\dot{Q}_T$ and $\dot{Q}_B$, thermal exergy input ($\dot{E}_T^Q$ and $\dot{E}_B^Q$) is greater for non-adiabatic distillation columns. Thus the non-adiabatic reversible columns operate, as was to be expected, with the minimum net thermal exergy input, which from (4.107) must be equal (with $\dot{I}=0$) to the minimum exergy of separation ($\dot{E}_T + \dot{E}_B - \dot{E}_M$).

The idealisation described above may be approached in a practical distillation column with intermediate condensers and reboilers[4.5]. Other arrangements using

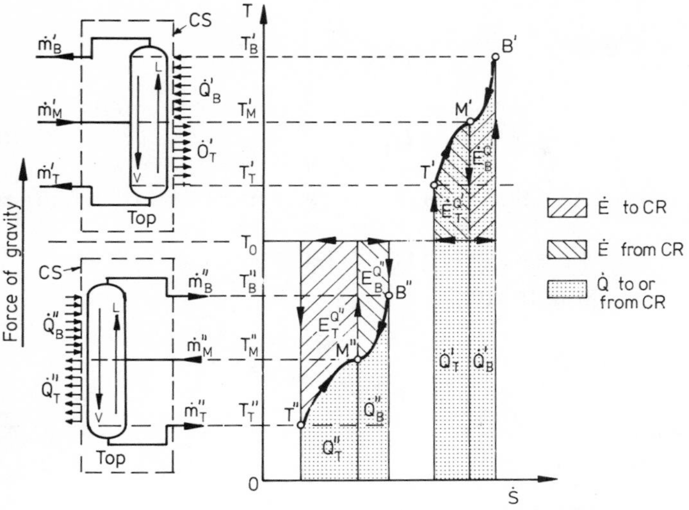

Fig 4.38 Heat and exergy transfers in ideal, non-adiabatic distillation columns for both $T > T_0$ and $T < T_0$.

vapour compressors have been proposed[4.6–4.8] to improve distillation column efficiency. All such improvements involve additional capital cost and increases in other forms of irreversibility, eg in increased pressure losses. A proper thermoeconomic analysis (Chapter 6) is required before the economic advantages of such systems can be proved.

Example of analysis of an adiabatic distillation column

As an example[4.3] of exergy analysis of a distillation column, consider an air separation process taking place at 0.1 MPa. Assume that air consists of 79% N_2 and 21% O_2 by mole and that the two products, pure N_2 and O_2, are in dry saturated vapour state. Under these conditions the sum of the enthalpies of the products is practically equal to the enthalpy of the mixture entering and hence, from (4.106), $Q_B = Q_T$.

Preliminary calculations show that minimum values of Q_B and Q_T, and consequently the minimum consumption of exergy, occurs when the reflux ratio is 0.402. The value of Q_B and Q_T obtained by a construction[4.4] on an enthalpy–concentration diagram was found to be 16 520 kJ/[kmol of O_2]. The temperatures corresponding to Q_B and Q_T for this ideal case are $T_B = 90.2$ K and $T_T = 77.3$ K. Also, $T_0 = 293.15$ K.

Adiabatic column with a minimum reflux ratio

The two previous sub-sections showed that the adiabatic distillation column is inherently irreversible. This intrinsic irreversibility, I_{INTR}, may be evaluated using the exergy balance (4.107), assuming that no other forms of irreversibilities are present. This expression can be adapted for the present purpose as:

$$I_{INTR} = \Delta E^{Q}_{NET} - \Delta E_{MIN} \tag{4.108}$$

where:

$$\Delta E_{MIN} = E_T + E_B - E_M \tag{4.109}$$

and:

$$\Delta E^{Q}_{NET} = E^{Q}_{T} + E^{Q}_{B}$$
$$= Q_T \tau_T + Q_B \tau_B$$
$$= Q_T \frac{T_T - T_0}{T_T} + Q_B \frac{T_B - T_0}{T_B} \tag{4.110}$$

The minimum exergy input as determined from the exergy–concentration diagram (see Appendix E) is:

$$\Delta E_{MIN} = 6\,500 \text{ kJ/[kmol } O_2]$$

The net thermal exergy input from (4.110) is:

$$\Delta E^{Q}_{NET} = -16\,520\left[\frac{77.3 - 293.15}{77.3} - \frac{90.2 - 293.15}{90.2}\right]$$
$$= 16\,520(2.79 - 2.25)$$
$$= 8\,820 \text{ kJ/[kmol } O_2]$$

Consequently the intrinsic irreversibility is:

$$I_{INTR} = 8\,820 - 6\,500 = 2\,320 \text{ kJ/[kmol } O_2]$$

If the necessary exergy input to the column is ΔE^{Q}_{NET} then, from (4.108), the output is ΔE_{MIN}. Hence the rational efficiency of the column is:

$$\psi = \frac{\Delta E_{MIN}}{\Delta E^{Q}_{NET}} = \frac{6\,500}{8\,820} = 0.737 \tag{4.111}$$

No air separation plant using an adiabatic column can have a rational efficiency greater than 0.737 even if the remaining elements of the plant operate reversibly.

Adiabatic column with a practical reflux ratio

In an actual distillation column operating with a limited number of stages the reflux ratio has to be higher than the minimum value by, say, about 20%. Thus:

$$Q_B = -Q_T = 16\,520 \times 1.2 = 19\,824 \text{ kJ/[kmol } O_2]$$

and the net thermal exergy input is:

$$\Delta E^{Q}_{NET} = 19\,824(2.79 - 2.25)$$
$$= 10\,705 \text{ kJ/[kmol } O_2]$$

The irreversibility of the adiabatic column increases to:

$$I = 10\,705 - 6\,500 = 4\,205 \text{ kJ/[kmol } O_2]$$

and the corresponding rational efficiency becomes:

$$\psi = \frac{6\,500}{10\,705} = 0.607$$

Irreversibilities in the condenser and the reboiler

Assuming that the heat transfers in both the condenser and the reboiler take place over a temperature difference of 3 K, the heating medium will have to be supplied at a correspondingly higher temperature and the cooling medium at a lower temperature. Hence the new required temperature will be $T_T^x = 74.3$ K and $T_B^x = 93.2$ K.

The irreversibility due to heat transfer over a constant finite temperature difference in the condenser is (see (3.70) and (3.71)):

$$I_T^{\Delta T} = Q_T T_0 \left[\frac{1}{T_T} - \frac{1}{T_T^x} \right]$$

$$= -19\,824 \times 293.15 \left[\frac{1}{77.3} - \frac{1}{74.3} \right]$$

$$= 3\,036 \text{ kJ/[kmol } O_2]$$

Similarly for the reboiler:

$$I_B^{\Delta T} = 19\,824 \times 293.15 \left[\frac{1}{90.2} - \frac{1}{93.2} \right]$$

$$= 2\,074 \text{ kJ/[kmol } O_2]$$

Hence the total increase in irreversibility is:

$$I^{\Delta T} = 3\,036 + 2\,074 = 5\,110 \text{ kJ/[kmol } O_2]$$

resulting in an increase in the net thermal exergy input to the column to:

$$\Delta E_{NET}^{Q} = 10\,705 + 5\,110 = 15\,815 \text{ kJ/[kmol } O_2]$$

with a corresponding rational efficiency of:

$$\psi = \frac{6\,500}{15\,815} = 0.411$$

Other forms of irreversibilities

Other forms of irreversibilities include:

Fluid friction: This type of irreversibility manifests itself as a pressure difference between the mixture and the products. Increases in this form of irreversibility require inceased power input to the pumps and compressors necessary to compensate for the pressure losses in the column.

Heat transfer between the column and the environment: The corresponding irreversibility can be evaluated in the way described for the heat exchangers in Section 4.3.

Lengthwise heat conduction in the structure of the column: As the two ends of the column are at substantially different temperatures the lengthwise temperature gradients established in the structure of the column can lead to some heat conduction which, clearly, is an irreversible phenomenon leading to degradation of energy.

4.5 Chemical processes including combustion

Before the exergy analysis, mass and energy balances on the system are required to determine the flow rates and energy transfer rates at the control surface. Exergy analysis then follows as the second stage.

Figure 4.39 shows an open system undergoing a steady-flow process involving one or more chemical reactions. The exergy balance for such a system in its basic form is

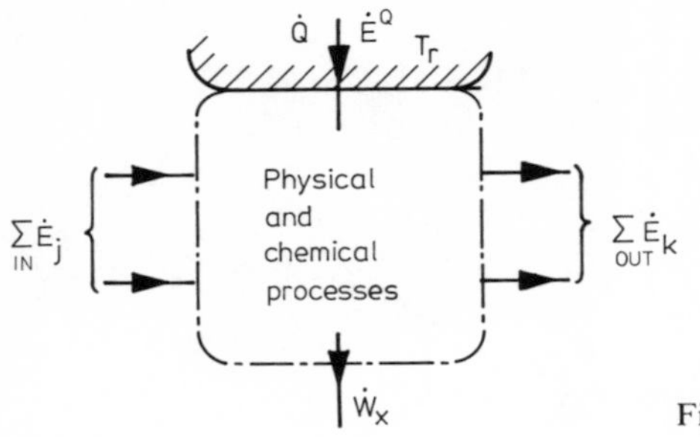

Fig 4.39 Control region for a reacting open system.

essentially the same as for an open system undergoing purely physical processes, and can be written conveniently in the form:

$$\sum_{IN} \dot{E}_j - \sum_{OUT} \dot{E}_k = -\dot{E}^Q + \dot{W}_x + \dot{I} \tag{4.112}$$

where subscripts j and k refer to streams entering and leaving the control region, respectively.

The elements of the RHS of (4.112) were covered in Chapters 2 and 3 and require no further discussion here.

In the open system under consideration the chemical composition of the streams entering the control region is different from those leaving it, so the exergy values for each substance must be based on the system of reference substances described in Section 2.5 and in Appendix A. Using this scheme, the exergy rate of a stream of substance (neglecting the potential and kinetic components) can be written in the form:

$$\dot{E} = \dot{E}_{ph} + \dot{E}_0 \tag{4.113}$$

Hence, using molar quantities, the LHS of (4.112) is:

$$\left(\sum_{IN} \dot{n}_j \tilde{\varepsilon}_{ph,j} - \sum_{OUT} \dot{n}_k \tilde{\varepsilon}_{ph,k}\right) + \left(\sum_{IN} \dot{n}_j \tilde{\varepsilon}_{0,j} - \sum_{OUT} \dot{n}_k \tilde{\varepsilon}_{0,k}\right) \tag{4.114}$$

allowing the chemical and the physical changes in the exergy of the streams of matter to be dealt with separately.

Chemical components of exergy

If the streams consist of pure substances their chemical exergies can be obtained from Tables A.3 or A.4 and used directly in (4.114).

If any stream crossing the control surface is a mixture, its chemical exergy can be evaluated from (2.21a) or (2.21b). In most cases the environmental state will be sufficiently close to the standard state for the standard values of chemical exergy to be used. Otherwise a correction for temperature difference may be carried out as shown in Example A.2.

A simple combustion process is an exothermic chemical reaction; the reactants are usually air and fuel and the products mainly of a mixture of common environmental substances. A method of calculating the chemical exergy of industrial fuels in liquid and solid form is given in Appendix C.

Physical components of exergy

The most general definition of physical exergy can be expressed in molar form as:

$$\tilde{\varepsilon}_{ph} = (\tilde{h} - \tilde{h}_0) - T_0(\tilde{s} - \tilde{s}_0) \tag{4.115}$$

which can be used with suitable property tables when considering, for example, steam, water, and other liquids and solids. For an ideal gas, (4.115) may be conveniently expressed (see (2.13), (2.14) and (2.15)) in terms of the thermal component $\tilde{\varepsilon}^{\Delta T}$ and the pressure component $\tilde{\varepsilon}^{\Delta P}$ as:

$$\tilde{\varepsilon}_{ph} = \tilde{c}_P^{\varepsilon}(T - T_0) + \tilde{R}T_0 \ln \frac{P}{P_0} \tag{4.116}$$

where:

$$\begin{aligned} \tilde{c}_P^{\varepsilon} &= \frac{\tilde{\varepsilon}^{\Delta T}}{T - T_0} \\ &= \frac{1}{T - T_0}\left[\int_{T_0}^{T} \tilde{c}_P \mathrm{d}T - T_0 \int_{T_0}^{T} \frac{\tilde{c}_P \mathrm{d}T}{T}\right] \end{aligned} \tag{4.117}$$

$\tilde{c}_P^{\varepsilon}$ may be looked upon as *mean, molar isobaric exergy capacity.*

If $\tilde{c}_P$ is obtainable in a polynomial form for a given ideal gas, $\tilde{c}_P^{\varepsilon}$ may be evaluated for the required temperature range, $(T - T_0)$. Calculation of the thermal component of physical exergy is simplified by using tabulated values of mean specific heat capacities and values of $\tilde{c}_P^{\varepsilon}$. Two forms of mean specific heat capacity may be defined, one for enthalpy and the other for entropy evaluation:

$$\tilde{c}_P^{h} = \frac{1}{T - T_0} \int_{T_0}^{T} \tilde{c}_P \,\mathrm{d}T \tag{4.118}$$

and:

$$\tilde{c}_P^{s} = \frac{1}{\ln(T/T_0)} \int_{T_0}^{T} \frac{\tilde{c}_P \,\mathrm{d}T}{T} \tag{4.119}$$

From (4.117), (4.118) and (4.119):

$$\tilde{\varepsilon}_{ph} = \tilde{c}_P^{h}(T - T_0) - T_0 \tilde{c}_P^{s} \ln(T/T_0) + \tilde{R}T_0 \ln(P/P_0) \tag{4.120}$$

Values of $\tilde{c}_P^{h}$ and $\tilde{c}_P^{s}$ are tabulated in Appendix D for the standard temperature T^0 over a range of values of T for several common gases. The tabulated values of $\tilde{c}_P^{h}$ are also useful in energy balances in conjunction with enthalpy of devaluation (see Appendix A). The difference between $\tilde{c}_P^{h}$ and $\tilde{c}_P^{s}$ is zero at $T = T^0$ and increases with increasing temperature. However, even at temperatures over 2 000 K this difference amounts to only a few per cent. In some calculations, particularly, those involving small values of $(T - T^0)$, using $\tilde{c}_P^{h}$ in calculating both enthalpy and entropy changes may give adequate accuracy.

Values of $\tilde{c}_P^{\varepsilon}$ are also tabulated for the same selection of gases in Appendix D. These permit rapid evaluation of the thermal component of exergy from (4.116).

Mixtures of ideal gases

It follows from the Gibbs–Dalton rules that the physical exergy of a mixture of N components can be evaluated from:

$$(\tilde{\varepsilon}_{ph})_M = \sum_{i=1}^{N} x_i \tilde{\varepsilon}_i^{\Delta T} + \tilde{R}T_0 \ln(P/P_0) \tag{4.121}$$

where P is the total pressure of the mixture. Using tabulated values of $\tilde{c}_P^{\mathrm{h}}$, $\tilde{c}_P^{\mathrm{s}}$ and $\tilde{c}_P^{\varepsilon}$, (4.121) may be written in the alternative forms:

$$(\tilde{\varepsilon}_{\mathrm{ph}})_{\mathrm{M}} = \sum_{i=1}^{\mathrm{N}} x_i[\tilde{c}_P^{\mathrm{h}}(T-T_0) - T_0\tilde{c}_P^{\mathrm{s}} \ln(T/T_0)]_i + \tilde{R}T_0 \ln(P/P_0) \tag{4.122}$$

$$(\tilde{\varepsilon}_{\mathrm{ph}})_{\mathrm{M}} = (T-T_0) \sum_{i=1}^{\mathrm{n}} x_i\tilde{c}_{P,i}^{\varepsilon} + \tilde{R}T_0 \ln(P/P_0) \tag{4.123}$$

Values of $\tilde{c}_P^{\mathrm{h}}$, $\tilde{c}_P^{\mathrm{s}}$ and $\tilde{c}_P^{\varepsilon}$ given for the common gases in Appendix D should ease these calculations. These tables can be easily extended, using, say, a programmable calculator, to include other gases.

Exothermic and endothermic reactions

Exothermic reactions are accompanied by release of thermal energy. The chemical potential of the reactants in exothermic reactions is always greater than that of the products and this difference acts as a driving force so that an exothermic reaction, once started, will proceed on its own, although sometimes a catalyst may be required. Processes which take place by virtue of finite driving forces, may be called spontaneous or uncontrolled, and are inherently irreversible. The irreversibility of exothermic reactions may also be looked upon as being due to degradation of chemical energy, a relatively ordered form, to thermal energy which is of highly disordered form.

Endothermic reactions require an input of thermal energy from an external source. Consequently endothermic reactions are, in principle, controlled although in practical reactions there are irreversibilities due to lack of homogeneity of composition and to temperature gradients in the mass of the reactants and products.

4.6 Combustion processes

Irreversibility of an adiabatic combustion process

Combustion processes are often accompanied by heat transfer as well as fluid friction and mixing so there is usually more than one form of irreversibility present. In principle it is impossible to evaluate in this case what part of the total irreversibility is due to any particular cause. The process of combustion can be examined, however, by assuming that it takes place under adiabatic conditions and that irreversibilities due to friction and mixing are negligible.

Figure 4.40 illustrates an isobaric adiabatic combustion process, corresponding to the open system shown in Fig 4.41 in T–$\dot{S}$ co-ordinates. It is assumed that the reactants entering the combustion chamber are in the environmental state, denoted as R0. Using the exergy balance, the irreversibility rate of this process is:

$$\dot{I} = \dot{\mathrm{E}}_{\mathrm{R}0} - \dot{\mathrm{E}}_{\mathrm{P}2} \tag{4.124a}$$

Alternatively using the Gouy–Stodola relation, with $\dot{Q}=0$:

$$\dot{I} = T_0\dot{\Pi} = T_0(\dot{S}_{\mathrm{P}2} - \dot{S}_{\mathrm{R}0}) \tag{4.124b}$$

The change in entropy, $(\dot{S}_{\mathrm{P}2} - \dot{S}_{\mathrm{R}0})$, is much larger than the entropy difference between products and reactant $(\dot{S}_{\mathrm{P}0} - \dot{S}_{\mathrm{R}0})$ at T_0 which corresponds to the calorimetric process. (4.124b) is interpreted graphically in Fig 4.40 where $\dot{I}$ is given as a rectangular area (shown dotted).

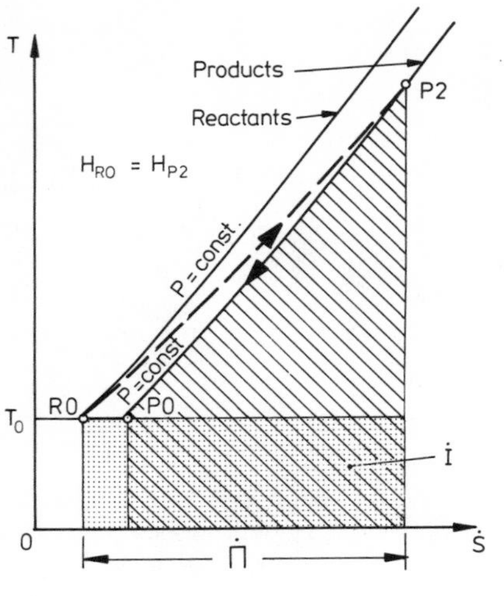

Fig 4.40 Irreversibility in an adiabatic, isobaric combustion process.

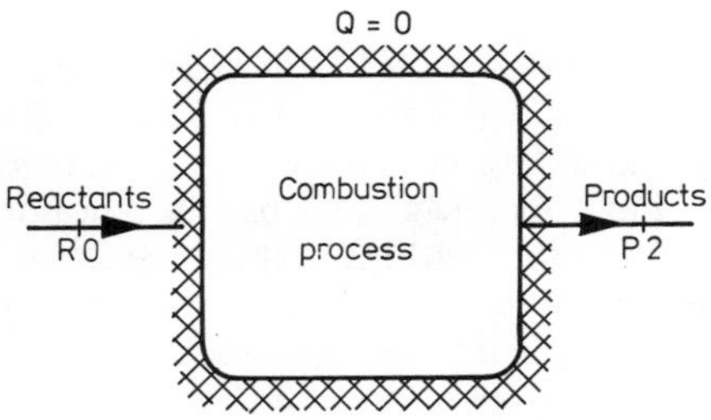

Fig 4.41 Adiabatic combustion chamber.

The cross-hatched area under the isobar between states P2 and P0 represents the calorific value of the fuel. The irreversibility inherent in the adiabatic combustion process corresponds to a large proportion of the original exergy of the fuel. To reduce this irreversibility requires a reduction in the rate of increase in entropy (entropy production rate) $\dot{\Pi}$, which is always associated with an increase in the maximum temperature of the products. Consider three ways of achieving this:

Isochoric combustion involves changing the character of the process. Isochors on a T–S diagram are steeper than isobars so, for given calorific value, the entropy increase, $\dot{\Pi}$, will be less and the final products temperature, T_{P2}, will be higher than in an isobaric process.

Oxygen enrichment of the reactants reduces N_2 dilution and thus also reduces the heat capacity of the reactants and products. This leads to a reduction in $\dot{\Pi}$ and an increase in T_{P2}. Although the reduction in irreversibility rate produced by oxygen enrichment is greater than the exergy of the oxygen used[4.9], this method is usually economical only when other considerations, such as the need for higher flame temperature, also favour the use of oxygen enrichment.

Preheating the reactants is the most common way of reducing the irreversibility of a combustion process. The preheating is usually carried out using products of combustion after they have performed their main heating duty and before they are

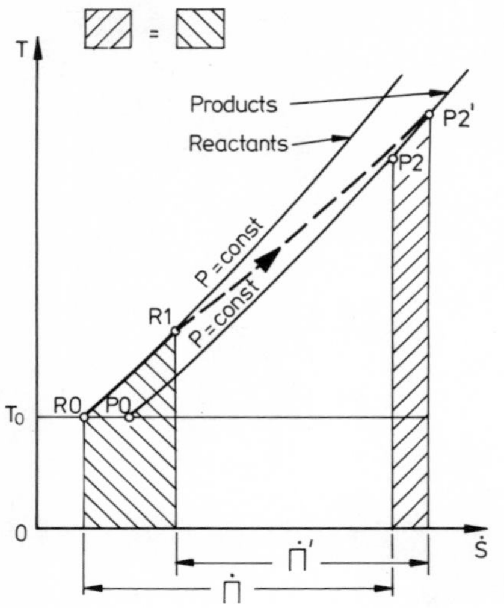

Fig 4.42 Effect of preheating of the reactants on the irreversibility rate in an adiabatic, isobaric combustion process.

discharged into the atmosphere. Figure 4.42 demonstrates the effect of preheating on the process irreversibility rate. The cross-hatched area under the isobar between R0 and R1 represents the heat transfer to the reactants and the consequent increase in their enthalpy; thus the final enthalpy of the products must also be greater by this amount. Hence the cross-hatched area under the isobar between P2 and P2′ is the same as that mentioned earlier. However, as these two areas are different in height, the entropy production rate, $\dot{\Pi}'$, for the process involving preheating will be less than that, $\dot{\Pi}$, for the process without preheating.

Other forms of irreversibility in combustion

In addition to the inherent irreversibility of the combustion process, important causes of irreversibilities include:

Heat loss: Because of the high temperatures involved in a combustion process, there are inevitably some heat losses. This form of irreversibility can be evaluated approximately from:

$$\dot{I}^{Q} = \dot{Q}_0 \frac{T_m - T_0}{T_m} \tag{4.125}$$

where T_m is the mean temperature of the combustion process, and $\dot{Q}_0$ is the rate of heat transfer to the environment.

Incomplete combustion: This form of irreversibility arises from the loss of small pieces of solid fuel through the grate of the furnace from where they are eventually removed with the ashes. If the rate of loss of fuel is $\dot{m}_{FL}$ and the specific exergy of the fuel ε_F, the irreversibility rate is:

$$\dot{I}_{FL} = \dot{m}_{FL}\,\varepsilon_F \tag{4.126}$$

Exhaust losses: These equal the exergy rate of the flue gases discharged to atmosphere. The flue gases consist principally of common environmental substances such as CO_2, H_2O and N_2 but may also contain, in smaller quantities, soot, CO, NO_x, SO_2 and, more rarely, CH_4 and H_2. Once the composition, pressure and temperature of the flue gases is established, the exergy rate of such a mixture $\dot{E}_{FG}$ can be calculated (Chapter 2). Hence, the loss is:

$$\dot{I}_{FL} = \dot{E}_{FG} \tag{4.127}$$

Note that $\dot{E}_{FG}$ includes contributions due to the chemical exergy of the unburnt components (soot, CO, *etc*), the physical exergy due to the temperature difference between the stream and the environment, difference in composition between the products of combustion and the environment (mainly the atmosphere) and possibly some kinetic exergy and pressure component of exergy.

A simple analysis of a steam boiler

Consider a steam boiler (Fig 4.43(a)) operating under steady conditions. The exergy balance for the control region is:

$$\dot{m}_F \varepsilon_F + \dot{m}_A \varepsilon_A - \dot{m}_P \varepsilon_P - \dot{m}_s(\varepsilon_2 - \varepsilon_1) = \dot{I} \tag{4.128}$$

If the air is delivered under atmospheric conditions (no preheating), and the control region is extended to include the mixing region of the products of combustion:

$$\dot{m}_A \varepsilon_A = \dot{m}_P \varepsilon_P = 0 \tag{4.129}$$

From (4.128) and (4.129):

$$\underbrace{\dot{m}_F \varepsilon_F}_{\text{input}} - \underbrace{\dot{m}_s(\varepsilon_2 - \varepsilon_1)}_{\text{output}} = \dot{I} \tag{4.130}$$

From (4.130), the rational efficiency of the boiler has the form:

$$\psi = \frac{\dot{m}_s(\varepsilon_2 - \varepsilon_1)}{\dot{m}_F \varepsilon_F} \tag{4.131}$$

The conventional combustion efficiency is:

$$\eta_{comb} = \frac{\dot{m}_s(h_2 - h_1)}{\dot{m}_F(\text{NCV})} \tag{4.132}$$

From (C.2) and (4.132):

$$\dot{m}_F \varepsilon_F = \frac{\dot{m}_s(h_2 - h_1)\varphi}{\eta_{comb}} \tag{4.133}$$

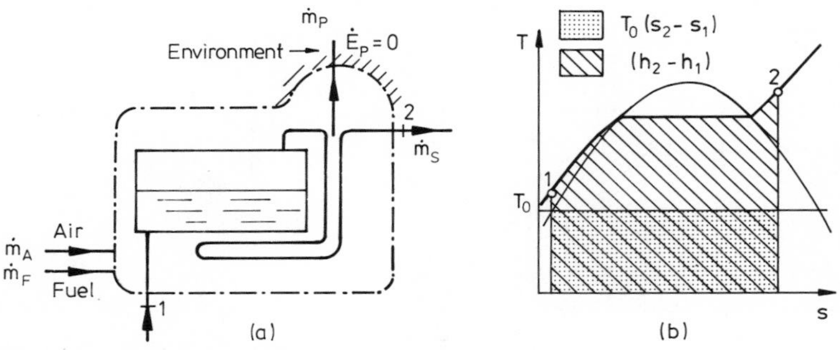

Fig 4.43 Steam boiler.

Also:

$$\varepsilon_2 - \varepsilon_1 = (h_2 - h_1) - T_0(s_2 - s_1) \tag{4.134}$$

(4.130), (4.133) and (4.134) yield the following expression for specific (per mass of steam) irreversibility of the boiler:

$$\dot{I}/\dot{m}_s = (h_2 - h_1)\left(\frac{\varphi}{\eta_{\text{comb}}} - 1\right) + T_0(s_2 - s_1) \tag{4.135}$$

The magnitude of the first term on the RHS of (4.135) is largely dependent on factors relating to the combustion process itself. In the special case when $\varphi = 1$, this term decreases as $\eta_{\text{comb}} \to 1$. In the limiting case the specific irreversibility is given by $T_0(s_2 - s_1)$.

(4.131), (4.133) and (4.134) give a relationship between the conventional combustion efficiency and the rational efficiency of the boiler:

$$\psi = \frac{\eta_{\text{comb}}}{\varphi}\left[1 - \frac{T_0(s_2 - s_1)}{h_2 - h_1}\right] \tag{4.136}$$

The quantities $T_0(s_2 - s_1)$ and $(h_2 - h_1)$ are represented by the dotted area and the cross-hatched area respectively in Fig 4.43(b). Therefore, the factor in the square brackets in (4.136) is always substantially less than unity and hence the numerical value of ψ is smaller than that of η_{comb}.

Clearly then, the processes taking place in a steam boiler always involve large irreversibilities, the main sources of which are the irreversible nature of the combustion process, the exhaust flue losses and the heat transfer between the high temperature gases and the relatively low temperature H_2O.

Example 4.3

Figure 4.44 is an idealised model of a methane reforming process. The control region corresponds to that part of the reformer furnace where the endothermic reaction takes

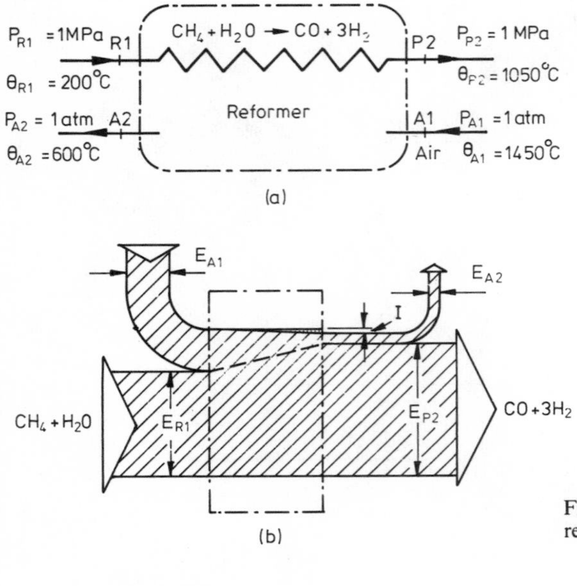

Fig 4.44 An idealised model of a methane reforming process.

place. To simplify the analysis, the products of combustion of the reformer furnace are replaced by a stream of air of corresponding parameters, as shown in the figure. The reactants consists of equal molar proportions of CH_4 and water vapour. It is assumed that the reforming process is fully completed so that the reformed gas consists of CO and H_2 in the molar proportions of 1 to 3. It is further assumed that pressure losses of both streams and heat losses are negligible. Calculate the irreversibility of the process per kmol of CH_4.

Solution: The basis of the calculation will be 1 kmol of CH_4.

Exergy balance

The required number of moles of air can be obtained from the energy balance. With $Q=0$, $W_x=0$, $\Delta \mathbf{E}_k=0$ and $\Delta \mathbf{E}_p=0$ (see Eq A.32):

$$n_A(h_{A1}-h_{A2})=(H_{ph.P2}-H_{ph.R1})+(H^0_{d.P2}-H^0_{d.R1}) \quad \text{(a)}$$

For air:

$$(h_{A1}-h_{A2})=(T_{A1}-T^0)\tilde{c}^{h}_{P,A1}-(T_{A1}-T^0)\tilde{c}^{h}_{P,A2} \quad \text{(b)}$$

From Table D.1 for air:

when $\theta=1\,450°C$ $\quad \tilde{c}^{h}_{P}=31.81$ kJ/kmolK

when $\theta=600°C$ $\quad \tilde{c}^{h}_{P}=30.55$ kJ/kmolK

Hence, using (b):

$$(h_{A1}-h_{A2})=1\,425\times 31.81-575\times 30.55$$
$$=27\,763 \text{ kJ/kmol}$$

Assuming the reactants and products to behave as ideal gases:

$$H_{\text{Mixture}}=\sum_i (n_i\tilde{h}_i) \quad \text{(c)}$$

Enthalpies of devaluation are obtained from Tables A.3 and A.4:

	CH_4	H_2O	CO	H_2
h^0_d/[kJ/kmol]	802 320	0	283 150	242 000

Hence from (c):

$$H^0_{d.P2}-H^0_{d.R1}=(283\,150+3\times 242\,000)-(802\,320+0)$$
$$=206\,830 \text{ kJ}$$

The change in the physical enthalpy can be expressed as:

$$H_{ph,P2}-H_{ph,R1}=(T_{P2}-T^0)(\tilde{c}^{h}_{P,CO}+3\tilde{c}^{h}_{P,H_2})-(T_{R1}-T^0)(\tilde{c}^{h}_{P,CH_4}+\tilde{c}^{h}_{P,H_2O}) \quad \text{(d)}$$

From Table D.1 values of $\tilde{c}^{h}_{P}$ are:

	200°C		1 050°C	
	CH_4	H_2O	CO	H_2
$\tilde{c}^{h}_{P}$/[kJ/kmol/K]	40.52	33.93	31.71	30.47

Substituting these numerical values in (d):

$$H_{ph.P2} - H_{ph.R1} = 1\,025(31.71 + 3 \times 30.47) - 175(40.52 + 33.93) = 113\,169 \text{ kJ}$$

Substituting the calculated values of enthalpy changes in (a):

$$n_A = \frac{113\,169 + 206\,830}{27\,763} = 11.51$$

The irreversibility occuring in the control region per kmol of CH_4 can be calculated now from the exergy balance:

$$I = (E_{A1} - E_{A2}) - (E_{P2} - E_{R1}) \quad \text{(e)}$$

Exergy of air

Since the air stream passes through the control region at the reference pressure of 1 atm only the isobaric components of exergy need be considered (see (4.116)). From Table D.3 when $\theta_{A1} = 1\,450°C$, $(\tilde{c}_P^{\varepsilon})_{A1} = 21.12$ kJ/kmolK. Hence for state A1:

$$E_{A1} = n_A(T - T^0)(\tilde{c}_P^{\varepsilon})_{A1} = 11.51 \times 1\,425 \times 21.12$$
$$= 346\,405 \text{ kJ}$$

At the air outlet state A2, $\theta_{A2} = 600°C$ and $(\tilde{c}_P^{\varepsilon})_{A2} = 13.68$ kJ/kmolK. Hence:

$$E_{A2} = 11.51 \times 575 \times 13.68 = 90\,538 \text{ kJ}$$

Exergy of the reactants and the products

Taking:

$$E_{R1} = n_R(\tilde{\varepsilon}_{ph.R1} + \tilde{\varepsilon}^0_{R1}) \quad \text{(f)}$$
$$E_{P2} = n_P(\tilde{\varepsilon}_{ph.P2} + \tilde{\varepsilon}^0_{P2}) \quad \text{(g)}$$

The molar standard chemical exergy from the reactants and the products is calculated from:

$$\tilde{\varepsilon}^0 = \sum x_i \tilde{\varepsilon}_i^0 + \tilde{R}T^0 \sum x_i \ln x_i \quad \text{(h)}$$

Values of $\tilde{\varepsilon}_i^0$ are obtained from Table A.3:

	CH_4	H_2O	CO	H_2
$\tilde{\varepsilon}^0$/[kJ/kmol]	836 930	11 710	275 430	238 490

For the reactants:

$$\tilde{\varepsilon}^0_{R1} = (0.5 \times 836\,930 + 0.5 \times 11\,710)$$
$$+ 8.3144 \times 298.15(0.5 \ln 0.5 + 0.5 \ln 0.5)$$
$$= 422\,603 \text{ kJ/kmol}$$

For the products:

$$\tilde{\varepsilon}^0_{P2} = (0.25 \times 275\,430 + 0.75 \times 238\,490)$$
$$+ 8.3144 \times 298.15(0.25 \ln 0.25 + 0.75 \ln 0.75)$$
$$= 246\,331 \text{ kJ/kmol}$$

When there is a change in the chemical composition of a stream then both thermal and pressure components of exergy must be taken into account as given by (4.123), even if there is no change in pressure. From Table D.3:

	200°C		1 050°C	
	CH_4	H_2O	CO	H_2
$\tilde{c}_P^{\varepsilon}$/[kJ/kmolK]	8.93	7.22	18.23	17.44

Using these values, for the reactants:

$$\tilde{\varepsilon}_{\mathrm{ph.R\,1}} = 175(0.5 \times 8.93 + 0.5 \times 7.22) + 8.3144 \times 298.15 \ln \frac{1}{0.101\,325}$$

$$= 1\,413 + 5\,672 = 7\,085 \text{ kJ/kmol}$$

Similarly, for the products:

$$\tilde{\varepsilon}_{\mathrm{ph.P\,2}} = 1\,025(0.25 \times 18.23 + 0.75 \times 17.44) + 5\,672$$
$$= 23\,750 \text{ kJ/kmol}$$

Substituting the calculated physical and chemical exergy components in (f) and (g):

$$E_{R\,1} = 2(7\,085 + 422\,603)$$
$$= 859\,376 \text{ kJ}$$
$$E_{P\,2} = 4(23\,750 + 246\,331)$$
$$= 1\,080\,324 \text{ kJ}$$

Substituting the calculated values of physical and chemical energy in (e):

$$I = (346\,405 - 90\,538) - (1\,080\,324 - 859\,376)$$
$$= \underline{34\,919 \text{ kJ per kmol of } CH_4}$$

Figure 4.44(b) shows the exergy balance for the control region in the form of a Grassmann diagram. Note that the exergy of the products, the reformed gas, shows a gain in relation to the reactant gases, at the expense of the exergy of the air stream. The irreversibility is mainly due to heat transfer over a finite temperature difference which takes place between the two gas streams. Although the irreversibility of this process is small, it must be appreciated that this is only one of several processes taking place in the plant. Other processes such as combustion, gas compression, and gas cooling are likely to be associated with much larger irreversibilities[4.10].

Example 4.4

In a coal fired steam boiler (Fig 4.45) steam is generated at the constant pressure of 20 MPa. The coal used is anthracite of the composition specified in Example C.1. Both the fuel and the air are delivered at the standard pressure and temperature. Air excess of 30 per cent over the stoichiometric requirement is provided. Assuming no heat losses and no pressure losses calculate the irreversibilities per 100 kg of the fuel burned occurring in:

Sub-region I, corresponding to adiabatic combustion.

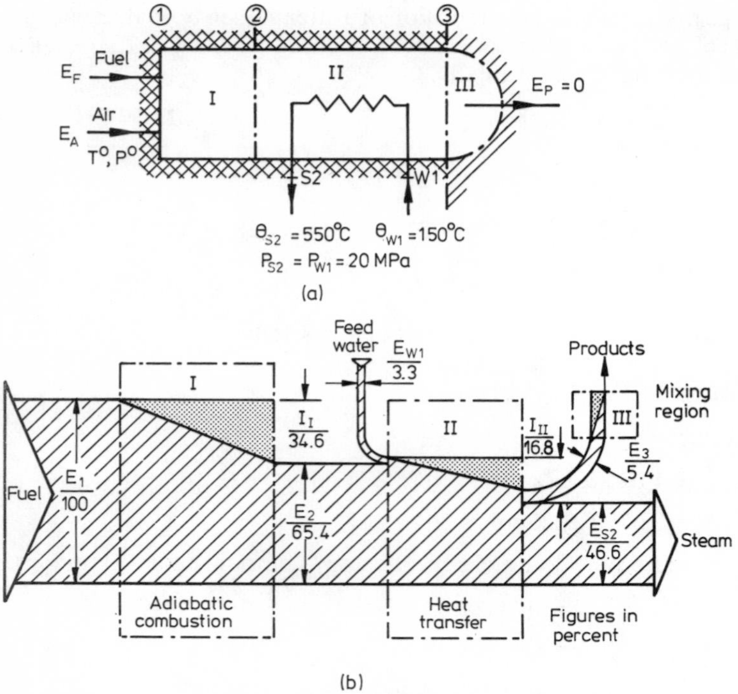

Fig 4.45 Coal fired steam boiler.

Sub-region II, corresponding to heat transfer to steam.
Sub-region III, region of mixing of the products of combustion with ambient air.

Also, calculate the rational efficiency of the boiler and construct the Grassmann diagram for the plant.

Solution: From Example C.1, for the fuel $(\text{NCV})^0 = 28\,940$ kJ/kg and $\varepsilon_F^0 = 30\,705$ kJ/kg. A mass balance shows the products of combustion to have the following composition:

	CO_2	H_2O	N_2	O_2	SO_2	Total
n_k/[kmol/100 kg fuel]	6.51	1.634	35.22	9.324	0.047	52.735
x_k	0.1234	0.0310	0.6679	0.1768	0.0009	1.000

Sub-region I, calculation of θ_2

As the coal and air enter the adiabatic combustion space at T^0 and P^0, the energy balance for the process (A.32 and A.33) may be written:

$$m_F(\text{NCV})^0 = \sum_k n_k \tilde{h}_{\text{ph},k}$$

$$= (\theta_2 - \theta^\circ) \sum_k n_k \tilde{c}_{P,k}^{\text{h}} \qquad \text{(a)}$$

The sum $\sum_k n_k \tilde{c}^{\text{h}}_{P,k}$ can be determined from values of $\tilde{c}^{\text{h}}_P$ extracted from Table C.1 for a trial value of θ_2. The iteration procedure gives the following values of $\tilde{c}^{\text{h}}_P$ corresponding to 1 550°C:

	CO_2	H_2O	N_2	O_2	SO_2
$\tilde{c}^{\text{h}}_P$/[kJ/kmolK]	53.24	40.78	32.69	34.35	52.57

From these values $\sum_k n_k \tilde{c}^{\text{h}}_{P,k} = 1\,887.3$ kJ/K. Substituting in (a):

$$\theta_2 = \theta^0 + \frac{m_{\text{F}}(\text{NCV})^0}{\sum_k n_k \tilde{c}^{\text{h}}_{P,k}} = 25 + \frac{100 \times 28\,940}{1\,887.3} = 1\,558°\text{C}$$

Since the calculated value of θ_2 is close enough to the previous trial value (1 550°C), no further iteration is required.

Irreversibility in Sub-region I

Since $\text{E}^{Q} = 0$, $W_x = 0$ and $\text{E}^0_{\text{air}} = 0$, the exergy balance for this case yields ((4.112) and (4.114)):

$$I_{\text{I}} = \text{E}_{\text{F}} - \text{E}_2 \tag{e}$$

where:

$$\text{E}_{\text{F}} = m_{\text{F}}\,\varepsilon^0_{\text{F}}$$

$$\text{E}_2 = n_{\text{P}2}\tilde{\varepsilon}^0_{\text{P}2} + \sum_k n_k \tilde{\varepsilon}_{\text{ph},k} \tag{g}$$

$$\tilde{\varepsilon}^0_{\text{P}2} = \sum_k x_k \tilde{\varepsilon}^0_k + \tilde{R}T^0 \sum_k x_k \ln x_k \tag{h}$$

From Table A.3:

	CO_2	H_2O	N_2	O_2	SO_2
$\tilde{\varepsilon}^0$/[kJ/kmol]	20.140	11 710	720	3 970	303 500

Using this data and the molar composition of the products given earlier in expression (h):

$$\tilde{\varepsilon}^0_{\text{P}2} = 1\,953.9 \text{ kJ/kmol}$$

The physical component of exergy of the products is evaluated using the data extracted from Table D.3 for $\theta = 1\,550$°C:

	CO_2	H_2O	N_2	O_2	SO_2
$\tilde{c}^{\varepsilon}_P$/[kJ/kmolK]	35.49	26.91	21.46	22.60	34.84

Hence for $P = P^0 = \text{const}$:

$$\sum_k n_k \tilde{\varepsilon}_{\text{ph},k} = (\theta_2 - \theta^0) \sum_k n_k \tilde{c}^{\varepsilon}_{P,k}$$

$$= (1\,558 - 25)1\,243.3 = 1\,905\,800 \text{ kJ}$$

and from (g):
$\text{E}_2 = 52.735 \times 1\,953.9 + 1\,905\,800 = \underline{2.009 \times 10^6 \text{ kJ}}$

Substituting the calculated values of the exergy terms in (e):

$$I_{\text{I}} = 100 \times 30\,705 - 2.009 \times 10^6$$
$$= \underline{1\,061\,600 \text{ kJ per 100 kg of fuel}}$$

Sub-region II, energy balance

The mass of H_2O, m_s, corresponding to the given operating conditions can be calculated from the energy balance:

$$H_2 - H_s = m_s(h_{s2} - h_{w1}) \tag{i}$$

From steam tables:

$$h_{s2} = 3\,394 \text{ kJ/kg}$$
$$h_{w1} = 644.5 \text{ kJ/kg}$$

and as calculated earlier:

$$H_2 = 1\,887.3 \times (1\,558 - 25)$$
$$= 2\,893\,200 \text{ kJ}$$

H_3 will be calculated from $\tilde{c}_P^{\text{h}}$ values extracted from Table C.1 for $\theta_3 = 200°\text{C}$:

	CO_2	H_2O	N_2	O_2	SO_2
$\tilde{c}_P^{\text{h}}/[kJ/kmolK]$	41.52	33.93	29.59	29.50	42.88

Hence:

$$H_3 = (\theta_3 - \theta^0) \sum_k n_k \tilde{c}_{P,k}^{\text{h}}$$
$$= 175 \times 1\,645 = 287\,880 \text{ kJ}$$

Substituting numerical values in (i):

$$m_s = \frac{2\,893\,200 - 287\,880}{3\,394 - 644.5} = 947.58 \text{ kg}$$

Irreversibility

Irreversibility in this sub-region is given by:

$$I_{\text{II}} = (\text{E}_2 - \text{E}_3) - (\text{E}_{s2} - \text{E}_{w1}) \tag{j}$$

For the water substance:

$$s_{s2} = 6.337 \text{ kJ/kgK}$$
$$s_{w1} = 1.821 \text{ kJ/kgK}$$

Hence:

$$\text{E}_{s2} - \text{E}_{w1} = m_s[(h_{s2} - h_{w1}) - T^0(s_{s2} - s_{w1})]$$
$$= 947.58[(3\,394 - 644.5) - 298.15(6.337 - 1.821)]$$
$$= 1\,329\,500 \text{ kJ}$$

E_3 will be calculated from the following data obtained from Table C.3 for $\theta_3 = 200°\text{C}$:

	CO_2	H_2O	N_2	O_2	SO_2
$\tilde{c}_P^\varepsilon/[\text{kJ/kmolK}]$	9.09	7.22	6.34	6.45	9.31

For $P=P^0=\text{const}$:

$$(\mathrm{E}_{\mathrm{ph}})_3=(\theta_3-\theta^0)\sum_k n_k\tilde{c}_{P,k}^\varepsilon$$

$$=175\times 354.85=62\ 098\ \text{kJ}$$

$$\mathrm{E}_3=\mathrm{E}_3^0+(\mathrm{E}_{\mathrm{ph}})_3$$

$$=52.735\times 1\ 953.9+62\ 098$$

$$=165\ 140\ \text{kJ}$$

Substituting in (j):

$$I_{\mathrm{II}}=(2\ 009\ 000-165\ 140)-(1\ 329\ 500)$$

$$=\underline{514\ 360\ \text{kJ}}$$

Sub-region III

In this sub-region all the exergy of the gases given by E_3 is lost through dissipation (mixing, cooling, *etc*). Hence:

$$I_{\mathrm{III}}=\mathrm{E}_3=\underline{165\ 140\ \text{kJ}}$$

Rational efficiency of the boiler

Since the function of the boiler is to increase the exergy of the stream of water substance and take the exergy of the fuel as the input:

$$\psi=\frac{\mathrm{E}_{\mathrm{s}2}-\mathrm{E}_{\mathrm{w}1}}{\mathrm{E}_{\mathrm{F}}}$$

$$=\frac{1\ 430\ 130-100\ 623}{3\ 070\ 500}=\underline{0.433}$$

This value may be compared with the conventional boiler efficiency known as combustion efficiency:

$$\eta_{\mathrm{comb}}=\frac{m_{\mathrm{s}}(h_{\mathrm{s}2}-h_{\mathrm{w}1})}{m_{\mathrm{F}}(\mathrm{NCV})}$$

$$=\frac{947.58(3\ 394-644.5)}{100\times 28\ 940}$$

$$=0.90$$

As will be seen from the Grassmann diagram, the rational efficiency of this plant takes into account three types of losses: intrinsic irreversibility of the adiabatic combustion process I_{I}; irreversibility due to heat transfer over a finite temperature difference I_{II}; and irreversibility due to dissipation of exergy of the products of combustion I_{III}. In the case of η_{comb}, the main loss considered is the loss of energy of the products of combustion.

Chapter 5 Examples of thermal and chemical plant analysis

This chapter demonstrates the application of the basic principles and techniques already presented in the analysis of whole thermal and chemical plants. Four typical plants have been selected: the Linde air liquefaction plant; a sulphuric acid plant; a gas turbine plant; and a refrigeration plant.

5.1 Linde air liquefaction plant

The Linde air liquefaction plant has strong interactions between its components and relatively small changes in the plant structure can give substantial improvement in plant performance. A simple Linde plant and then a modified plant with auxiliary refrigeration will be analysed.

Simple Linde process

Figure 5.1 is a plant schematic and Fig 5.2 shows the thermodynamic processes in *T–s* co-ordinates. The state given by points 0–6 correspond to the simple Linde process*. Compression (0–1) is represented symbolically in both figures, without specifying the number of stages or intercoolers. Compressor operation will be specified adequately by its isothermal and mechanical efficiency. In the plant arrangement used here, all air entering the compressor is fresh atmospheric air; the air leaving the heat exchanger, state 6, is discharged into the atmosphere.

Basis: In the analysis which follows all the values of exergy and irreversibility will be specified per kg of liquefied air.

Given operating parameters

Environmental state, $P_0 = 0.1$ MPa, $T_0 = 293$ K.
Compressor, air exit temperature, $T_1 = 293$ K.
Maximum air pressure, $P_1 = 20$ MPa.
Air discharge temperature, $T_6 = 288$ K.
Electric motor efficiency, $\eta_{el} = 0.95$.
Compressor mechanical efficiency, $\eta_m = 0.90$.
Isothermal efficiency, $\eta_{iso} = 0.70$.

* Points 7–9 relate to the process with auxiliary refrigeration.

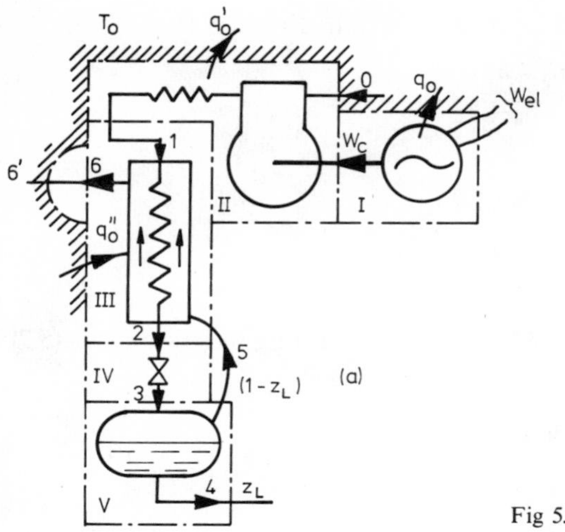

Fig 5.1 Plant diagram for the simple Linde process.

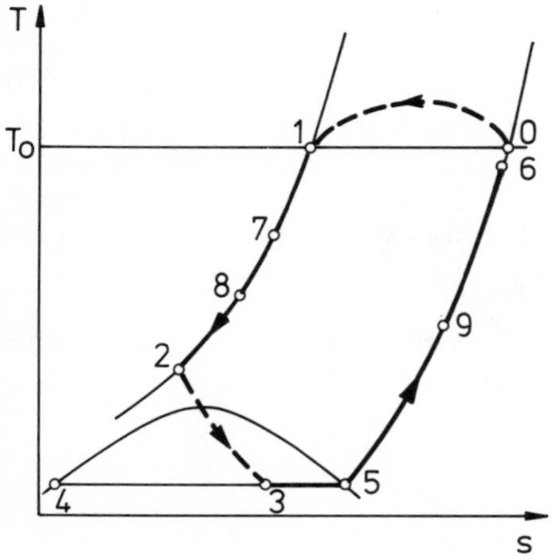

Fig 5.2 Temperature–entropy diagram for the Linde processes.

Principal assumptions

(i) Pressure losses in the heat exchanger are negligible.
(ii) A 'heat leak' to the heat exchanger of 2.1 kJ per kg of fresh air is assumed. Other plant components, except for the electric motor and the compressor, will be assumed to operate adiabatically.
(iii) Kinetic and potential components of energy (and exergy) of the air stream are negligible.

Energy analysis of the plant

From the energy balance for sub-regions III, IV and V:

$$h_1 + q_0'' = z_L h_4 + (1 - z_L) h_6 \tag{5.1}$$

per kg of fresh air, where z_L is liquefied fraction of the air. From (5.1):

$$z_L = \frac{h_6 - h_1 - q_0''}{h_6 - h_4} \quad (5.2)$$

From Fig E.1, for the process:

$$h_1 = 406 \text{ kJ/kg}$$
$$h_4 = 25 \text{ kJ/kg}$$
$$h_6 = 437 \text{ kJ/kg}$$

Also, as stated above, $q_0'' = 2.1$ kJ per kg of fresh air. Substituting these values in (5.2):

$$z_L = 0.070$$

Hence, the mass of fresh air per kg of liquid air is:

$$m_a = 1/z_L = \underline{14.25 \text{ kg}}$$

For adiabatic throttling process:

$$h_2 = h_3$$

Since h_3 lies in the two phase region:

$$h_3 = h_4 + (h_5 - h_4)(1 - z_L) \quad (5.3)$$

The value of h_5 from Fig E.2 is:

$$h_5 = 230 \text{ kJ/kg}$$

Substituting the numerical values of h_4, h_5 and z_L in (5.3):

$$h_2 = h_3 = 215.7 \text{ kJ/kg}$$

The thermodynamic properties for the different states of the process are given in Table 5.1. The reversible isothermal work of the compressor is given by:

$$W_{iso} = m_a R T_0 \ln \frac{P_1}{P_0}$$

$$= 14.25 \times 0.287 \times 293 \ln \frac{20}{0.1}$$

$$= 6349 \text{ kJ} = \underline{6.349 \text{ MJ}}$$

Table 5.1

	State						
Quantity	0	1	2	3	4	5	6
m/kg	14.25	14.25	14.25	14.25	1.0	13.25	13.25
h/[kJ/kg]	443	406	215.7	215.7	25	230	437
T/K	293	293	174	82	82	82	288
ε/[kJ/kg]	0.0	443	502	203	702	166	0.12
E/MJ	0.0	6.313	7.154	2.893	0.702	2.200	0.002

From the definitions of isothermal efficiency and mechanical efficiency:

$$W_{ind} = W_{iso}/\eta_{iso} = 6349/0.70 \text{ kJ} = 9.070 \text{ MJ}$$
$$\text{and}: W_c = W_{ind}/\eta_m = 9070/0.9 \text{ kJ} = 10.078 \text{ MJ}$$

The required input of electrical energy is:

$$W_{el} = W_c/\eta_{el} = 10078/0.95 \text{ kJ} = \underline{10.608 \text{ MJ}}$$

which represents the exergy input to this plant.

Exergy analysis

Construction of the Grassmann diagram for the plant requires the irreversibilities corresponding to the different sub-regions (Fig 5.1). These will be calculated from the exergy balance. Specific exergies for the different states will be obtained from the exergy-enthalpy chart in Fig E.2.

Sub-region I: electric motor

The only exergy terms in the exergy balance for this sub-region are the work quantities W_{el} and W_c. Hence:

$$I_I = W_{el} - W_c = 10.608 - 10.078 = 0.530 \text{ MJ}$$

shown on the Grassmann diagram in Fig 5.3.

Sub-region II: compressor

As all the thermal energy is rejected to the environment, $E^Q = 0$. Also, since the air entering the compressor is in the environmental state, $E_0 = 0$. With $W_x = -W_c$ the exergy balance for this case reduces to:

$$I_{II} = W_c - E_1 \tag{5.4}$$

Taking the value of E_1 from Table 5.1, the irreversibility for the compressor is:

$$I_{II} = 10.078 - 6.313 = \underline{3.765 \text{ MJ}}$$

which may be considered to comprise two components, one associated with mechanical losses and equal to the difference between the shaft work and the actual work done on the gas (ie the indicated work), and the other, called *internal irreversibility*, is equal to the difference between the indicated work and the reversible work of compression given by E_1. Hence:

$$I_{II} = (I_{II})_{mech} + (I_{II})_{int} \tag{5.5}$$

where:

$$(I_{II})_{mech} = W_c - W_{ind} \tag{5.6}$$

$$(I_{II})_{int} = W_{ind} - E_1 \tag{5.7}$$

Substituting numerical values:

$$(I_{II})_{mech} = 10.078 - 9.070 = 1.008 \text{ MJ}$$

$$(I_{II})_{int} = 9.070 - 6.313 = 2.757 \text{ MJ}$$

Figure 5.3 shows the calculated values of exergy and the two components of irreversibility for this sub-region.

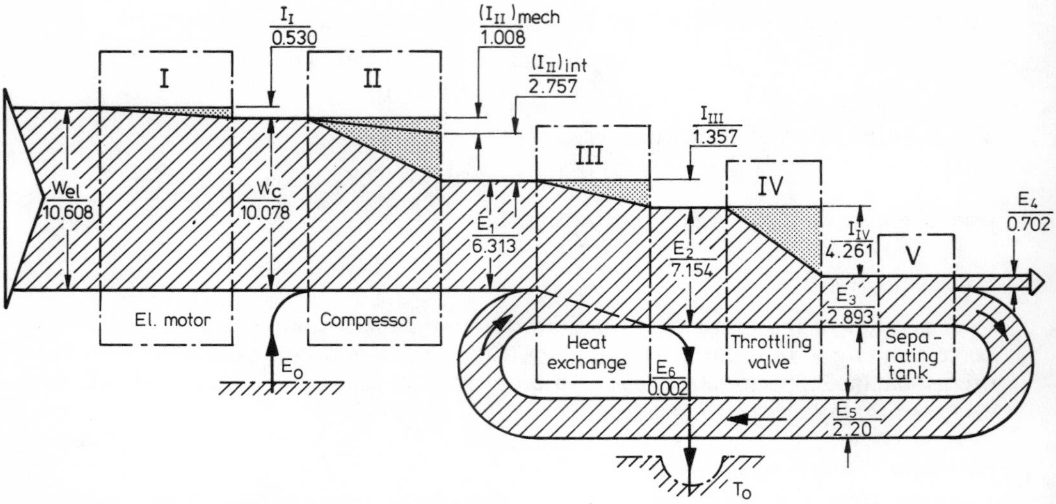

Fig 5.3 Grassmann diagram for the simple Linde process. Values of I and E in MJ per kg of liquefied air.

Sub-region III: heat exchanger

The exergy balance for the heat exchanger takes the form:

$$I_{III} = (E_5 - E_6) - (E_2 - E_1) \quad (5.8)$$

Taking the values of exergy from Table 5.1, the irreversibility for heat exchanger, from (5.8), is:

$$I_{III} = \underline{1.357 \text{ MJ}}$$

Although this is a counter-flow heat exchanger, the exergy stream E_5 is shown in Fig 5.3 entering from the same direction as E_1. This is justified because both E_5 and E_1 represent flows of exergy into the sub-region. The exergy flux E_6, which leaves the sub-region, has only a very small value. As shown symbolically in the diagram, it is, ultimately, dissipated in the environment.

Sub-region IV: throttling valve

In this case the exergy balance is:

$$I_{IV} = E_2 - E_3 \quad (5.9)$$

Substituting E_2 and E_3 from Table 5.1:

$$I_{IV} = \underline{4.261 \text{ MJ}}$$

Sub-region V: separating tank

Here liquid is separated from the vapour under saturation conditions, using gravity. The process is virtually reversible, hence:

$$I_V = 0$$

The exergies of the separated liquid and vapour streams, E_4 and E_5 respectively, are given in Table 5.1 and shown graphically in Fig 5.3. The exergy flux E_4 represents the exergy output of the plant. The air stream associated with E_5 is taken back to the regenerative heat exchanger.

Plant rational efficiency

From the general principles of formulation of rational efficiency, for this plant:

$$\psi = \frac{E_4}{W_{el}} \tag{5.10}$$

Substituting the numerical values in (5.10):

$$\psi = \frac{0.702}{10.608} = \underline{0.0662}$$

Thus the efficiency of this plant is very low. Figure 5.3 shows that there are large irreversibilities associated with processes occurring in the compressor, the heat exchanger and the throttling valve. Efficiency of throttling valves in cryogenic applications has been discussed in Section 4.1 where it was shown that, as the temperature of the stream entering the valve approaches the saturation temperature, T_3, of the vapour–liquid mixture which emerges from the valve, the irreversibility of the process gets smaller and the liquid fraction, z_L, gets larger. This clearly points to the need for a reduction in T_2. As the temperature difference at the 'warm end' of the heat exchanger is already quite small, no appreciable reduction in T_2 can be achieved through an improvement in the heat transfer conditions, eg increasing the size of the heat exchanger. The large temperature difference $T_2 - T_5$ at the 'cold end' of the heat exchanger, and hence the resulting irreversibility, is due to the difference in the heat capacities of the two air streams.

One solution to this problem is auxiliary refrigeration of the forward air stream between states 1 and 2.

Linde process with auxiliary refrigeration

Figure 5.4 shows the plant layout for this process and the additional thermodynamic states which must be specified for its description are shown as 7, 8 and 9 in Fig 5.2. All the operating parameters and simplifying assumptions used for the simple Linde plant will be taken to apply to this version of the plant. With regard to 'heat leak' it will be assumed that in this plant the value used previously of 2.1 kJ per kg of fresh air will be divided, as follows, among the three heat exchangers:

$$q_0^{III} = 0.4 \text{ kJ}$$
$$q_0^{IV} = 0.7 \text{ kJ}$$
$$q_0^{V} = 1.0 \text{ kJ}$$

Temperatures corresponding to states 8 and 9 should not be chosen arbitrarily, but determined in a thermoeconomic optimisation. For this example, the temperatures adopted are:

$$T_8 = 200 \text{ K}, \qquad T_9 = 195 \text{ K}$$

The corresponding values of specific enthalpy (Fig E.2) are

$$h_8 = 267 \text{ kJ/kg}, \qquad h_9 = 346 \text{ kJ/kg}$$

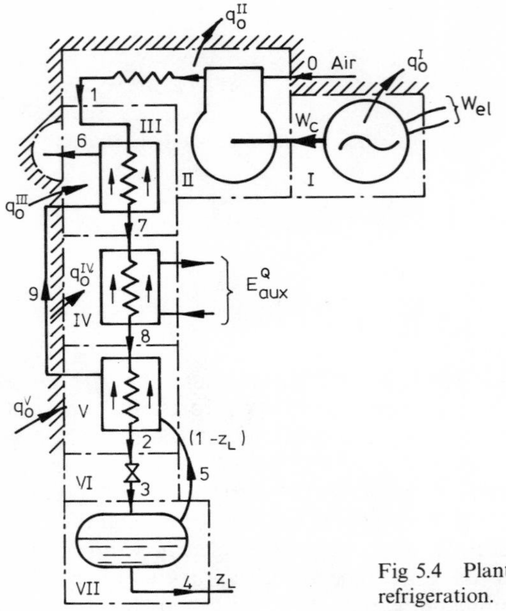

Fig 5.4 Plant diagram from the Linde process with auxiliary refrigeration.

The new value of z_L can be obtained from an energy balance for an open system consisting of sub-regions V, VI and VII which leads to

$$z_L = \frac{h_9 - h_8 - q_0^V}{h_9 - h_4}$$

Substituting the numerical values:

$$z_L = 0.243$$

The corresponding mass of fresh air is:

$$m_a = 1/z_L = 4.12 \text{ kg}$$

This mass of fresh air drawn per kg of liquified air is considerably less than the previous value of 14.25 kg. As the efficiencies of the components in sub-regions I and II are, in this case, the same as in the simple Linde process, the exergy values and irreversibilities will be smaller proportionately to the reduction in m_a. Hence, scaling down the values calculated for the simple Linde process, for the electric motor:

$$W_{el} = 3.052 \text{ MJ} \qquad I_I = 0.152 \text{ MJ}$$

and for the compressor:

$$W_c = 2.900 \text{ MJ} \qquad E_1 = 1.816 \text{ MJ}$$

$$(I_{II})_{mech} = 0.290 \text{ MJ} \qquad (I_{II})_{int} = 0.794 \text{ MJ}$$

Applying the energy balance to sub-region III:

$$h_7 = 337.8 \text{ kJ/kg}$$

and hence from Fig E.2, $T_7 = 242$ K.

Table 5.2

	State									
Quantity	0	1	2	3	4	5	6	7	8	9
m/kg	4.12	4.12	4.12	4.12	1	3.12	3.12	4.12	4.12	3.12
h/[kJ/kg]	443	406	180.2	180.2	25	230	437	337.8	269	345
T/K	293	293	155	82	82	82	288	242	200	195
ε/[kJ/kg]	0.0	443	529	300	702	166	0.12	450	472	23
E/kJ	0.0	1 816	2 169	1 230	702	515	0.37	1 845	1 935	71.3

The new value of h_3 can be calculated using (5.3). Thus:

$$h_3 = h_2 = 180.2 \text{ kJ/kg}$$

Using the calculated values of thermodynamic properties and the ε–h chart (Fig E.2), all the thermodynamic properties relevant to the process can be obtained (Table 5.2). The auxiliary refrigeration required corresponds to an isobaric process between states 7 and 8. Thus:

$$Q_{\text{aux}} = m_{\text{a}}(h_8 - h_7)$$
$$= -282.1 \text{ kJ/kg}$$

Assuming that the refrigerant brought into the heat exchanger performs this duty while evaporating at the constant temperature of $T_{\text{ref}} = 195$ K, the value of the auxiliary thermal exergy can be calculated from:

$$\mathrm{E}^{\mathrm{Q}}_{\text{aux}} = \mathrm{Q}_{\text{aux}} \frac{T_{\text{ref}} - T_0}{T_{\text{ref}}} = 0.142 \text{ MJ}$$

The remaining values of exergy and irreversibilities are calculated as shown earlier for the simple Linde process. From these values the Grassmann diagram in Fig 5.5 has been constructed. Since all the values of exergy and irreversibility are presented in both Figs 5.3 and 5.5 in MJ per kg of liquified air, the results of the analysis of the two plants can be compared easily.

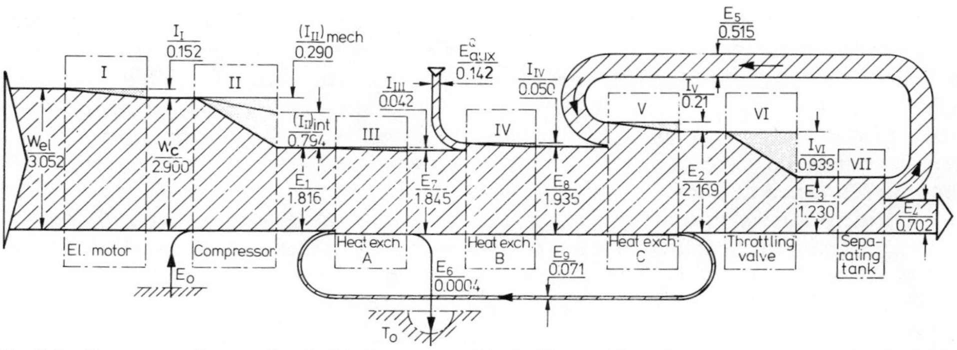

Fig 5.5 Grassmann diagram for the Linde process with auxiliary refrigeration. Values of I and E in MJ per kg of liquefied air.

Discussion

Clearly, from comparison of the Grassmann diagrams for the two versions of the plant, the introduction of auxiliary refrigeration has substantially improved process efficiency. In formulating the rational efficiency for the modified plant, the thermal exergy corresponding to the auxiliary refrigeration must be included as a part of the input. Thus:

$$\psi' = \frac{\mathrm{E}_4}{W_{el} + \mathrm{E}^{Q}_{aux}} = 0.2198$$

The rational efficiency of the process increases in the ratio of 0.2198/0.0662 = 3.32. However, the plant which would be necessary to produce the auxiliary refrigeration was excluded from the control region, so this ratio is unduly optimistic. For a fair comparison, consider the amount of electrical energy $(W_{el})_{aux}$, which would be required to generate the necessary auxiliary refrigeration. Assuming that $(W_{el})_{aux}/\mathrm{E}^{Q}_{aux} = 10$, ie the rational efficiency of the auxiliary refrigeration plant is 0.1, the rational efficiency formulated on this basis is:

$$\psi'' = \frac{\mathrm{E}_4}{W_{el} + (W_{el})_{aux}} = 0.1570$$

Thus, even in this case the improvement in the efficiency of the process is 0.1570/0.0662 = 2.37.

The success of this modification in improving the efficiency is due to the strong interaction between the components of this plant. A reduction in the mean temperature difference between two streams exchanging heat normally reduces the specific irreversibility of the process. Here, this results in the reduction in T_2, leading to an increase in z_L, which, for given plant output, reduces fresh air intake, m_a. Lower m_a yields a corresponding decrease in the irreversibilities of all the components, even without any improvement in their specific irreversibilities or rational efficiencies. This has been shown to be the case for the electric motor and the compressor. The irreversibility of the heat-exchange process is therefore doubly affected, by reducing mean temperature difference and by reducing m_a. This is shown[2.2] graphically in τ–ΔH co-ordinates in Fig 5.6. As explained in Section 3.6, the shaded area between the process curves represents irreversibility due to heat transfer over a finite temperature difference, $I^{\Delta T}$. Note that the area corresponding to the process with auxiliary refrigeration is much smaller than that for the simple Linde process. The effect of a reduction in $\Delta\tau$, which has its minimum value (a pinch point) at points 8 and 9, and the effect of a reduction in the overall change in enthalpy, ΔH, can be seen from the diagram. The subject of mutual interactions, known as structural bonds, between plant components and its relevance to plant optimization will be discussed in Chapter 6.

In principle, irreversibility due to 'heat leaks' can be shown on the τ–ΔH diagram if information on the heat flux and the corresponding surface temperature is available for the heat exchanger surface. This was not the case in this example since the 'heat leaks' were specified as lumped values for each heat exchanger. However, as these values were relatively small in relation to the enthalpy changes of the air streams, the omission of this form of irreversibility from the diagram does not affect materially the conclusions regarding the effect of auxiliary refrigeration on plant performance.

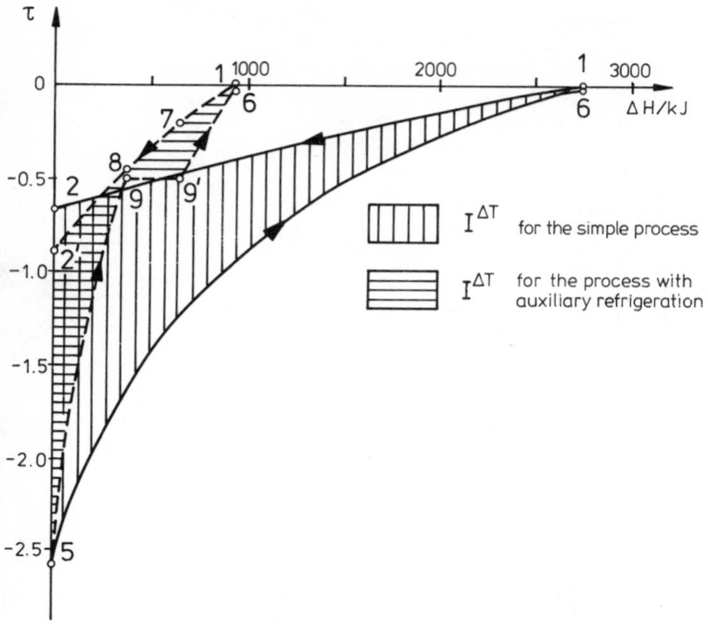

Fig 5.6 Dimensionless exergetic temperature–enthalpy change diagram.

5.2 Sulphuric acid plant

The sulphuric acid plant considered here contains a number of elements common to many large-scale chemical plants. The special characteristics of this plant are two types of exothermic reactions releasing large quantities of thermal energy; this by-product of the process must be used in the best possible way. In the present plant, the thermal energy is used partly for internal purposes (sulphur melting) but mostly for steam raising for external applications as process steam and for power generation. Some of the electric power produced is used for driving pumps, blowers and other auxiliary equipment. Since, as will be seen later, the magnitude of the two forms of output, in exergy terms, are comparable, their production should be properly integrated for high overall plant rational efficiency.

Description of the plant

Figure 5.7(a) is a simplified version of the sulphur-burning contact plant. Secondary effects such as 'heat losses' and pressure losses are neglected. As the process is dominated by transformation from chemical exergy to the thermal component of exergy and by heat transfer between the streams, the pressure component of exergy, $\varepsilon^{\Delta P}$, plays a relatively small part in the exergy balance.

Sulphur is fed in the environmental state to the sulphur-melting tank shown in Sub-region I (S-R I). Molten sulphur and clean, dehumidified atmospheric air are introduced into the sulphur burner where SO_2 is produced. The products of combustion are used in Waste-Heat Boiler A, S-R III, to raise process steam and to provide dry saturated steam for melting sulphur. The gases are then filtered and passed through catalytic

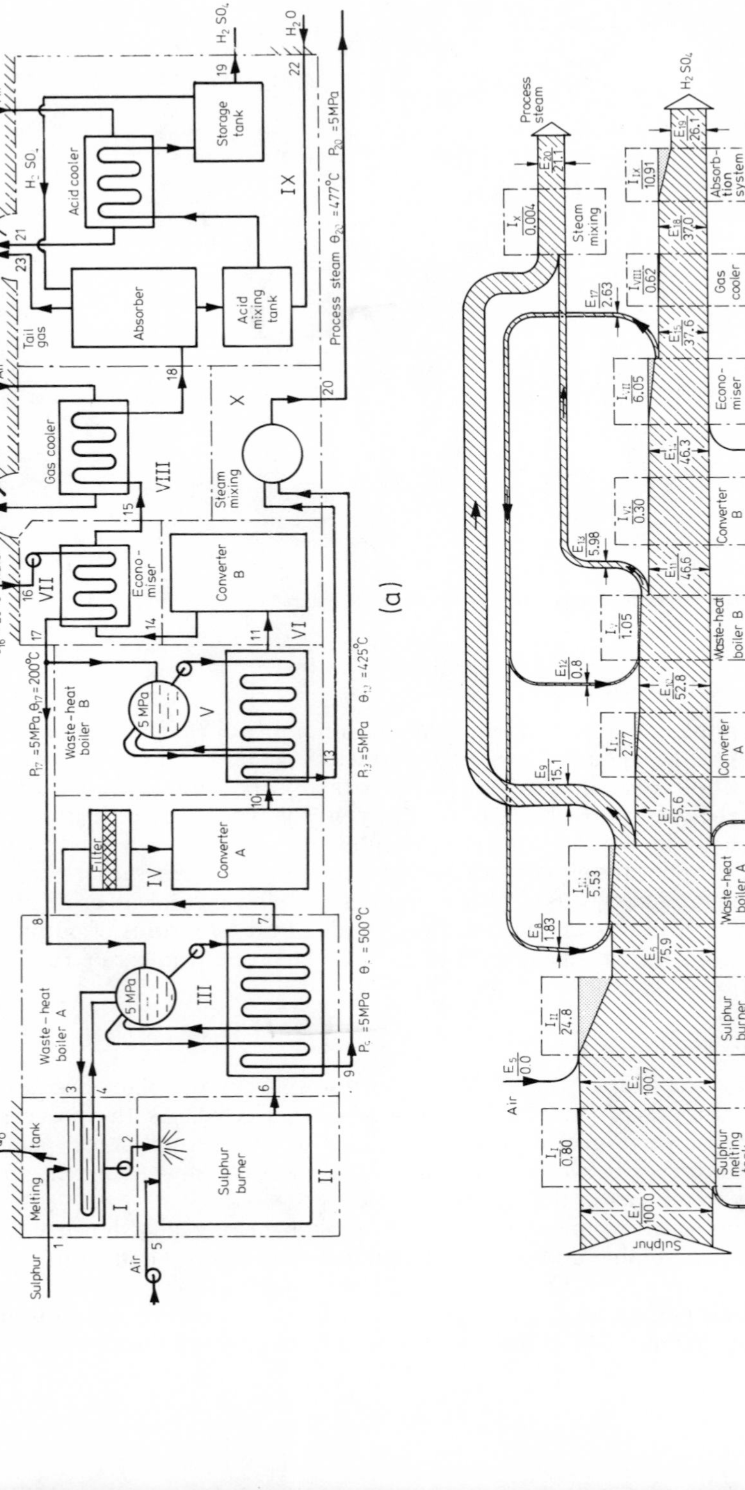

Fig 5.7 Sulphuric acid plant, (a) plant diagram, (b) Grassmann diagram. Values of *I* and E are given as percentages of plant input.

Converter A, S-R IV, which converts 60% of SO_2 to SO_3. The increase in the temperature of the gases from the exothermic reaction generates more process steam in Waste-Heat Boiler B, S-R V, while reducing the temperature to the required level for the second converter. At the exit from Converter B, (S-R VI), 97% of the original SO_2 has been converted to SO_3. The temperature increase resulting from the reaction in Converter B is used in the economiser, S-R VII, for pre-heating feed water for both the waste-heat boilers. The gases are then cooled by a stream of air (S-R VIII) before passing to the absorption system, S-R IX. To simplify the calculations, all the sulphuric acid output is taken as 100% H_2SO_4. The tail gas from the absorber, stream 23, is discharged into the atmosphere. The usual arrangement for gas cleaning before its discharge is not shown.

Steam is generated in both the waste-heat boilers at the same pressure of 5 MPa but is superheated to different temperatures. The two streams are therefore mixed (S-R X) before delivery to the outside user.

Most of the operating parameters necessary for the exergy analysis of the plant are shown in Fig 5.7(a). Additional parameters required are: $\theta_2 = 180°C$; $\theta_7 = 325°C$; $\theta_{11} = 300°C$; $\theta_{18} = 50°C$; and $\theta_{19} = 25°C$.

Principal assumptions

1. Pressure losses are negligible.
2. 'Heat losses', except where indicated, are negligible.
3. For all gaseous streams, except steam, ideal gas behaviour will be assumed.
4. Changes in kinetic and potential energy are negligible.

Plant analysis

The analysis of the plant shown below will be given in full for Sub-regions I, II, III, IV, IX and X. The remaining sub-regions are offered to the reader as exercises.

Basis: 1 kmol of sulphur.

Sub-region I: melting tank

The melting tank process, 1–2, can be considered conveniently in three stages (Fig 5.8):

(i) Heating of solid sulphur, process $1 - s$. Change of enthalpy:

$$Q_{1,s} = H_s - H_1 = \tilde{c}_{P,\mathrm{sol}}(T_{\mathrm{sf}} - T_1) \tag{5.12}$$

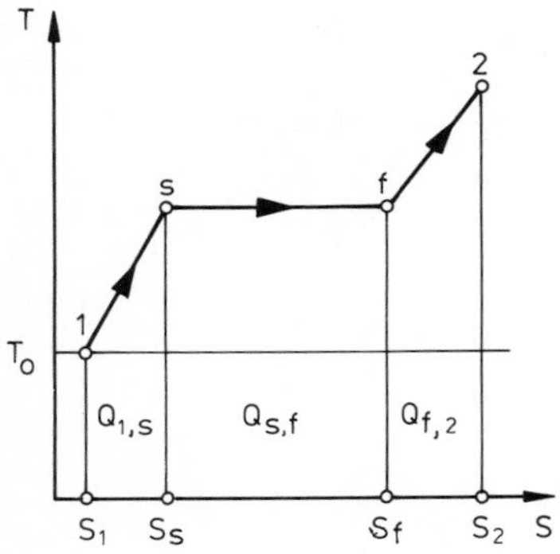

Fig 5.8 Sulphur melting process in T–S co-ordinates.

For change of entropy, the approximate relation:

$$S_s - S_1 = \frac{H_s - H_1}{(T_{sf} + T_1)/2} \tag{5.13}$$

(ii) Melting of sulphur; change of enthalpy:

$$Q_{s,f} = H_f - H_s = \tilde{h}_{sf} \tag{5.14}$$

Change of entropy:

$$S_f - S_s = \tilde{h}_{sf}/T_{sf} \tag{5.15}$$

(iii) Heating of liquid sulphur. Following from (5.12) and (5.13), change of enthalpy for the liquid is:

$$Q_{f,2} = H_2 - H_f = \tilde{c}_{P,\mathrm{liq}}(T_2 - T_{sf}) \tag{5.16}$$

Change of entropy:

$$S_2 - S_f = \frac{H_2 - H_f}{(T_{sf} + T_2)/2} \tag{5.17}$$

Assuming the 'heat losses' from S-R I are 10% of the thermal energy of the heating steam supplied, the energy balance can be expressed as:

$$\tilde{c}_{P,\mathrm{sol}}(T_{sf} - T_1) + \tilde{h}_{sf} + \tilde{c}_{P,\mathrm{liq}}(T_2 - T_{sf}) = 0.9\, m_3 h_{fg} \tag{5.18}$$

The melting point for sulphur is $\theta_{sf} = 119°C$. The mean values of molar heat capacities for the solid and the liquid phases of sulphur are[5.3] respectively:

$$\tilde{c}_{P,\mathrm{sol}} = 21.8 \text{ kJ/kmol K}$$

$$\tilde{c}_{P,\mathrm{liq}} = 31.5 \text{ kJ/kmol K}$$

and the molar enthalpy of fusion of sulphur:

$$\tilde{h}_{sf} = 14\,200 \text{ kJ/kmol}$$

At 5 MPa the enthalpy of evaporation of steam is

$$h_{fg} = 1639 \text{ kJ/kg}$$

and:

$$T_{SAT} = 537.1 \text{ K}$$

Substituting these values in (5.18), the mass of steam, m_3, required per kmol of S is:

$$m_3 = 12.39 \text{ kg}$$

Hence, the thermal energy supplied by steam to (S-R I) is:

$$Q_{3,4} = m_3 h_{fg}$$
$$= 20\,307 \text{ kJ}$$

With $W_x = 0$ and $E_0^Q \equiv 0$ the exergy balance for (S-R I) is:

$$I_I = E_{3,4}^Q - (E_2 - E_1) \tag{5.19}$$

where:

$$E_{3,4}^Q = E_3 - E_4$$
$$= Q_{3,4} \frac{T_{SAT} - T_0}{T_{SAT}}$$

and with $E_1 = E_0$:

$$E_2 = E_0 + (H_2 - H_1) - T_0(S_2 - S_1) \tag{5.21}$$

From Table A.3, for sulphur:

$$E_1 = E_0 = \tilde{\varepsilon}_0$$
$$= 598\,850 \text{ kJ/kmol}$$

Using (5.12) to (5.21) and substituting the numerical values:

$$E^Q_{3,4} = 20\,307 \frac{537.1 - 298}{537.1}$$
$$= 9\,040 \text{ kJ}$$
$$H_2 - H_1 = 1\,921.5 + 14\,200 + 2\,158.2$$
$$= 18\,280 \text{ kJ}$$
$$S_2 - S_1 = 6.26 + 36.22 + 4.55$$
$$= 47.03 \text{ kJ/K}$$
$$E_2 = 598\,850 - 18\,280 - 298 \times 47.03$$
$$= 603\,108 \text{ kJ}$$

From (5.19):

$$I_1 = 9\,040 - (603\,108 - 598\,850)$$
$$= 4\,782 \text{ kJ}$$

In this case the proportion of I_{I} due to the heat loss is easily evaluated as:

$$I^Q_{\text{I}} = 0.1\ E^Q_{3,4}$$

or about 19% of I_{I}.

In the Grassmann diagram in Fig 5.7(b), all values of exergy and irreversibility are quoted as percentages of the exergy input, ie the exergy sulphur, E_1. On this basis:

E_1—100.00%

E_2—100.70%

I_{I}— 0.80%

$E^Q_{3,4}$— 1.51%

Sub-region II: sulphur burner

The reaction taking place is:

$$S_{(1)} + O_{2(g)} \rightarrow SO_{2(g)}$$

With a 150% air excess and assuming the molar composition of air to be $O_2 = 0.21$, $N_2 = 0.79$, the molar balance for the process is:

1 kmol S + 2.5 kmol O^2 + 9.59 kmol N_2 → 1 kmol SO_2 + 1.5 kmol O_2 + 9.59 kmol N_2

Hence $n_{SO_2} = 1$ kmol, $n_{O_2} = 1.5$ kmol, $n_{N_2} = 9.59$ kmol, the total mass, $n_6 = 12.09$ kmol, and the fractions for the product stream are $x_{SO_2}=0.083$, $x_{O_2}=0.124$, and $x_{N_2}=0.793$.

The exit gas temperature T_6 will be calculated from the energy balance assuming the process to be adiabatic. As we have decided to neglect pressure losses, we will assume the air to be drawn into the sulphur burner in its environmental state. From the concept of enthalpy of devaluation (Appendix A), the enthalpy of atmospheric air is zero. Hence, $H_5=0$ and with $Q=0$ and $W_x=0$, the energy balance reduces to:

$$H_2=H_6 \tag{5.22}$$

Using the concept of enthalpy of devaluation, for sulphur:

$$H_2=(H_d^0)_S+(H_2-H^0) \tag{5.23}$$

From Table A.3:

$$(H_d^0)_S=(\tilde{h}_d^0)_S=724\ 580\ \text{kJ/kmol}$$

Now, since $H_1=H^0$ (see S-R I) the second term on the RHS of (5.23) has already been calculated above. Substituting in (5.23):

$$H_2=724\ 580+18\ 280=742\ 860\ \text{kJ}$$

The enthalpy of the products of combustion can be expressed in terms of its chemical and physical components (see Section 4.5):

$$H_6=\sum_k n_k\tilde{h}_{d,k}^0+\sum_k n_k\tilde{h}_{ph,k} \tag{5.24}$$

Values of $\tilde{h}_d^0$ for the different components of the mixture can be obtained from Table A.3.

The second term on the RHS can be expressed as:

$$\sum_k n_k\tilde{h}_{ph,k}=(T_6-T^0)\sum_k n_k\tilde{c}_{P,k}^{h} \tag{5.25}$$

Values of $\tilde{c}_P^h$ extracted from Table D1 for a trial value of T_6 can be used in the energy balance to calculate T_6 by iteration. The following table gives values of $\tilde{c}_P^h$ for $T_6=$ 1 073.15 K and values of $\tilde{h}_d^0$ for the components of the mixture.

	SO_2	O_2	N_2
n_k/kmol	1	1.5	9.59
$\tilde{h}_d^0$/[kJ/kmol]	427 480	0	0
$\tilde{c}_P^h$/[kJ/kmol K]	49.35	32.93	31.11

From the tabulated values:

$$\sum_k n_k\tilde{h}_{d,k}^0=427\ 480\ \text{kJ}$$

and:

$$\sum_k n_k\tilde{c}_{P,k}^h=397.09\ \text{kJ/K}$$

Using (5.22) to (5.25) gives:

$$742\ 860=427\ 480+(T_6-T^0)\ 397.09$$

$$\therefore\ T_6=1\ 067\ \text{K}$$

The irreversibility of the process can now be calculated using the exergy balance for the sub-region:

$$I_{II} = E_2 - E_6 \tag{5.26}$$

From (S-R I), $E_2 = 603\ 108$ kJ. E_6 can be evaluated from the expression:

$$E_6 = n_6 \left[\tilde{\varepsilon}_6^0 + (T_6 - T^0) \sum_k x_k \tilde{c}_{P,k}^{\varepsilon} \right] \tag{5.27}$$

To calculate E_6 requires the following data on $\tilde{\varepsilon}^0$ from Table A.3 and on $\tilde{c}_P^{\varepsilon}$ (for $T_6 =$ 1 067 K) from Table D.3:

	SO_2	O_2	N_2
x_k	0.083	0.124	0.793
$\tilde{c}_P^{\varepsilon}$/[kJ/kmol K]	25.72	17.02	15.92
$\tilde{\varepsilon}^0$/[kJ/kmol]	303 500	3 970	690

Hence, for the particular mixture of gases under consideration:

$$\tilde{\varepsilon}_6^0 = \sum_k x_k \tilde{\varepsilon}_k^0 + \tilde{R}T^0 \sum_k x_k \ln x_k$$
$$= 24\ 620 \text{ kJ/kmol}$$

Also, using the tabulated data:

$$\sum_k x_k \tilde{c}_{P,k}^{\varepsilon} = 16.87 \text{ kJ/kmol K}$$

Substituting in (5.27):

$$E_6 = 12.09[24\ 620 + (794 - 25)16.87]$$
$$= \underline{454\ 502 \text{ kJ}}$$

Using (5.26):

$$I_{II} = 603\ 108 - 454\ 502 = 148\ 606 \text{ kJ}$$

This represents a substantial loss of the original exergy of sulphur due to the uncontrolled nature of the chemical reaction.

Expressing the calculated values of E_6 and I_{II} as a percentage of the original exergy of sulphur:

E_6—75.90%
I_{II}—24.82%

Sub-region III: Waste-Heat Boiler A

Since the processes taking place in this sub-region are all physical in nature (ie no changes in composition occur), in principle, physical enthalpy and physical exergy could be used in the energy and exergy balances. However, as total values will be required in the next sub-region and in the construction of the Grassmann diagram, they will be used in all the calculations.

To calculate m_8, use the energy balance for this sub-region:

$$m_8(h_9 - h_8) + Q_{3,4} = H_6 - H_7 \tag{5.28}$$

From S-R II:

$$H_6 = H_2 = 742\,860 \text{ kJ}$$

To calculate $(H_7)_{ph}$, we use the following data from Table D.1, for $T_7 = 598.15$ K:

	SO_2	O_2	N_2
n_k/kmol	1	1.5	9.59
$\tilde{c}_P^h$/[kJ/kmol K]	44.67	30.73	29.93

Hence:

$$\sum_k n_k \tilde{c}_P^h = 377.79 \text{ kJ/K}$$

Using the value of $H_d^0 = 427\,480$ kJ calculated for S-R II:

$$H_7 = 427\,480 + 300 \times 377.79$$

$$= 540\,818 \text{ kJ}$$

For the water substance for states 8 and 9:

$$h_8 = 853.6 \text{ kJ/kg}$$

$$h_9 = 3\,434 \text{ kJ/kg}$$

Also, as calculated for S-R I:

$$Q_{3,4} = 20\,307 \text{ kJ}$$

Substituting these values in (5.28):

$$m_8 = \underline{66.55 \text{ kg}}$$

The irreversibility for this sub-region will be calculated from the exergy balance, which can be expressed in the form:

$$I_{III} = (\mathrm{E}_6 - \mathrm{E}_7) - (\mathrm{E}_9 - \mathrm{E}_8) - \mathrm{E}_{3,4}^Q \tag{5.29}$$

To calculate $(\mathrm{E}_7)_{ph}$, use the following data for $T_7 = 598.15$ K from Table D.3:

	SO_2	O_2	N_2
n_k/kmol	1	1.5	9.59
$\tilde{c}_P^\varepsilon$/[kJ/kmol K]	14.11	9.73	9.28

Hence, $\sum_k n_k \tilde{c}_P^\varepsilon = 117.61$ kJ/k.

Using from S-R II the value of $\tilde{\varepsilon}_6^0 = \tilde{\varepsilon}_7^0$:

$$\mathrm{E}_7 = 12.09 \times 24\,620 + 300 \times 117.61 = \underline{332\,939 \text{ kJ}}$$

To calculate E_8 and E_9, from steam tables:

$$s_8 = 2.325 \text{ kJ/kgK}$$

$$s_9 = 6.977 \text{ kJ/kgK}$$

The specific exergy function for water in the standard state (Appendix A) is:

$$\beta^0 = h^0 - T^0 s^0 = -4.7 \text{ kJ/kg}$$

Thus:

$$\mathrm{E}_8 = m_8[h_8 - T^0 s_8 - \beta^0]$$
$$= 10\,980\ \mathrm{kJ}$$
$$\mathrm{E}_9 = m_8[h_9 - T^0 s_9 - \beta^0]$$
$$= \underline{90\,402\ \mathrm{kJ}}$$

Also, from S-R II:

$$\mathrm{E}_6 = 454\,502\ \mathrm{kJ}$$

Substituting the calculated values in (5.29):

$$I_{\mathrm{III}} = 156\,844 - 35\,282 - 90\,402 + 10\,980 - 9\,040$$
$$= \underline{33\,100\ \mathrm{kJ}}$$

This irreversibility is due to the heat transfer over a finite temperature difference between the hot gases and the water substance. Expressing the energy values as percentages of the plant input:

E_7—55.60%
E_8— 1.83%
E_9—15.10%
I_{III}— 5.53%

Sub-region IV: Converter A

This sub-region contains the gas filter and the converter. As we have assumed that both pressure and 'heat losses' are negligible, the former of the two components does not incur any irreversibility. The reaction in the converter is:

$$SO_2 + \tfrac{1}{2}O_2 \rightarrow SO_3$$

leading to 60% conversion of SO_2 to SO_3 by the end of the process.

Using the energy balance, the exit temperature, T_{10}, will be determined. For this sub-region:

$$H_7 = H_{10} \tag{5.30}$$

where H_{10} can be expressed with the aid of (5.24) and (5.25) used earlier for H_6. For an initial trial value of T_{10} followed by an iteration, the table below gives values of $\tilde{c}_P^{\mathrm{h}}$ extracted from Table D.1 corresponding to $T = 748.15$ K. A set of values of $\tilde{h}_{\mathrm{d}}^0$ for the components of the gas mixture and its molar composition in state 10 are also given.

	SO_2	SO_3	O_2	N_2
n_k/kmol	0.4	0.6	1.2	9.58
$\tilde{h}_{\mathrm{d}}^0$/[kJ/kmol]	427 480	329 140	0	0
$\tilde{c}_P^{\mathrm{h}}$/[kJ/kmol K]	46.48	65.66	31.69	30.32

The total amount of the gas is $n_{10} = 11.78$ kmol. From the table:

$$\sum_k n_k \tilde{h}_{\mathrm{d},k}^0 = 368\,476\ \mathrm{kJ}$$

and:

$$\sum_k n_k \tilde{c}_{P,k}^{\mathrm{h}} = 386.48 \text{ kJ/K}$$

Also, as calculated for S-R II:

$$H_7 = 540\,818 \text{ kJ}$$

Using (5.30):

$$540\,818 = 368\,476 + (T_{10} - 298.15)386.48$$
$$T_{10} = 744 \text{ K}$$

The irreversibility of the sub-region can now be calculated from the exergy balance:

$$I_{\mathrm{IV}} = \mathrm{E}_7 - \mathrm{E}_{10} \qquad (5.31)$$

E_{10} will be calculated by a procedure similar to that used earlier in the calculation of E_6, and using values of $\tilde{\varepsilon}^0$ from Table A.3 and $\tilde{c}_P^{\varepsilon}$ for $T = 744$ K from Table D.3, as given in the table below:

	SO_2	SO_3	O_2	N_2
x_k	0.0340	0.0509	0.1019	0.8132
$\tilde{\varepsilon}^0$/[kJ/kmol]	303 500	225 070	3 970	690
$\tilde{c}_P^{\varepsilon}$/[kJ/kmol K]	18.58	26.63	12.64	11.88

Hence:

$$\tilde{\varepsilon}_{10}^0 = \sum_k x_k \tilde{\varepsilon}_k^0 + \tilde{R}T^0 \sum_k x_k \ln x_k = 21\,086 \text{ kJ/kmol}$$

and:

$$\sum_k x_k \tilde{c}_P^{\varepsilon} = 12.936 \text{ kJ/kmol K}$$

Using these values:

$$\mathrm{E}_{10} = 11.78[21\,086 + (471 - 25)12.936]$$
$$= \underline{316\,357 \text{ kJ}}$$

As calculated for S-R III:

$$\mathrm{E}_7 = 332\,939 \text{ kJ}$$

From (5.31):

$$I_{\mathrm{IV}} = 332\,939 - 316\,357$$
$$= \underline{16\,582 \text{ kJ}}$$

This is an unavoidable, intrinsic irreversibility of the process.

Expressing the calculated values of E_{10} and I_{IV} as percentages of the plant input, E_1:

E_{10}—52.8%
I_{IV}— 2.77%

Sub-region V: Waste-Heat Boiler B

Using procedures similar to those applied to Waste-Heat Boiler A the following values have been obtained:

$m_{13} = 28.59$ kg
$E_{11} = 278\,922$ kJ
$E_{12} = 4\,717$ kJ
$E_{13} = 35\,822$ kJ
$I_V = 6\,260$ kJ

and as percentages of plant input, E_1:

E_{11}—46.6%
E_{12}— 0.79%
E_{13}— 5.98%
I_V— 1.05%

The irreversibility occurring in this sub-region is mainly due to the heat transfer over a finite temperature difference.

Sub-region VI: Converter B

In this converter further conversion of SO_2 to SO_3 takes place. At the exit from the converter, state 14, the gas mixture has the composition:

	SO_2	SO_3	O_2	N_2
n_k/kmol	0.03	0.97	1.015	9.58
x_k	0.0026	0.0837	0.0875	0.8262

The total amount of mixture, $n_{14} = 11.595$ kmol. The enthalpy of devaluation for the mixture is:

$$(H_d^0)_{14} = \sum_k n_k \tilde{h}_{d,k}^0 = 332\,090 \text{ kJ}$$

and its chemical exergy:

$$E_{14}^0 = 11.595 \times 19\,073 = 221\,146 \text{ kJ}$$

From the energy balance the exit temperature is:

$$T_{14} = \underline{692 \text{ K}}$$

Using this temperature:

$$E_{14} = \underline{277\,168 \text{ kJ}}$$

Hence, from the exergy balance:

$$I_{VI} = E_{11} - E_{14}$$
$$= \underline{1\,824 \text{ kJ}}$$

The last two calculated values, as percentages of the input, are

E_{14}—46.3%
I_{VI}— 0.30%

The irreversibility of this reaction is intrinsic and is, clearly, very small.

Sub-region VII: economiser

From the calculated values obtained for the two waste-heat boilers, the total mass of water passing through the economiser is:

$$m_{16} = m_{17} = 66.55 + 28.59$$
$$= 95.14 \text{ kg}$$

Using the energy balance, the exit gas temperature is:

$$T_{15} = \underline{388 \text{ K}}$$

The exergy values for the streams of gas and water are

$E_{15} = 225\,206$ kJ
$E_{16} = 0$
$E_{17} = 15\,741$ kJ

Hence, using the exergy balance:

$I_{VII} = 36\,221$ kJ

As percentages of the input these values become

E_{15}—37.6%
E_{16}= 0
E_{17}— 2.63%
I_{VII}— 6.05%

Because of the large temperature differences between the two streams exchanging heat, the irreversibility of this process is quite high.

Sub-region VIII: gas cooler

The process occurring here is similar to that in the previous sub-region, except that the exergy of the coolant stream, air, is not used for any useful purpose. For the exit gases, $\theta_{18} = 50°C$, the exergy of the stream is:

$E_{18} = 221\,502$ kJ

From the exergy balance (with $E_0^Q \equiv 0$):

$$I_{VIII} = E_{15} - E_{18}$$
$$= 3\,704 \text{ kJ}$$

As the percentage of the plant input these values are:

E_{18}—37.0%
I_{VIII}— 0.62%

As the temperature range over which the gas stream is cooled is close to the environmental temperature, the irreversibility due to dissipation of its energy is not very great.

Sub-region IX: absorption system

The principal process here can be described by the exothermic reaction:

$$SO_3 + H_2O \rightarrow H_2SO_4$$

Since the amount of SO_3 is 0.97 kmol, that of H_2SO_4 produced is also 0.97 kmol. From Table A.3:

$$\tilde{\varepsilon}^0_{H_2SO_4} = 161\ 010\ \text{kJ/kmol}$$

Hence, the exergy of the acid stream is:

$$E_{19} = 0.97 \times 161\ 010$$
$$= \underline{156\ 170\ \text{kJ}}$$

As the exergy of the streams 23 and 21 is ultimately dissipated in the environment, $(E_{23})_0 = 0$ and $(E_{21})_0 = 0$. Also, the mixing water is supplied in the environmental state and thus $E_{22} = 0$. The exergy balance for this sub-region reduces to:

$$I_{IX} = E_{18} - E_{19}$$
$$= \underline{65\ 332\ \text{kJ}}$$

Thie represents a large loss of exergy, partly due to dissipation of the thermal energy released in the exothermic reaction and partly due to dissipation of the chemical exergy of the tail gases (stream 23). To evaluate the latter loss, use the following table which gives the necessary data relating to the tail gas:

	SO_2	SO_3	O_2	N_2
n_k/kmol	0.03	0	1.015	9.58
x_k	0.0028	0	0.0955	0.9016
$\tilde{\varepsilon}^0$/[kJ/kmol]	303 500	—	3 970	690

The total quantity of the tail gas is $n_{23} = 10.63$ kmol. Using the expression for the chemical exergy of a mixture of ideal gases gives that part of I_{IX} which arises from the differences in composition between the tail gas and the environment. Thus:

$$E^0_{23} = 1\ 023\ \text{kJ}$$

This is clearly only a small fraction (about 1.6%) of I_{IX}. Thus, most of the loss of exergy in this sub-region is due to the failure to utilise the large amounts of thermal energy evolved in the various processes, in this sub-region.

As percentages of the input, E_{19} and I_{IX} are:

E_{19}—26.1%
I_{IX}—10.91%

Sub-region X: steam mixing

Steam is generated in the two waste heat boilers at the same pressure of 5 MPa but is superheated to different temperatures. To obtain the temperature of the stream emerging from the steam mixer, apply the energy balance:

$$m_{20}h_{20} = m_9h_9 + m_{13}h_{13} \tag{5.32}$$

Hence, $h_{20} = 3\ 381$ kJ/kg which corresponds to $\theta_{20} = 477°C$. The irreversibility of the mixing process is obtained from:

$$I_X = E_9 + E_{13} - E_{20}$$

with $E_{20}=126\ 203$ kJ:

$$I_X=90\ 402+35\ 822-126\ 203$$
$$=\underline{21 \text{ kJ}}$$

Thus, the irreversibility arising from the mixing of the two streams at the same pressure but at different temperatures ($\theta_9=500°C$ and $\theta_{13}=425°C$) is very small.

As percentages of the plant input E_{20} and I_X are:

E_{20}—21.1%
I_X— 0.004%

Rational efficiency of the plant

The exergy input to the plant has been taken to be that corresponding to one kmol of sulphur. The exergy of the corresponding amount of H_2SO_4, together with the corresponding exergy of the process steam produced represent the exergy output of the plant. Consequently the rational efficiency of the plant can be expressed as:

$$\psi=\frac{E_{19}+E_{20}}{E_1} \tag{5.33}$$

Substituting the numerical values we get

$$\underline{\psi=0.472}$$

Discussion

The total exergy output of the plant comprises two components, the exergy of H_2SO_4 and the exergy of steam. The former does not leave much scope for improvement since it depends only on materials balance. In the best case 1 mole of S leads to 1 mole of H_2SO_4. Any shortfall in the production of the acid, in this case 3%, is due to incomplete conversion of SO_2 to SO_3. Consequently, virtually the only possible means of improving the overall performance of the plant is in the generation of steam. An increase in the steam output and its specific exergy can be realised by making a better use of the available thermal energy evolved at various stages in the production of H_2SO_4. To achieve this, it is necessary to integrate the two parallel production processes of H_2SO_4 and steam, so that the heat transfers occur over as low temperature differences as can be arranged within the prevailing economic constraints. The subject of process integration and thermo-economics is discussed in Chapter 6.

The Grassmann diagram, in Fig 5.7(b), shows at a glance the distribution of irreversibilities in the plant. The largest, in the sulphur burner, S-R I, is an intrinsic irreversibility associated with the uncontrolled nature of the process. Some reduction in this form of irreversibility could be achieved by preheating the air supplied to it, eg by using the gas cooler, S-R VIII. The irreversibilities occurring in the two converters, S-R IV and S-R VI, are also intrinsic. However, they are relatively small and there is not much that can be done to reduce them. The irreversibilities in the remaining sub-regions, except S-R X, are associated with heat transfer, and their reduction can be achieved through better process integration.

The analysis presented does not include the effect of pressure losses. To assess the effect of this simplification on the analysis presented here, assume that all the electric power supplied to the plant, mainly to operate pumps and blowers, is ultimately dissipated through viscous friction. The figure quoted[5.4] for the consumption of

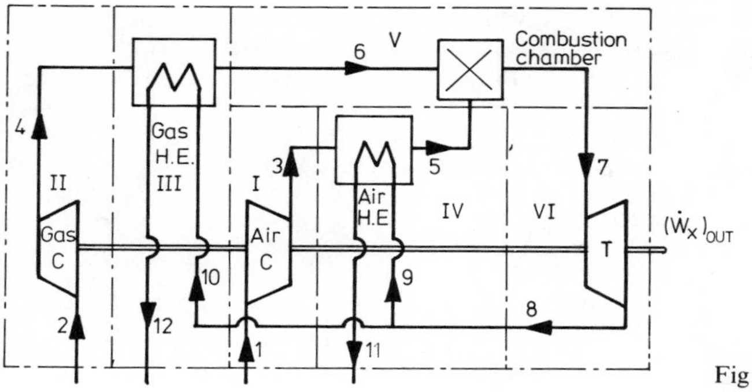

Fig 5.9 Gas turbine plant diagram.

electric energy, per tonne of H_2SO_4, produced in this type of plant is about 25 kWh. Hence, on the average, the irreversibilities in the various components would be about 3% greater if pressure losses were not neglected.

5.3 Gas turbine plant

The gas turbine plant* considered here is shown in Fig 5.9. The plant uses a gaseous fuel, blast furnace gas. Fuel and air are compressed in separate single-stage turbines and then pre-heated by the products of combustion in separate heat exchangers. The two streams are assumed to enter the combustion chamber at the same pressure and temperature, where combustion takes place at constant pressure. The products of combustion are expanded in the turbine from which part of the power output is diverted to drive the two compressors.

Plant parameters

Plant net power output $(\dot{W}_x)_{out} = 3\,000$ kW. Mechanical efficiency of the compressors and the turbine $\eta_m = 0.98$.

Both calculated and assumed pressures, temperatures and flow rates are given in Table 5.3.

Environmental parameters

The environmental pressure and temperature are $P_0 = 98$ kPa and $T_0 = 291$ K. The partial pressures of the mean constituents of air are taken as:

$$(P_{00})_{N_2} = 76.26 \text{ kPa} \qquad (P_{00})_{CO_2} = 0.03 \text{ kPa}$$

$$(P_{00})_{O_2} = 20.27 \text{ kPa} \qquad (P_{00})_{H_2O} = 1.44 \text{ kPa}$$

Specification of the fuel gas

The molar composition of the fuel is taken as:

CO—0.2888	CO_2—0.1028
H_2—0.0166	N_2—0.5682
CH_4—0.0029	H_2O—0.0206

* This example is reproduced from Ref [3.3], by permission.

Table 5.3 Thermodynamic parameters of the working fluids in the gas turbine plant

State (Fig 5.9)	Working fluid	Pressure, P/kPa	Temp, °C	Flow rate, $\dot{n}$/[kmol/s]
0	Environmental air	98	18	—
1	Air	98	18	1.2523
2	Fuel gas	98	18	0.1376
3	Air	353	167	1.2523
4	Fuel gas	353	167	0.1376
5	Air	343	357.4	1.2523
6	Fuel gas	343	355	0.1376
7	Combustion products	343	627	1.3713
8	Combustion products	103	417	1.3713
9	Combustion products	103	417	1.2283
10	Combustion products	103	417	0.1430
11	Combustion products	98	230	1.2283
12	Combustion products	98	230	0.1430

The net calorific value of the fuel is (NCV) = 90 050 kJ/kmol. The fuel is delivered to the turbine in the environmental state.

Principal assumptions

1. The 'heat loss' from the combustion chamber is taken to be 25 120 kJ/kmol of the dry fuel gas.
2. 'Heat losses' from all other plant components are neglected.
3. Kinetic and potential components of exergy are negligible.
4. The combustion process is complete.
5. All gases are treated as ideal.

Gas turbine plant calculations

Using the normal methods of gas turbine plant analysis[5.5] the following quantities were calculated:

The air–fuel ratio is 8.966 mol of dry air per mol dry fuel gas

The composition of the products of combustion in kmol/kmol of dry fuel gas is:

$$n_{CO_2} = 0.403 \qquad n_{O_2} = 1.721$$

$$n_{H_2O} = 0.178 \qquad n_{N_2} = 7.663$$

Gas consumption:

$$\dot{n}_F = 0.1376 \text{ kmol (dry fuel gas)/s}$$

Power generated by the gases in the turbine (internal power):

$\dot{W}_{7,8} = 9\,445$ kW.

Power requirement for gas compression (internal power):

$\dot{W}_{2,4} = 638.3$ kW

Power requirement for air compression (internal power):

$$\dot{W}_{1,3} = 5\,492 \text{ kW}$$

Irreversibilities of mechanical type

Irreversibilities due to mechanical inefficiency:

Turbine:

$$\dot{I}_{\text{m,VI}} = (1-\eta_{\text{m}})\dot{W}_{7,8} = 188.8 \text{ kW}$$

Hence, the effective power output of the turbine is:

$$(\dot{W}_{\text{x}})_{\text{VI}} = \dot{W}_{7,8} - \dot{I}_{\text{m,VI}} = 9\,256 \text{ kW}$$

Fuel gas compressor:

$$\dot{I}_{\text{m,II}} = \left(\frac{1}{\eta_{\text{m}}} - 1\right)\dot{W}_{2,4} = 13 \text{ kW}$$

Hence, the gross power input to the fuel-gas compressor is:

$$(\dot{W}_{\text{x}})_{\text{II}} = \dot{W}_{2,4} + \dot{I}_{\text{m,II}} = 651 \text{ kW}$$

Air compressor:

$$\dot{I}_{\text{m,I}} = \left(\frac{1}{\eta_{\text{m}}} - 1\right)\dot{W}_{1,3} = 112 \text{ kW}$$

Hence, the gross power input to the air compressor is:

$$(\dot{W}_{\text{x}})_{\text{I}} = \dot{W}_{1,3} + \dot{I}_{\text{m,I}} = 5\,604 \text{ kW}$$

Calculation of exergy of the gas streams

As the specified environmental parameters differ somewhat from standard conditions (Table A.1) some adjustment is necessary in the calculation of the exergy of the gas streams. The general expression used for calculating the exergy rate of the fuel gas is:

$$\dot{E} = \dot{n} \sum_i x_i \tilde{\varepsilon}_i^{\Delta T} + \dot{n}\tilde{R}T_0 \ln (P/P_0)$$
$$+ \dot{n}\left[\sum_i x_i \tilde{\varepsilon}_{0,i} + \tilde{R}T_0 \sum_i x_i \ln x_i\right] + \dot{n}_{\text{H}_2\text{O}}\tilde{R}T_0 \ln\left(\frac{P^{00}}{P_{00}}\right)_{\text{H}_2\text{O}} \quad (5.34)$$

$\tilde{\varepsilon}_i^{\Delta T}$ can be calculated from (4.122). Negligible errors will be incurred by using uncorrected values of $\tilde{c}_p^{\text{h}}$ and $\tilde{c}_p^{\text{s}}$ from Tables D1 and D2 despite the difference here between T_0 and T^0.

The values of $\tilde{\varepsilon}_{0,i}$ are obtained by correcting standard values, $\tilde{\varepsilon}_i^0$, using (A.36).

The last term in (5.34) corrects $\dot{E}$ for the difference between the actual and the standard partial pressures (Table A.1) of water vapour in air. Here this correction is a little over 1% of the chemical exergy of the fuel.

In principle (5.34) can be used also for calculating the exergy rate of the products of combustion, but:

$$(\dot{E}_{\text{RS}})_{\text{M}} = \dot{n} \sum_i x_i \tilde{\varepsilon}_i^{\Delta T} + \dot{n}\tilde{R}T_0 \sum_i x_i \ln \frac{x_i P}{P_{00,i}} \quad (5.35)$$

is more convenient and more directly applicable to mixtures consisting entirely of gaseous reference substances.

The first term on the RHS of (5.35) represents the thermal component of the mixture and is identical to the corresponding term in (5.34). The second term represetns the sum of the pressure component of exergy and the chemical exergy of the mixture. It may also be looked upon as a sum of work terms arising from the reversible isothermal expansion of the components of the mixture at $T = T_0 = \text{const}$ from its initial partial pressure $P_i = x_i P$ to the partial pressure $P_{00,i}$ at which the particular gaseous reference substance exists in the environment. The ideal model which could be used for determining this work would correspond to the model shown in Fig 2.13 except that it would operate between the states indicated above and not those shown on that diagram. As in this case $P_{00,i}$ and T_0 define the actual dead state of the reference substance rather than the standard one, no correction for any differences between the two is required.

The scheme used for computing exergy rates of the gas streams at various points of the plant is shown in Table 5.4. First, the exergy values per kmol of dry fuel gas for air, fuel gas and products of combustion in selected states, 1, 2 and 12 respectively*, have been computed using (5.34) and (5.35). These values are given in column 4 as 'known exergy'. As the air in state 1 is in the environmental state, its exergy is zero. These values are then used to calculate the exergy for other states of the particular working media by evaluating differences in enthalpy and entropy between the selected state (state z) and the state under consideration (state x).

Calculation of irreversibilities and criteria of performance

The specific chemical exergy of the fuel calculated from (5.34) is $\tilde{\varepsilon}_F = 87\,550$ kJ/kmol. This is somewhat smaller than (NCV), with $\varphi = 0.972$. Hence the exergy input to the plant is:

$$\mathrm{E}_2 = \dot{n}_F \tilde{\varepsilon}_p = 12\,077 \text{ kW}$$

The output of the plant being $(\dot{W}_x)_{OUT}$, the rational efficiency of the plant is:

$$\psi = \frac{3\,000}{12\,077} = \underline{0.2484}$$

The exergy losses of the plant comprise internal irreversibilities of the various plant components and external irreversibilities, primarily that due to discharge of the hot products of combustion into the environment, given by:

$$\dot{I}_{EXT} = \dot{\mathrm{E}}_{11} + \dot{\mathrm{E}}_{12}$$
$$= \dot{n}_F \times 19\,377 = \underline{2\,666 \text{ kW}}$$

This is 22.07% of the plant input. The external irreversibility is shown as the width of the dotted band in Fig 5.10.

The irreversibilities of the various plant components can be calculated with aid of the exergy balance, using the exergy values given in Table 5.4. As an example of such calculations, for the combustion chamber (S-R V):

$$\dot{I}_V = \dot{\mathrm{E}}_6 + \dot{\mathrm{E}}_5 - \dot{\mathrm{E}}_7$$

* As will be observed from Table 5.3, the thermodynamic states corresponding to points 11 and 12 on the plant diagram are identical. The exergy value $\nu\varepsilon_z$ in column 5 is for the two product streams combined.

Table 5.4 Calculation of exergy of the working fluids at various points in the gas turbine plant

Working fluid	Amount of working fluid per mole of dry fuel, ν $\frac{\text{kmol}}{\left[\frac{\text{kmol}}{\text{dry fuel}}\right]}$	State under considera-tion, x	State with 'known' exergy, z	A Known exergy, $\frac{\nu\varepsilon_Z}{\text{kJ}} \left[\frac{\text{kmol}}{\text{dry fuel}}\right]$	B Difference in enthalpy $\nu\Delta h]_0^x$ $\frac{\text{kJ}}{\text{kmol dry fuel}}$	C Difference in enthalpy $\nu\Delta h]_0^z$ $\frac{\text{kJ}}{\text{kmol dry fuel}}$	$\frac{p_X}{p_Z}$	D Difference in entropy $\frac{\nu\Delta s]_z^x}{\text{kJ}} \left[\frac{\text{kmol}}{\text{dry fuel}}\right]$	A+B−C−DT_0 Exergy $\frac{\nu\varepsilon_X}{\text{kJ}} \left[\frac{\text{kmol}}{\text{dry fuel}}\right]$	Exergy rate $\dot{E}_X = \dot{n}_F \cdot \nu\varepsilon_X$ kW	Exergy rate $\frac{\dot{E}_X}{\dot{E}_2} \times 100$ %
1	2	3	4	5	6	7	8	9	10	11	12
Air	9.1	3	1	0	39 915	0	3.6	12.37	36 316	4 997	41.36
Air	9.1	5	1	0	92 121	0	3.5	113.88	58 983	8 116	67.18
Fuel gas	1.0215	4	2	87 793	4 639	0	3.6	1.80	91 909	12 650	104.69
Fuel gas	1.0215	6	2	87 793	10 718	0	3.5	13.65	94 537	13 010	107.68
Combustion products	9.965	7	12+11	19 377	190 379	63 459	3.5	79.94	123 034	16 930	140.14
Combustion products	9.965	8	12+11	19 377	121 745	63 459	1.05	92.69	50 688	6 975	57.74

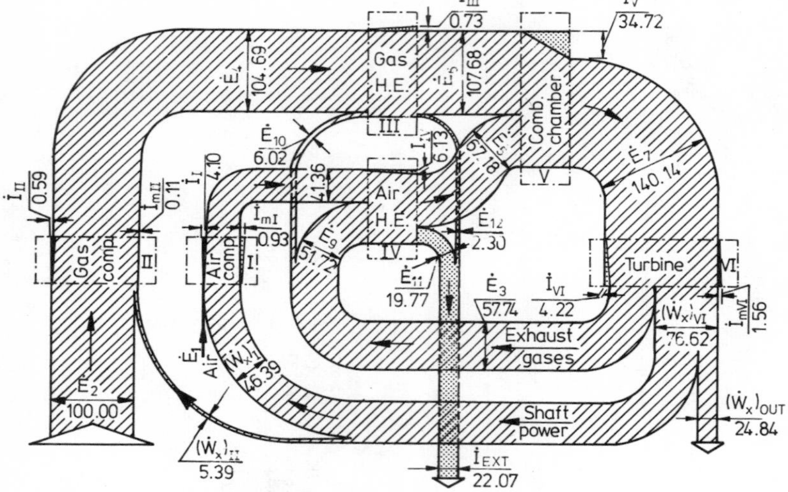

Fig 5.10 Grassmann diagram for the gas turbine plant. Values of I and E are given as percentages of plant input. (Reproduced from Ref [3.3], by premission.)

Taking the numerical values from column 11 of Table 5.4:

$$\dot{I}_V = 13\,010 + 8\,116 - 16\,930$$
$$= \underline{4\,196\text{ kW}}$$

$\dot{E}_2$ is 34.72% of the exergy input.

Discussion

As will be seen from the Grassmann diagram (Fig 5.10), the largest contributions fo the total plant efficiency defect, $(1-\psi)$, are from the combustion chamber, $\dot{I}_V$, and from the external irreversibility, $\dot{I}_{EXT}$. Although the irreversibility rate of the combustion chamber is large by comparison with those of other plant components, its performance as a combustion system can be independently assessed on the basis of its rational efficiency. For this present case:

$$\psi_V = \frac{E_7}{E_6 + E_5}$$
$$= 0.80$$

Compared with the value calculated for a coal-fired boiler ($\psi = 0.433$) in Example 4.4, it is clear that the present combustion chamber is much more efficient despite the use of high air–fuel ratio and the heat losses. The main reason for this is the high degree of preheating of the reactants (Section 4.5).

The external losses, $\dot{I}_{EXT}$, are due to the products streams rejected to the environment from the fuel gas pre-heater and the air pre-heater. The loss associated with the former is much smaller than the latter owing to the relatively smaller heating duty. For the same reason the irreversibility $\dot{I}_{III}$ is much smaller than $\dot{I}_{IV}$, despite the fact that the two heat exchangers operate with approximately the same temperature differences and pressure losses.

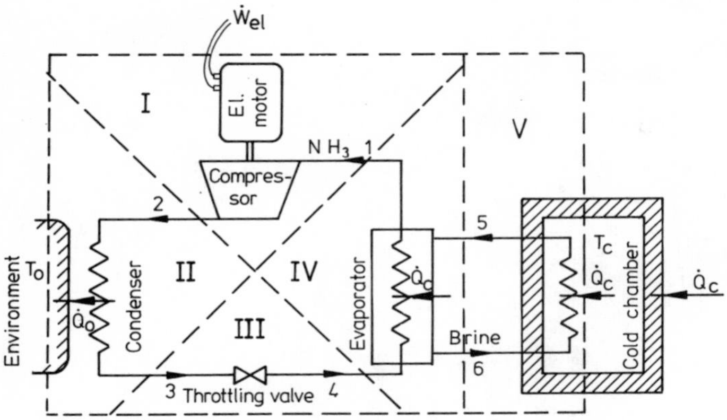

Fig 5.11 Refrigeration plant diagram. (Reproduced from Ref [3.3], by permission.)

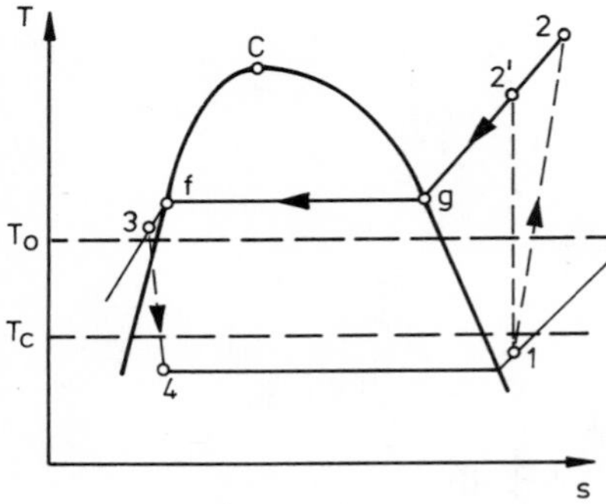

Fig 5.12 Temperature–entropy diagram for the refrigeration cycle.

5.4 Refrigeration plant

The vapour-compression refrigeration plant* shown in Fig 5.11 uses ammonia as the working fluid. The plant refrigerates a cold chamber by circulating brine, which is used as the heat transfer medium, between the evaporator and the cold chamber. The refrigeration cycle is shown in *T–s* co-ordinates in Fig 5.12 and in *ε–h* co-ordinates[5.6] in Fig 5.13.

Plant operating parameters

Plant refrigeration duty $\dot{Q}_c = 93.03$ kW
Environmental temperature $\theta_0 = 20°C$
Temperature of the cold chamber $\theta_c = -1°C$
Evaporator saturation temperature $\theta_{ev} = -12°C$
Compressor inlet temperature $\theta_1 = -10°C$
Compressor outlet temperature $\theta_2 = 119°C$
Condenser saturation temperature $\theta_{con} = 28°C$

* This example is based on one in Ref [3.3] and is reproduced by permission.

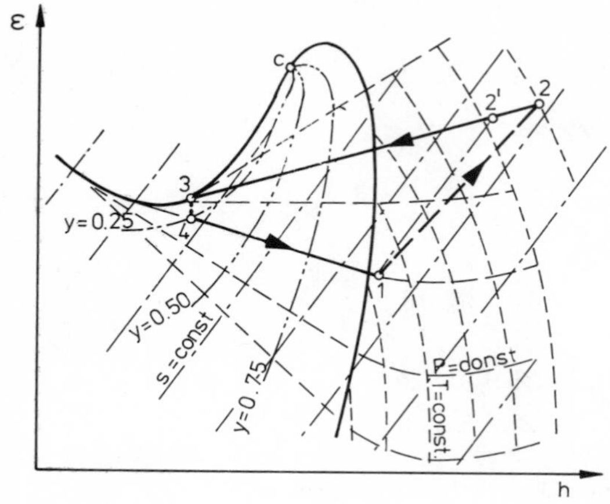

Fig 5.13 Exergy–enthalpy diagram for the refrigeration cycle.

Condenser outlet temperature $\theta_3 = 25°C$
Compressor mechanical efficiency $\eta_m = 0.83$
Electric motor efficiency $\eta_{el} = 0.90$
Brine temperature at the inlet to the evaporator $\theta_5 = -5°C$
Brine temperature at the outlet from the evaporator $\theta_6 = -7°C$
Specific isobaric heat capacity of brine $c_{P,b} = 2.85$ kJ/kgK

Principal assumptions

1. 'Heat losses' except in Sub-region I and 'heat leaks' are negligible.
2. Pressure losses are negligible.
3. Kinetic and potential components of exergy are negligible.
4. Power input to pump and other auxiliary equipment is negligible.

Calculations

All the processes taking place in the plant are physical in character and consequently the chemical component of exergy is not considered. As the plant operates on a closed cycle, the exergy of the working fluid can be calculated (see (2.22)) with reference to a convenient reference state for which the exergy of the fluid is allocated zero value. Here this reference state, indicated by the letter *a* in Fig 5.12, was chosen for ammonia as:

$$P_a = 0.27 \text{ MPa} \qquad \theta_a = \theta_0 = 20°C$$

For brine, zero exergy was taken to coincide with $\theta_b = -5°C$. The specific exergy for ammonia may be calculated from (2.22). In the case of brine, which has been assumed to undergo no pressure changes and to have a constant value of isobaric heat capacity:

$$\varepsilon_b = c_{P,b}\left[(T - T_b) - T_0 \ln \frac{T}{T_b}\right] \tag{5.36}$$

may be used to calculate its 'relative' specific exergy.

As an alternative to calculating values of exergy of ammonia as outlined above, one can interpolate for the specified states on the ε–*h* diagram given in Fig E.4. The resulting values of 'relative' specific exergy for ammonia and brine are listed in Table 5.5.

Table 5.5 Thermodynamic parameters of ammonia and brine in the refrigeration plant shown in Fig 5.11

State	Flow rate, $\dot{m}$/[kg/s]	Pressure, P/MPa	Temperature, °C	Specific enthalpy, h/[kJ/kg]	'Relative' specific exergy, ε/[kJ/kg]	Exergy rate, $\dot{E}$/kW
1	0.082	0.27	−10	1 671	2.8	0.23
2	0.082	1.1	119	1 947	222.9	18.28
3	0.082	1.1	25	536	156.6	12.84
4	0.082	0.27	−12	536	142.6	11.69
5	16.33	—	−5	19.95	0	0
6	16.33	—	−7	14.25	0.554	9.05

From the given refrigeration duty $\dot{Q}_c$ and the listed values of enthalpy, the mass flow rate of ammonia is:

$$\dot{m}_A = \frac{\dot{Q}_c}{h_1 - h_4} = 0.082 \text{ kg/s}$$

Similarly for brine:

$$\dot{m}_B = \frac{\dot{Q}_c}{c_{P,b}(T_6 - T_7)} = 16.33 \text{ kg/s}$$

The heat transfer rate in the condenser is given by:

$$\dot{Q}_0 = \dot{m}_A(h_2 - h_3) = 115.6 \text{ kW}$$

Assuming the ammonia compressor to operate adiabatically, the electric power necessary to drive it is:

$$\dot{W}_{el} = \frac{\dot{m}_A(h_2 - h_1)}{\eta_m \eta_{el}} = 30.22 \text{ kW}$$

Because the power input to the auxiliary equipment was neglected, this power input must be regarded as the exergy input to the plant.

The irreversibility rates corresponding to the various sub-regions of the plant, indicated in Fig 5.11, can be calculated using the exergy balance or the Gouy–Stodola relation. Since the exergy rates for the various points in the plant had to be calculated in any case in order to construct the Grassmann diagram, the former of the two methods was adopted.

Sub-region I: electric motor and compressor

The exergy balance for the sub-region is:

$$\dot{I}_I = \dot{W}_{el} + \dot{E}_1 - \dot{E}_2$$

Using 'relative' exergy rate values from Table 5.5:

$$\dot{I}_I = 12.17 \text{ kW}$$

The mechanical-electrical losses can be obtained as a separate item from:

$$(\dot{I}_{\mathrm{I}})_{\mathrm{m,el}} = \dot{W}_{\mathrm{el}}(1 - \eta_{\mathrm{m}}\eta_{\mathrm{el}})$$
$$= 7.65 \text{ kW}$$

The difference between the two values corresponds to 'internal' irreversibility, due to fluid friction:

$$(\dot{I}_{\mathrm{I}})_{\mathrm{int}} = \underline{4.52 \text{ kW}}$$

Sub-region II: condenser

With $\dot{E}_0^Q = 0$, the exergy balance in this case is:

$$\dot{I}_{\mathrm{II}} = \dot{E}_2 - \dot{E}_3 = \underline{5.44 \text{ kW}}$$

Sub-region III: throttling valve

Here:

$$\dot{I}_{\mathrm{III}} = \dot{E}_3 - \dot{E}_4 = \underline{1.15 \text{ kW}}$$

Sub-region IV: evaporator

In this case two streams, ammonia and brine, exchanging heat must be considered. From the exergy balance:

$$\dot{I}_{\mathrm{IV}} = (\dot{E}_4 - \dot{E}_1) - (\dot{E}_6 - \dot{E}_5) = \underline{2.41 \text{ kW}}$$

Sub-region V: cold chamber

The thermal exergy rate corresponding to the refrigerator duty $\dot{Q}_c$, at the temperature $T_c = 272.15$ K is:

$$\dot{E}_c^Q = -93.03 \frac{272.15 - 293.15}{272.15}$$

$$= 7.18 \text{ kW}$$

From the exergy balance for this sub-region:

$$\dot{I}_{\mathrm{V}} = (\dot{E}_6 - \dot{E}_5) - \dot{E}_c^Q = \underline{1.87 \text{ kW}}$$

Rational efficiency of the plant

Considering the plant as a whole, the plant exergy output is $\dot{E}_c^Q$. Hence, the overall rational efficiency is:

$$\psi_{\mathrm{ov}} = \frac{\dot{E}_c^Q}{\dot{W}_{\mathrm{el}}} = 0.238$$

This value includes the effect of the irreversibilities arising from the two heat transfer processes involving the brine stream; one in the cold chamber and the other in the evaporator. The performance of the refrigeration plant itself may be assessed using the 'net' rational efficiency, drawn up for Sub-regions I, II and III only. This criterion of performance has the form:

$$\psi_{\mathrm{NET}} = \frac{\dot{E}_4 - \dot{E}_1}{\dot{W}_{\mathrm{el}}}$$

Hence:

$$\psi_{NET} = \underline{0.379}$$

Discussion

In Fig. 5.14 a graphical representation of both the energy balance, ie the Sankey diagram, and the exergy balance, ie the Grassmann diagram, is given for the refrigeration plant. The only information which the Sankey diagram provides relates to energy transfers to or from the control region under consideration. It gives no information about changes in the quality of energy and it gives equal weight to both the electric energy supplied to the plant and the low-grade thermal energy rejected by the condenser to the environment.

The Grassmann diagram, on the other hand, gives quantitative information regarding the proportion of the exergy input to the plant which is dissipated in the

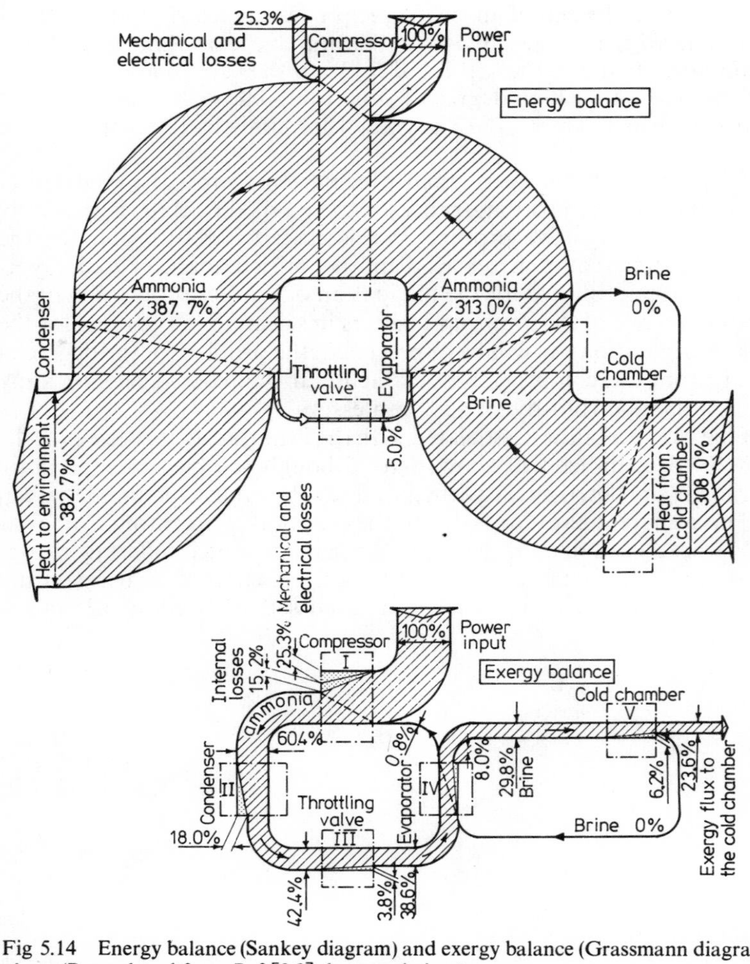

Fig 5.14 Energy balance (Sankey diagram) and exergy balance (Grassmann diagram) for the refrigeration plant. (Reproduced from Ref [2.3], by permission.)

different plant components and, in some cases, in what form this dissipation occurred. For example, in Sub-region I the losses occurring as a result of mechanical friction, electrical dissipation, and fluid friction inside the compressor can be listed as separate items.

The largest irreversibilities occur in Sub-region I, the motor-compressor sub-assembly. The magnitude of these losses, over 40% of the plant input, is associated with the electrical, mechanical and isentropic efficiencies which are low because of the relatively small size of plant considered here. These large losses emphasise the need for paying close attention to the selection of this type of equipment, since components of inferior performance can considerably reduce the overall performance of the plant.

The second-largest irreversibility occurs in the condenser. This is partly due to the large degree of superheat achieved at the end of the compression process, leading to large temperature differences associated with the initial phase of heat transfer (process 2-g in Fig 5.12). This is an interesting example of a case where the cause of an irreversibility in one component must be sought in another.

The degree of superheat at the end of an isentropic process depends on the slope of the saturated vapour line for the refrigerant used in relation to the slope of an isentrope in the same co-ordinate system (see Figs 5.12 and 5.13). In the case of ammonia, this difference in the slopes leads to a high degree of superheat. The degree of superheat would have been smaller if, for example, R-12 instead of ammonia had been used in the refrigerator.

A similar kind of dependence of the irreversibility incurred in a process on the type of refrigerant used is found in the throttling process. Although throttling is generally regarded as an inherently dissipative process, in this case it is responsible for the smallest irreversibility of all the components. Had another refrigerant been used, say R-12, this process would have accounted for the dissipation of a larger proportion of the plant input. This points to one of the important factors in selecting refrigerants for the type of plant under consideration. As will be appreciated, the problem of the degree of superheat at the exit from the compressor is less acute in well-cooled reciprocating compressors than in adiabatic roto-dynamic compressors.

A comparison of the irreversibilities associated with the heat transfer processes in the evaporator and the cold chamber will show that, although the mean temperature difference in the former is smaller, the relative magnitudes of their irreversibilities are in the reverse order. This apparent paradox is due to the evaporator operating at a lower temperature than the heat exchanger in the cold chamber and hence suffering greater irreversibility with a smaller mean temperature difference. The effect of the temperature range on heat exchanger performance has been discussed in Section 4.3 and is illustrated in Fig 4.26.

Chapter 6 Thermoeconomic applications of exergy

The preceding chapters of this book deal mainly with applications of the Exergy Method to the *analysis* of thermal plants. This type of analysis reveals the relative magnitudes and the nature of component irreversibilities thus exposing, qualitatively, potential for improvements in the plant's thermodynamic and economic performance. To make a good use of such results, it is necessary to make a transition from analysis to the optimal synthesis of a new system.

The main objectives of optimal synthesis are:

- attainment of an optimal system structure, ie an optimal arrangement of components of the system;
- optimisation of the geometrical parameters of the components for maximum component efficiency; and
- assessment of economically justified costs of these components through thermoeconomic optimisation.

The first objective is the most difficult. There are no general formal techniques for utilising the findings of system analysis for determining an optimal system structure. Some progress has been made in the special areas of heat exchanger networks[6.1–6.3] and distillation plants[6.4,6.5] but, in other areas, success depends heavily on the practical experience and the inventive skills of the engineer. Given the necessary insight into the nature of irreversibilities, an appreciation of the practical limits to their reduction, and an understanding of the interrelations inherent in system structures, the optimal structure of a plant can be arrived at through an empirical process of successive improvements. The concept of structural coefficients, introduced in this chapter, may be of some help in this process.

The second objective, the optimisation of the geometric parameters of plant components will be discussed in Section 6.4.

The third objective, the thermoeconomic optimisation of plant components has been dealt with by a number of authors; some of the techniques developed will be presented.

6.1 Structural coefficients

The distribution of irreversibilities and the associated exergy flows among the various components of a steadily operating plant as given by, say, a Grassmann diagram, can be of great value in the analysis of a plant. Of further interest, however, is the way in which the local irreversibility rates and exergy fluxes alter in relation to the overall plant irreversibility rate, or the exergy input to the plant, with changes in a selected operating parameter. These relative changes can be expressed conveniently by two types of *structural coefficients*. The concept of structural coefficients was originally put forward by Beyer, who also developed some of the associated techniques[6.6–6.10]. A number of contributions to the development of this subject has been made by Brodyanskii and his co-workers[2.2,6.11,6.12].

Structural coefficients are used in the study of system structure, optimisation of plant components and product pricing in multi-product plants.

The coefficient of structural bonds

The coefficient of structural bonds (CSB) is defined by:

$$\sigma_{k,i}=\left(\frac{\partial \dot{I}_T}{\partial x_i}\right)\Big/\left(\frac{\partial \dot{I}_k}{\partial x_i}\right) \tag{6.1a}$$

where $\dot{I}_T$ is the irreversibility rate of the system under consideration, $\dot{I}_k$ the irreversibility rate of the k-th component of the system, and x_i the parameter of the system which produces the changes. Alternatively:

$$\sigma_{k,i}=\left(\frac{\partial \dot{I}_T}{\partial \dot{I}_k}\right)_{x_i=\text{var}} \tag{6.1b}$$

Here, the system, a thermal or chemical plant, is assumed to operate in a steady fashion and the input to the system is assumed to be of invariable quality, a condition satisfied by an input of fuel, electric energy, or steam supply at a fixed pressure and temperature. The effect of a change in x_i on the system would be to alter the rate of exergy input while leaving the output constant. This assumption conforms with the usual practice of specifying a plant in terms of its output rather than its input. From the exergy balance of the system:

$$\dot{E}_{IN}=\dot{E}_{OUT}+\dot{I}_T \tag{6.2}$$

it is clear that with $\dot{E}_{OUT}=\text{const}$:

$$\Delta\dot{E}_{IN}=\Delta\dot{I}_T \tag{6.3}$$

or, in other words, changes in the irreversibility of the system are equivalent to changes in the input.

Now, consider the significance of the different possible ranges of values of the CSB:

(i) If:

$$\sigma_{k,i}>1 \tag{6.4}$$

the reduction in the input to the system (saving in primary exergy) is greater than the reduction in the irreversibility of the element under consideration (k-th element). Clearly, the change in x_i improves not only the k-th element but, owing to the bonds between them, other elements. In this situation it is particularly advantageous to optimise the k-th element because of the potential favourable impact on the overall plant efficiency.

(ii) When:

$$\sigma_{k,i}<1 \tag{6.5}$$

the reduction in the input to the system is less than the reduction in the irreversibility of the k-th element. Clearly, here a reduction in the irreversibility rate in the k-th element is accompanied by increases in the irreversibility rates in other elements of the plant. This range of values of $\sigma_{k,i}$ demonstrates an unfavourable structure of the system.

(iii) A special case of case (ii) is:

$$\sigma_{k,i}=0 \tag{6.6}$$

where improvement in the performance of the k-th element is counterbalanced by an equal reduction in performance of other elements (as measured by $\dot{I}$) so that there is no effect on the overall plant efficiency. This case demonstrates a rigid system structure which does not permit the benefits of a local improvement in performance to be passed on to the plant as a whole.

(iv) It is possible to have:

$$\sigma_{k,i}<0 \tag{6.7}$$

indicating that x_i affects other elements more strongly than the k-th element, and in the opposite sense, ie when the irreversibility rate in the k-th element *decreases*, it *increases* in other plant elements by a greater amount. This value of $\sigma_{k,i}$ characterises a very unfavourable system structure which, subject to economic constraints, should be changed through plant modifications.

The CSB is useful in investigating the structure of a system, and in thermoeconomic optimisation of components of a plant.

The coefficient of external bonds

When dealing with a large and complicated system it may be advantageous to begin the analysis by examining its sub-systems. Such sub-systems differ from the system to which they belong in that the quality of their exergy inputs and the quantity and the quality of their exergy outputs cannot, in general, be taken to be constant. This makes it necessary, when analysing a sub-system, to relate the exergy fluxes of the sub-system to the exergy input to the system. This relationship can be expressed in terms of a coefficient which is known as the coefficient of external bonds (CEB):

$$\chi_{j,i}=\left(\frac{\partial \dot{E}_{IN}}{\partial \dot{E}_j}\right)_{x_i=\text{var}} \tag{6.9}$$

where $\dot{E}_{IN}$ is the exergy input to the system and $\dot{E}_j$ the exergy flux entering or leaving the sub-system under consideration.

CEB in the optimisation of autonomous sub-systems

As an example of application of the CEB, consider the case when the plant under examination is divisible into sub-systems which behave in an *autonomous* fashion. A sub-system can be regarded as autonomous when the thermodynamic bonds between it and other sub-systems are non-existent or very weak. This happens when exergy transfers between it and the other sub-systems are characterised by constant quality although not necessarily by constant exergy rates. Under these conditions each autonomous sub-system can be investigated and optimised independently.

The plant shown in Fig 6.1 consists of two such autonomous sub-systems, a process plant (PP) supplied with all its energy requirements (electric energy and heating steam of constant pressure and temperature (τ = const)) by a combined heating and power (CHP)

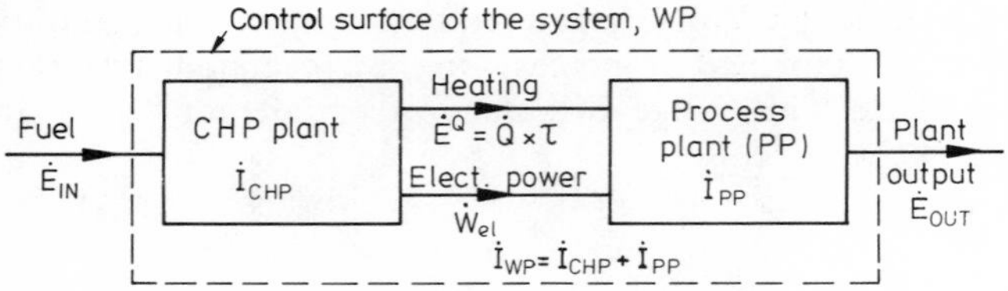

Fig 6.1 A system made up of a CHP plant and a process plant (PP) as its two sub-systems.

plant. The CSB for the k-th element of the process plant can be written as:

$$\sigma_{k,i}^{PP} = \left(\frac{\partial \dot{I}_{PP}}{\partial \dot{I}_k}\right)_{x_i = \mathrm{var}} \tag{6.10}$$

This relates the change in the irreversibility rate of the k-th component to that of the process plant resulting from a change in the x_i. If as a result of these changes the heat transfer rate $\dot{Q}$, which forms part of the input to the process plant, also changes, the corresponding CEB can be written:

$$\chi_{E^Q,Q} = \left(\frac{\partial \dot{E}_{IN}}{\partial \dot{E}^Q}\right)_{Q = \mathrm{var}} \tag{6.11}$$

Differentiating the exergy balance for the process plant with respect to x_i while keeping the plant output constant:

$$\frac{\partial \dot{E}^Q}{\partial x_i} = \frac{\partial \dot{I}_{PP}}{\partial x_i} \tag{6.12}$$

Similarly for the whole plant (WP):

$$\frac{\partial \dot{E}_{IN}}{\partial x_i} = \frac{\partial \dot{I}_{WP}}{\partial x_i} \tag{6.13}$$

The CSB for these changes, which is related to the whole plant, is defined as:

$$\sigma_{k,i}^{WP} = \left(\frac{\partial \dot{I}_{WP}}{\partial \dot{I}_k}\right)_{x_i = \mathrm{var}} \tag{6.14}$$

Consequently, from (6.10) to (6.14):

$$\sigma_{k,i}^{WP} = \chi_{E^Q,Q}\, \sigma_{k,i}^{PP} \tag{6.15}$$

This form of the CSB relates the change in the irreversibility rate $\dot{I}_k$ to changes in $\dot{I}_{WP}$ and hence to changes in the input of primary exergy $\dot{E}_{IN}$ ((6.13)). Such a relationship is essential in optimisation of plant components since the unit cost of primary exergy is, generally, known with more certainty than that of its derived forms.

Further examples of application of both the CSB and CEB will be given later.

Relationship between the CSB and the CEB

Consider the k-th element of a system (Fig 6.2) with a number of exergy input streams represented as $\sum_{IN,k} \dot{E}_j$ and output streams represented as $\sum_{OUT,k} \dot{E}_j$. With the irreversibility rate for the element denoted by $\dot{I}_k$, the exergy balance for the element is:

$$\dot{I}_k = \sum_{IN,k} \dot{E}_j - \sum_{OUT,k} \dot{E}_j \tag{6.16}$$

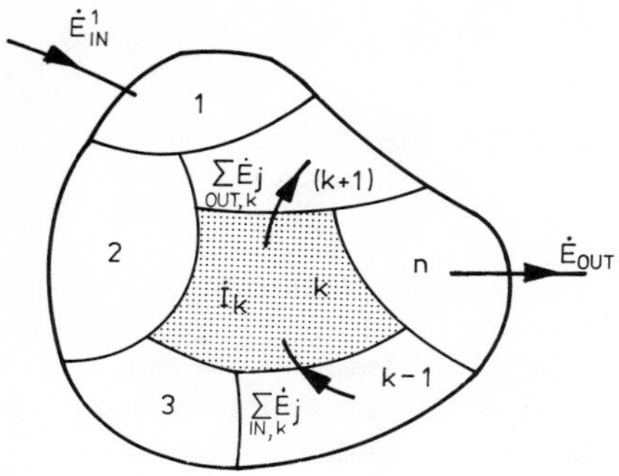

Fig 6.2 A system and its elements with the associated exergy fluxes.

Differentiating (6.16) with respect to x_i:

$$\frac{\partial \dot{I}_k}{\partial x_i} = \sum_{IN,k} \frac{\partial \dot{E}_j}{\partial x_i} - \sum_{OUT,k} \frac{\partial \dot{E}_j}{\partial x_i} \tag{6.17}$$

Similarly, with the irreversibility rate for the system denoted by $\dot{I}_T$ and with $\dot{E}_{OUT}$ = const, for the system:

$$\frac{\partial \dot{I}_T}{\partial x_i} = \frac{\partial \dot{E}_{IN}}{\partial x_i} \tag{6.18}$$

Dividing (6.17) by (6.18):

$$\frac{1}{\sigma_{k,i}} = \sum_{IN,k} \frac{1}{\chi_{j,i}} - \sum_{OUT,k} \frac{1}{\chi_{j,i}} \tag{6.19}$$

(6.19) will be used now to consider some special cases.

If one of the exergy streams of the k-th element is independent of x_i then, from (6.9):

$$\frac{1}{\chi_{j,i}} = 0 \tag{6.20}$$

Now restrict consideration to the case when, in the k-th element under consideration, there is only one exergy input stream and one output stream. Thus:

(i) When:

$$\frac{1}{(\chi_i)_{OUT,k}} = 0 \qquad \text{then} \qquad \sigma_{k,i} = (\chi_i)_{IN,k} \tag{6.21}$$

Clearly, the CSB for the element is equal to the CEB for its input stream.

(ii) When:

$$\frac{1}{(\chi_i)_{IN,k}} = 0 \qquad \text{then} \qquad \sigma_{k,i} = -(\chi_i)_{OUT,k} \qquad 6.22)$$

Here the CSB is equal to minus the CEB for the output stream.

(iii) When the exergy input to the plant, $\dot{E}_{IN}$, is independent of an exergy flux $\dot{E}_j$:

$$\chi_{j,i} = 0 \tag{6.23}$$

This indicates that there is a 'rigid bond'. If this can be proved for *any one* exergy stream crossing the boundary of the *k*-th element, then, from (6.19):

$$\sigma_{k,i}=0 \tag{6.24}$$

and all the other CEBs for the element need not be determined.

Determination of the CSB and the CEB

The most direct method of calculation of the CSB is to calculate and plot values of $\dot{I}_k$ and $\dot{I}_T$ for a number of values (at least two) of x_i, and hence determine the CSB from the slope of the plotted curve. A similar procedure can be used for calculating the CEB.

For large, complex plants a more formalised, and systematic approach involving suitable computer programs is required. The preparation of such programs can be greatly facilitated by the use of the graph-theoretic approach. This technique is outside the scope of this book and the reader is referred to the original publications[6.13,6.14].

6.2 Thermodynamic non-equivalence of exergy and exergy losses

The CSB may be said to express the non-equivalence of exergy losses (irreversibilities) in the various elements of a system. If the change in the irreversibility rate in the *k*-th element is given by $\Delta\dot{I}_k$ the corresponding change in the exergy input to the plant, assuming the output to be fixed, is:

$$\Delta\dot{E}_{IN}=\sigma_{k,i}\Delta\dot{I}_k \tag{6.25}$$

Hence the larger is $\sigma_{k,i}$ the larger will be the effect of a given change in the irreversibility rate on the plant exergy input.

For example, consider a system, shown in Fig 6.3, consisting of a chain of elements in

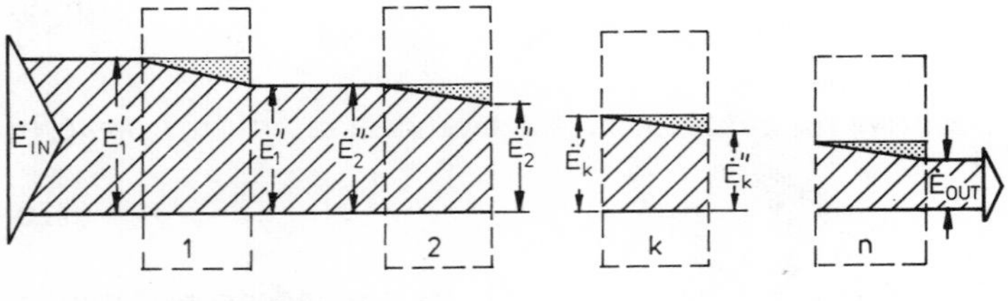

Fig 6.3 Grassmann diagram.

which the exergy input to the 1st element is processed sequentially, the output being delivered by the *n*-th element without withdrawal or input of exergy at intermediate elements.

Define a ratio for the *i*-th element:

$$\omega_i\equiv\frac{\dot{E}''_i}{\dot{E}'_i} \tag{6.26}$$

For some elements, this ratio will be identical to the rational efficiency, but not for others. For example, considering the compressor plant shown in Fig 3.20, $\omega_i=\psi_i$, for the electric motor and the compressor but not for the aftercooler or the transmission which are both purely dissipative elements.

From (6.26) the input to the k-th element can be written:

$$\dot{E}'_k = \dot{E}_{IN} \prod_{i=1}^{k-1} \omega_i \tag{6.27}$$

where $\dot{E}_{IN} = \dot{E}'_1$.

Now, consider a change in the irreversibility rate of the k-th element due to the effect of a local change. Assuming the plant output to remain constant and all the ratios ω_i to remain unchanged, the output of the k-th element, $\dot{E}''_k$, should also remain constant. Consequently from (6.27):

$$\Delta\dot{E}_{IN}/\Delta\dot{E}''_k = 1 \Big/ \prod_{i=1}^{k-1} \omega_i \tag{6.28a}$$

and with:

$$\left.\begin{aligned} \dot{E}_{OUT} &= \text{const} \\ \dot{E}''_k &= \text{const} \end{aligned}\right\} \tag{6.28b}$$

$$\Delta\dot{I}_T/\Delta\dot{I}_k = 1 \Big/ \prod_{i=1}^{k-1} \omega_i \tag{6.29}$$

The LHSS of (6.28a) and (6.29) are the CEB and the CSB respectively for the k-th element. Consequently for a system in which elements are arranged in chain form and with the restricting conditions stated above:

$$(\chi_k)_{\text{chain}} = (\sigma_k)_{\text{chain}} = 1 \Big/ \prod_{i=1}^{k-1} \omega_i \tag{6.30}$$

Because of the conditions specified by (6.28b) the CEB is equal to the CSB. The magnitude of these coefficients is equal to the reciprocal of the product of all the values of ω_i corresponding to elements preceding the element under consideration. Since for all these coefficients:

$$\omega_i < 1$$

therefore:

$$1 \Big/ \prod_{i=1}^{k-1} \omega_i > 1 \tag{6.31}$$

Thus, for this type of special arrangement of elements within the plant and with the restricting conditions specified, both the CSB and the CEB have values greater than unity and increase with the value of k, the index of the element under consideration. Therefore, from (6.25), a reduction in the irreversibility rate of a certain magnitude in an element near the end of the chain of processes leads to a greater saving in the exergy input to the plant than the same reduction occurring in an element at the beginning of the chain. The values of ω_i for the successive elements will always be less than unity provided there is no supplementary exergy input at any of the elements lying on this path. Hence for each of these elements the CSB will be greater than unity.

These conclusions can be extended to a system with an arbitrary arrangement of elements. Considering a 'path' through the plant followed in the direction of exergy flow starting at the point of exergy input to the plant, then for any element lying on this path the condition $\omega_i < 1$ will apply provided there is no supplementary exergy input to the element. Hence, for each of these elements the CSB will be greater than unity and (6.30) applies.

6.3 Structural coefficients for a CHP plant — an illustrative example

A CHP plant with an extraction turbine (Fig 6.4) operates on the steam cycle shown in Fig 6.5. The Grassmann diagram for the plant (Fig 6.6) illustrates the mode of heating which depends exclusively on extraction steam, ie without the use of fresh steam. For this reason the parts of the plant corresponding to the use of fresh steam have been drawn in Fig 6.4 using dotted lines.

In order not to obscure the techniques which are to be demonstrated with extraneous details, the plant selected for this purpose is reduced to its essential aspects and its behaviour subject to a number of simplifying assumptions.

The operating parameters for the plant are:

Boiler pressure = 40 bar
Steam temperature = 400°C
Extraction pressure = 4 bar
Condenser pressure = 0.05 bar
Environmental temperature = 20°C

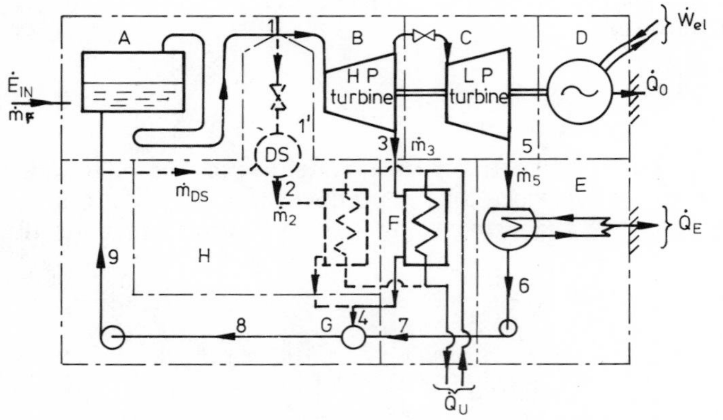

Fig 6.4 A CHP plant with an extraction turbine, DS—desuperheater.

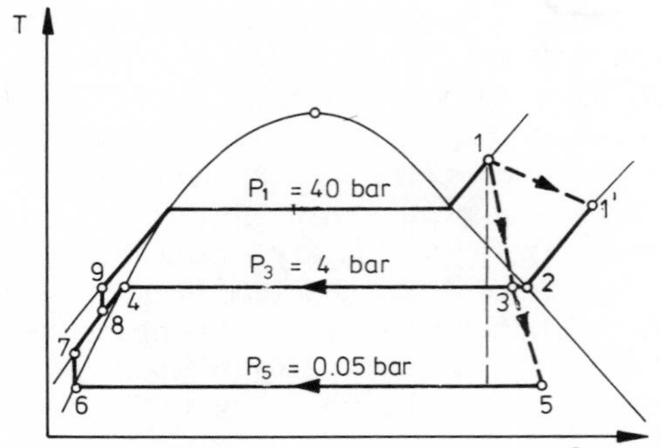

Fig 6.5 *T–s* diagram for processes in the CHP plant.

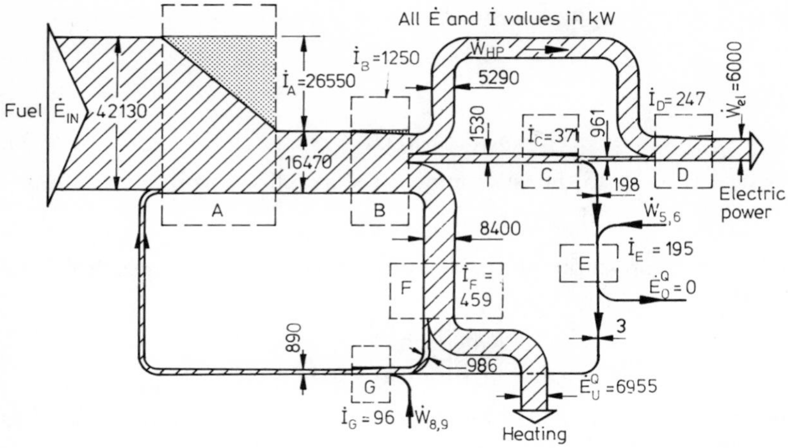

Fig 6.6 Grassmann diagram for the CHP plant for the case when $R_Q = 1$.

Efficiencies

Boiler efficiency $\eta_{\text{comb}} = 0.85$
Isentropic efficiency, for both HP and LP turbines $\eta_s = 0.75$
Mechanical efficiency $\eta_m = 0.98$
Electrical efficiency $\eta_{el} = 0.97$
Electric power output $\dot{W}_{el} = 6\,000$ kW
Useful heating, output $\dot{Q}_u = 25\,000$ kW
Heating, mean delivery temperature $T_u = 406.15$ K

The assumptions made in the calculations are:

1. All pressure losses are neglected.
2. All stray heat transfer losses are neglected except those due to dissipation of electrical and mechanical type which are taken to be confined to Sub-region D (denoted in Fig 6.4 by $\dot{Q}_0$).
3. Power input to the feed pumps and other auxiliary equipment is neglected.
4. The coefficient φ for the fuel in the relation:

$$\varepsilon_F = \varphi(\text{NCV})$$

is unity. This is a good approximation for some gaseous fuels (see Appendix C).
5. The heat transfer $\dot{Q}_u$ in the heat exchangers takes place at constant temperature T_u. This may mean that the heat transfer medium undergoes evaporation at constant pressure, or alternatively, that it undergoes only a small temperature change in the heat exchanger and hence a negligible error would be introduced by taking T_u to be the mean temperature of heat transfer.

Plant analysis

From Fig 6.4, the energy balance for the plant* is:

$$\dot{m}_F(\text{NCV})\eta_{\text{comb}} = \dot{W}_{el} + \dot{Q}_E + \dot{Q}_u + \dot{Q}_0 \qquad (6.32)$$

* The combustion efficiency is based here on the net calorific value, NCV.

Assuming that a part of the heating output, $\dot{Q}_u$, comes from fresh steam ($\dot{m}_2 \neq 0$) then:

$$\dot{Q}_u = \dot{m}_2(h_2 - h_4) + \dot{m}_3(h_3 - h_4) \tag{6.33}$$

Also, for the condenser:

$$\dot{Q}_E = \dot{m}_5(h_5 - h_6) \tag{6.34}$$

and for the stray heat transfer to the environment $\dot{Q}_0$, as follows from item 2 above:

$$\dot{Q}_0 = \dot{W}_{el}[1/(\eta_m \eta_{el}) - 1] \tag{6.35}$$

A ratio specifying the proportion of the heating output which comes from the extraction steam is:

$$R_Q = \frac{\dot{m}_3(h_3 - h_4)}{\dot{m}_2(h_2 - h_4) + \dot{m}_3(h_3 - h_4)} \tag{6.36}$$

The heating output can be expressed in terms of thermal exergy using the dimensionless exergetic temperature τ_u:

$$\dot{E}_u^Q = \dot{Q}_u \tau_u \tag{6.37}$$

where:

$$\tau_u = \frac{T_u - T_0}{T_u} \tag{6.38}$$

Finally, using the coefficient φ, the exergy input can be expressed in terms of combustion quantities:

$$\dot{m}_F \varphi(\text{NCV}) = \dot{E}_{IN} \tag{6.39}$$

An expression for $\dot{E}_{IN}$ can now be obtained in the form:

$$\dot{E}_{IN} = \frac{\varphi \dot{W}_{el}}{\eta_m \eta_{el} \eta_{comb}}\left[1 + \frac{\Delta h_{5,6}}{\Delta h_{1,5}}\right] + \frac{\varphi \dot{E}_u^Q}{\tau_u \eta_{comb}}\left[1 - R_Q \frac{\Delta h_{1,3}}{\Delta h_{3,4}} \frac{\Delta h_{5,6}}{\Delta h_{1,5}}\right] \tag{6.40}$$

The specific enthalpy changes in (6.40) are obtained by subtracting specific enthalpies in states given by the subscripts and indicated in Fig 6.4. Note that, although (6.40) relates exergy input to the two forms of the exergy output, it is not an exergy balance. The exergy balance for the plant is:

$$\dot{E}_{IN} = \dot{W}_{el} + \dot{E}_u^Q + \dot{I}_T \tag{6.41}$$

The exergy of a stream of water substance is calculated using specific enthalpies and entropies of H_2O from steam tables[2.4], from:

$$\dot{E} = \dot{m}[(h - T_0 s) - \beta_0] \tag{6.42}$$

where $\beta_0 = h_0 - T_0 s_0$ is calculated for the saturated liquid state of H_2O at 20°C. Its value was found to be -2.9 kJ/kg. Although β_0 was chosen to correspond to the environmental state, any other convenient state could have been chosen since, in a closed cycle, it is *changes in exergy* rather than absolute values that matter.

The Grassmann diagram (Fig 6.6) was constructed using the above relations for the case when all heating output is provided by extraction steam, ie $R_Q = 1$. The exergies of the different streams given by the widths of the corresponding bands are specified in kW. The power output from the HP turbine, $\dot{W}_{HP}$, is shown, for the sake of greater clarity, by-passing Sub-region C. The actual path of the transmission of power from the turbines to the alternator is immaterial for thermodynamic analysis.

Calculation of the CEBS for the CHP plant

(6.40) showed that the two forms of the plant output $\dot{W}_{el}$ and $\dot{E}_u^Q$ can be varied, independently, within certain limits, by simply changing the mass flow rates $\dot{m}_3$ and $\dot{m}_5$. Assuming the cycle operating parameters and the values of the different efficiencies remain constant, the CEBS obtained from (6.40) are:

$$\chi_W = \left(\frac{\partial \dot{E}_{IN}}{\partial \dot{W}_{el}}\right)_{\dot{E}_u^Q} = \frac{\varphi}{\eta_m \eta_{el} \eta_{comb}} \left[1 + \frac{\Delta h_{5,6}}{\Delta h_{1,5}}\right] \tag{6.43}$$

$$\chi_Q = \left(\frac{\partial \dot{E}_{IN}}{\partial \dot{E}_u^Q}\right)_{\dot{W}_{el}} = \frac{\varphi}{\tau_u \eta_{comb}} \left[1 - R_Q \frac{\Delta h_{1,3}}{\Delta h_{3,4}} \frac{\Delta h_{5,6}}{\Delta h_{1,5}}\right] \tag{6.44}$$

Since the two forms of plant output can be varied independently, input of primary exergy can be written as a total differential:

$$d\dot{E}_{IN} = \left(\frac{\partial \dot{E}_{IN}}{\partial \dot{W}_{el}}\right)_{\dot{E}_u^Q} d\dot{W}_{el} + \left(\frac{\partial \dot{E}_{IN}}{\partial \dot{E}_u^Q}\right)_{\dot{W}_{el}} d\dot{E}_u^Q \tag{6.45}$$

and hence:

$$d\dot{E}_{IN} = \chi_W \, d\dot{W}_{el} + \chi_Q \, d\dot{E}_u^Q \tag{6.46}$$

The effect of an increment in either $\dot{W}_{el}$ or $\dot{E}_u^Q$ on the consumption of primary exergy can be assessed from (6.43), (6.44) and (6.46). Clearly, the increase in the consumption of primary exergy caused by an increase in $\dot{W}_{el}$ is given by the first term on the RHS of (6.46), while that due to an increase in $\dot{E}^Q$ by the second term. Consequently the CEBS may be looked upon as reciprocals of *partial incremental rational efficiencies* for the two plant products and an alternative form of (6.46) is:

$$d\dot{E}_{IN} = \frac{d\dot{W}_{el}}{\psi'_W} + \frac{d\dot{E}_u^Q}{\psi'_Q} \tag{6.47}$$

where:

$$\left.\begin{aligned} \psi'_W &= 1/\chi_W \\ \psi'_Q &= 1/\chi_Q \end{aligned}\right\} \tag{6.48}$$

From the data given above for the plant, for the first CEB:

$$\chi_W = 4.3688$$

or:

$$\psi'_W = 0.2289$$

and for the second CEB:

$$\chi_Q = 4.2273[1 - 0.4591 R_Q]$$

The variation of χ_Q with R_Q is given in Table 6.1

Table 6.1

R_Q	χ_Q	ψ'_Q	Remarks
1	2.2866	0.4373	All extraction steam
0.75	2.7718	0.3608	
0.50	3.2570	0.3070	
0.25	3.7422	0.2672	
0.00	4.2273	0.2366	All fresh steam

The calculated value of χ_W shows that an increase in the electric power output results in 4.3688 times larger increase in the primary exergy consumption, ie the additional electric power generation takes place with a partial incremental rational efficiency, ψ'_W, of 0.2289. Note that these values are independent of the heating output or the fraction R_Q.

The CEB for $\dot{E}_u^Q$ increases as the proportion of heating from fresh steam increases, or as R_Q decreases. When all the heating output comes from extraction steam, the extra output of $\dot{E}_u^Q$ can be provided with a relatively low increase of primary exergy consumption, only 52% of that necessary for $\dot{W}_{el}$. At the other end of the range, when all the heating output comes from fresh steam the increase of primary exergy consumption is considerably higher, nearly the same as for an additional generation of electric power. In other words, the partial incremental rational efficiency of generation of thermal exergy, $\dot{E}_u^Q$, increases with R_Q. Thus, the use of fresh steam for heating is very inefficient and should be used only during peak demand. Although this conclusion is well known, the technique is of wider applicability and should prove of value in cases where plant behaviour is not so well understood.

Calculation of the CSB for a heat exchanger in the CHP plant

As an example of calculation of the CSB, consider the heat exchanger in Sub-region F (Fig 6.4) under conditions when $R_Q = 1$, ie when no fresh steam is used. As we have assumed that pressure and stray heat transfer losses in the plant are negligible, the only form of irreversibility in the heat exchanger is that due to heat transfer over a finite temperature difference. Since the temperature of the heat transfer medium is constant, $\theta_u = 133°C$, this temperature difference can be varied by varying the pressure, P_3, and the corresponding saturation temperature T_3 of the extraction steam. Thus the CSB for the heat exchanger can be expressed as:

$$\sigma_{F,P_3} = \left(\frac{\partial \dot{I}_T}{\partial \dot{I}_F}\right)_{P_3 = \text{var}} \tag{6.49}$$

Under conditions of constant temperature heat exchange, the irreversibility rate of the heat exchanger is:

$$\dot{I}_F = \dot{Q}_u T_0 \left(\frac{1}{T_u} - \frac{1}{T_3}\right) \tag{6.50}$$

The irreversibility rate of the plant as a whole, $\dot{I}_T$, can be calculated from (6.40) and (6.41). Results of calculations of the effect of variation in P_3 on the partial incremental rational efficiency ψ'_Q as given by (6.48) and (6.44) are given in Table 6.2.

Table 6.2

P_3/bar	$\dot{I}_F$/kW	$\dot{I}_T$/kW	ψ'_Q
3.5	258	28 550	0.4548
4.0	459	29 120	0.4373
4.5	639	29 670	0.4237
5.0	798	30 160	0.4116
5.0	947	30 610	0.4008

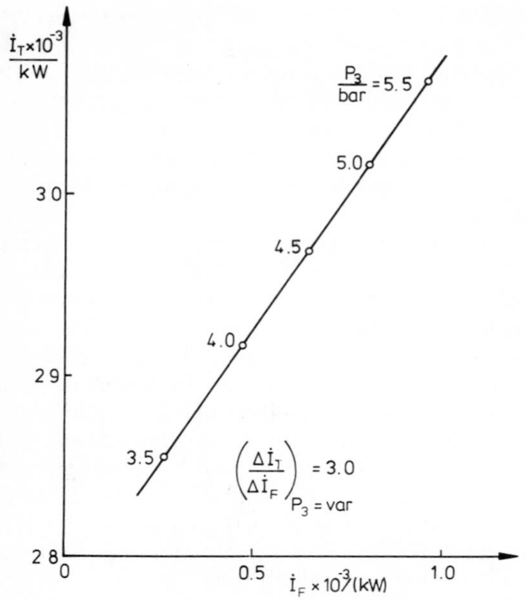

Fig 6.7 Determination of the CSB for the heat exchanger in the CHP plant, with $P_3 = \text{var}$.

A plot of $\dot{I}_T$ against $\dot{I}_F$ in Fig 6.7 gives a linear relationship. From the slope of the line:

$$\sigma_{F,P_3} = 3.0$$

Clearly, any reduction in the irreversibility rate of the heat exchanger gives a three times greater reduction in the irreversibility rate of the plant. From the values of ψ'_Q tabulated in Table 6.2 we can see that the effect of increasing extraction pressure P_3, is to reduce the partial rational efficiency of generation of thermal exergy. The calculated value of the CSB of the heat exchanger will be used in its thermoeconomic optimisation in Section 6.6.

6.4 Optimisation of component geometry

Optimisation of a thermal plant is a complex procedure generally involving many geometric and economic variables. Reducing the complexity of this task by breaking up the procedure into a number of relatively simple optimisation processes is usually helpful. One aspect of the overall problem which can be often treated separately, before the main thermoeconomic optimisation, is optimisation of the geometrical parameters of the plant components.

The different components comprising a thermal plant can be categorised as:

1. 'Ready made' components selected from a manufacturer's catalogue, eg pumps, compressors, turbines, *etc.*
2. Components specially designed, or 'tailor-made' for the plant, eg heat exchangers, pressure vessels, pipe and duct systems, *etc.*

Geometric parameters of 'ready made' components may be assumed to have been optimised by the manufacturer; at the thermoeconomic stage of optimisation, the plant designer need only select a component which fits the technical specification and whose capital cost is justified by its thermodynamic efficiency.

With the second type of component, it is convenient, from the point of view of the thermoeconomic stage of optimisation, to optimise the particular component geometry for a range of components of different prices. In other words, this optimisation process aims to offer the designer a range of components of maximum efficiencies for given prices. Thus, the two component types would be optimised thermoeconomically in basically the same way.

Ideally, component geometry should be optimised in the context of the plant using the annual plant operating cost as the objective function. Under these conditions, the exergy streams entering and leaving the component would automatically be given their proper economic values. However, such a procedure would be complex and time consuming. This difficulty can be overcome by optimising individual components, provided that, at least approximately, correct economic values of the exergy streams are used in the optimisation. Now, in a general case, the objective function would be the annual component operating cost, comprising principally the investment and running costs. However, when optimising for a component of a fixed capital cost, as suggested above, the cost of the investment is invariable and hence the running cost becomes, in effect, the objective function. Now, the running cost (cost of the exergy consumed) depends on the efficiency of the component, so the objective function can be some suitably modified form of a criterion of performance which takes into account the relative values of the exergy streams involved.

As an example of this optimisation technique, consider a case for which the cost of compensation for the losses of different forms of exergy can be assumed to be the same. Thus the total irreversibility rate of the component can be the objective function. The component geometry can be optimised if its total irreversibility rate, $\dot{I}_T$, consists of two components $\dot{I}_A$ and $\dot{I}_B$ which are both functions of some geometric parameters $x_1, x_2 \ldots x_N$, to be optimised, and perhaps a third component $\dot{I}_C$ which is independent of these parameters. This can be expressed as:

$$\dot{I}_T = \dot{I}_A(x_1, x_2 \ldots x_N) + \dot{I}_B(x_1, x_2 \ldots x_N) + \dot{I}_C$$

Under these conditions there may be a trade-off between $\dot{I}_A$ and $\dot{I}_B$ as a result of a change in any x_i, with the possibility of a minimum value of $\dot{I}_T$ at some value of x_i. Thus, the optimum condition with respect to all the relevant geometric parameters will correspond to:

$$\dot{I}_A(x_1, x_2 \ldots x_N) + \dot{I}_B(x_1, x_2 \ldots x_N) \rightarrow \min$$

For example, consider a single pass shell-and-tube air heater with condensing steam, shell side, as the heating medium[6.15].

In heat exchangers involving gas streams, the irreversibility rate due to pressure losses $\dot{I}^{\Delta P}$ is, in general, a substantial fraction of the total irreversibility rate. Consequently the trade-off in this case is between $\dot{I}^{\Delta P}$ and $\dot{I}^{\Delta T}$ (see Section 4.3).

For this type of heat exchanger the capital cost is given by the manufacturer as a function of the heat transfer area. Thus, keeping the heat transfer area constant satisfies the fixed investment condition. Essentially, then, this optimisation consists of spacial distribution of the available heat transfer area into the most efficient geometric configuration. Within certain limits, the number of tubes does not have a pronounced effect on heat exchanger performance so the only parameter used as an independent variable is the tube diameter D. Thus, the optimum condition is given in this case by:

$$\dot{I}^{\Delta T}(D)+\dot{I}^{\Delta P}(D)\rightarrow \min$$

or:

$$\frac{d\dot{I}^{\Delta P}}{dD}=-\frac{d\dot{I}^{\Delta T}}{dD}$$

For a perfect gas stream of mass flow rate $\dot{m}$ heated from a TER at the constant temperature T_S, the two components of irreversibility rate are:

$$\dot{I}^{\Delta T}=\dot{m}c_P T_0\left[\ln\frac{T_2}{T_1}-\frac{T_2-T_1}{T_S}\right]$$

$$\dot{I}^{\Delta P}=\dot{m}c_P T_0\frac{\gamma-1}{\gamma}\ln\frac{P_1}{P_2}$$

where the subscripts 1 and 2 refer to the inlet and outlet state, respectively, of the gas stream. The pressure ratio P_1/P_2 was obtained using the usual[6.16] empirical expression for the friction factor and allowing for pressure drops at the tube inlets and outlets. The friction factor was related to the film heat transfer coefficient through the Colburn analogy. Hence, the variation of $\dot{I}_T$ with tube diameter for different total heat transfer areas was plotted (Fig 6.8). Note that each of the curves plotted is the result of two conflicting tendencies. For small tube diameters, the contribution of $\dot{I}^{\Delta P}$ high due to large pressure losses while at large diameters $\dot{I}^{\Delta T}$ dominates irreversibility. As will be seen from Fig 6.8, the smaller the heat transfer area under consideration the smaller is the optimum tube diameter, D_{OPT}. Also, the value of D_{OPT} is more sharply defined for small heat transfer areas. For larger heat transfer areas, substantial departures from D_{OPT} are possible without paying too high a price in terms of an increased irreversibility rate.

The results of the optimisation are summarised in Fig 6.9, where the optimum tube length to diameter ratio and the minimum value of the irreversibility rate are plotted against the heat transfer area. Also the value of the intrinsic irreversibility rate (Section 3.3) based on zero pressure losses and zero value of minimum temperature difference, ie

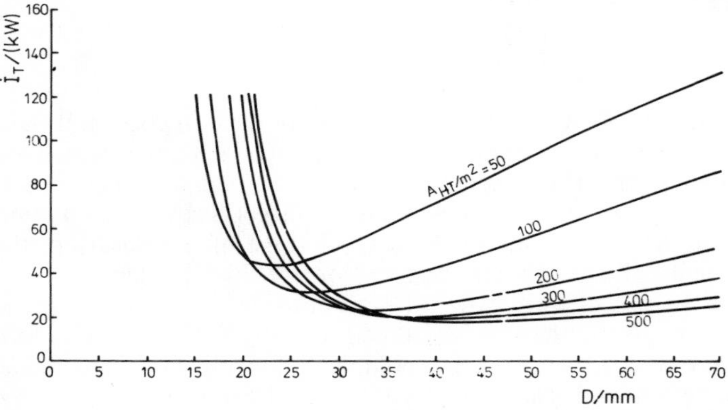

Fig 6.8 Dependence of the irreversibility rate of the heat exchanger on the tube diameter for different heat transfer areas.

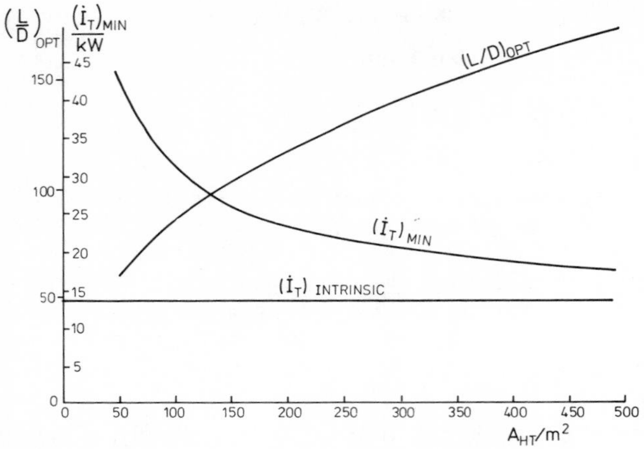

Fig 6.9 Optimum operating parameters of the heat exchanger.

$T_s - T_2 = 0$, is shown for comparison. The actual irreversibility rate $(\dot{I}_T)_{MIN}$ of the optimised heat exchanger tends asymptotically to $(\dot{I}_T)_{INTRINSIC}$ as the heat transfer area increases.

The assumption of equal cost of compensation for $\dot{I}^{\Delta P}$ and $\dot{I}^{\Delta T}$ in heat exchangers may be found in some cases too inaccurate, when separate unit costs of compensation must be allocated. Denoting them by $c^{I}_{\Delta P}$ and $c^{I}_{\Delta T}$ respectively, the cost rate of exergy consumption (running cost rate) can be expressed as:

$$\dot{C}^{I} = c^{I}_{\Delta P}\dot{I}^{\Delta P} + c^{I}_{\Delta T}\dot{I}^{\Delta T}$$

The condition of minimum running cost which determines the optimum diameter is given by:

$$\frac{d\dot{I}^{\Delta T}}{dD} = -F_w \frac{d\dot{I}^{\Delta P}}{dD}$$

where, F_w, a form of a weighting factor, is given by:

$$F_w = \frac{c^{I}_{\Delta P}}{c^{I}_{\Delta T}}$$

Now the condition for optimum geometry depends on ratio of the unit costs, not their absolute values. F_w can be calculated using the thermoeconomic techniques which are outlined in the following sections of this chapter. The effect of F_w here would be to shift the optimum diameters, for given heat transfer areas, to larger sizes. The function $(\dot{I}_T)_{MIN}(A_{HT})$ plotted in Fig 6.9 is relevant to the second stage of optimisation, the thermoeconomic stage, dealt with in the next two sections of this chapter.

Another interesting example of this type of optimisation is minimising the irreversibility caused by throttling in the valves of a reciprocating expander. To reduce throttling losses, the areas of the valve apertures available to flow must be increased as much as possible within the confined space of the cylinder head. So the problem is how to divide the available space between the two valves. This problem has been solved[6.17] by a technique which is similar to the one described above for the heat exchanger, except that in this case the trade-off was between irreversibility rate occurring in the inlet and outlet valves.

6.5 Thermoeconomic optimisation of thermal systems

Thermoeconomics is a discipline which combines concepts of the Exergy Method with those belonging to Economic Analysis. The purpose of thermoeconomic optimisation is to achieve, within a given system structure, a balance between expenditure on capital costs and exergy costs which will give a minimum cost of the plant product. The complexity of thermal systems often makes their thermoeconomic optimisation difficult. Ways are being sought, therefore, to simplify the optimisation procedure while ensuring that the results obtained are within acceptable limits of accuracy.

Thermoeconomic analysis must be treated as the final stage of optimisation following:

1. Thermodynamic examination of the system to devise the most effective system structure, a procedure known as system synthesis.
2. Thermodynamic optimisation of the geometric parameters of the elements of the system to maximise component efficiency for a given capital cost.

There are cost optimisation procedures which make no use of the exergy concept so the effectiveness of every change carried out on a plant component must be assessed in terms of the overall system parameters, eg its effect on the input to the system. This makes optimisation complex and time consuming. The advantage of using the Exergy Method of thermoeconomic optimisation is that the various elements of the system can be optimised on their own, the effect of the interaction between the given element and the whole system being taken into account by local unit costs of exergy fluxes or those of exergy losses. The individual optimisation of system elements is made possible by the universality of exergy as a standard of quality of energy and by using irreversibility rate as a measure of process imperfection.

Currently, there are basically two different methods which make use of the exergy concepts. The *structural method* based on concepts introduced by Beyer[6.8–6.10] depends on the use of structural coefficients and related concepts for the evaluation of local unit costs of irreversibilities. The other method, due to Tribus, Evans and their co-workers[6.18,6.19] allows the autonomous thermoeconomic optimisation of system elements by using local unit costs of exergy fluxes entering and leaving the elements under consideration. This method will be called the *autonomous method.*

The structural method of thermoeconomic optimisation

The purpose of this optimisation is to determine for a selected component (system element) the capital cost corresponding to the minimum annual operating cost of the plant for a given plant output and thus, by implication, to the minimum unit cost of the product.

Assume that there is a plant parameter x_i affecting the performance of the k-th element of the system and thus, in most cases, also indirectly affecting the performance of the system. Any variation in x_i will also, in general, cause changes in the irreversibility rates of the other elements of the system (Section 6.1), and necessitate changes in the capital costs of the different elements. The exergy balance for the system as a whole can be written:

$$\dot{I}_{\mathrm{T}}(x_i) = \dot{\mathrm{E}}_{\mathrm{IN}}(x_i) - \dot{\mathrm{E}}_{\mathrm{OUT}} \tag{6.51}$$

As shown, the term $\dot{E}_{OUT}$ which represents the joint exergy of the plant products is taken to be independent of x_i. The irreversibility rate $\dot{I}_T(x_i)$ may be looked upon as the *consumption of exergy* in the system, necessary to generate the product exergy $\dot{E}_{OUT}$. Any increase in exergy consumption will necessitate corresponding additional exergy input, $\Delta\dot{E}_{IN}(x_i)$.

The nature of the technique requires that the exergy input to the plant should have a single fixed unit cost. This condition can be satisfied by a single form of exergy input of invariable quality, eg fuel or electric energy. Alternatively, the input could be made up of more than one form of exergy of invariable quality in fixed proportions.

For this optimisation, take the objective function to be the annual cost of plant operation, C_T, which can be expressed as:

$$C_T(x_i) = t_{op} c_{IN}^{\varepsilon} \dot{E}_{IN}(x_i) + a^c \sum_{l=1}^{n} C_l^c(x_i) + b^c \tag{6.52}$$

where:

t_{op}—period of operation per year;
c_{IN}^{ε}—unit cost of input exergy to the system;
a^{ε}—capital-recovery factor, which when multiplied by the total investment gives the annual repayment necessary to pay back the investment after a specified period;
C_l^c—capital cost of the l-th element of the system consisting of n elements;
b^c—the part of the annual cost which is not affected by the optimisation.

Subject to the usual mathematical conditions being fulfilled, the objective function will be differentiated with respect to x_i. From (6.51):

$$\frac{\partial \dot{E}_{IN}}{\partial x_i} = \frac{\partial \dot{I}_T}{\partial x_i}$$

so:

$$\frac{\partial C_T}{\partial x_i} = t_{op} c_{IN}^{\varepsilon} \frac{\partial \dot{I}_T}{\partial x_i} + a^c \sum_{l=1}^{n} \frac{\partial C_l^c}{\partial x_i} \tag{6.53}$$

The second term on the RHS of (6.53) may be rearranged conveniently as:

$$a^c \sum_{l=1}^{n} \frac{\partial C_l^c}{\partial x_i} = a^c \sum_{l'=1}^{n} \frac{\partial C_{l'}^c}{\partial x_i} + a^c \frac{\partial C_k^c}{\partial x_i} \tag{6.54}$$

where $l' \neq k$, ie subscript l' marks any of the elements of the system except that one which is subject to the optimisation. Also, it will be convenient to make the rearrangement:

$$\sum_{l'=1}^{n} \frac{\partial C_{l'}^c}{\partial x_i} = \frac{\partial \dot{I}_k}{\partial x_i} \sum_{l'=1}^{n} \left(\frac{\partial C_{l'}^c}{\partial \dot{I}_k}\right) = \frac{\partial \dot{I}_k}{\partial x_i} \zeta_{k,i} \tag{6.55}$$

where:

$$\zeta_{k,i} = \sum_{l'=1}^{n} \left(\frac{\partial C_{l'}^c}{\partial \dot{I}_k}\right)_{x_i=\text{var}, l' \neq k} \tag{6.56}$$

$\zeta_{k,i}$ is the *capital cost coefficient*.

From (6.1a):

$$\frac{\partial \dot{I}_T}{\partial x_i} = \sigma_{k,i} \frac{\partial \dot{I}_k}{\partial x_i} \tag{6.57}$$

Using (6.54)–(6.57), (6.53) can be modified to:

$$\frac{\partial C_T}{\partial x_i}=t_{op}\,c^{I}_{k,i}\,\frac{\partial \dot{I}_k}{\partial x_i}+a^{c}\,\frac{\partial C^{c}_{k}}{\partial x_i} \tag{6.58}$$

where:

$$c^{I}_{k,i}=c^{\varepsilon}_{IN}\,\sigma_{k,i}+\frac{a^{c}}{t_{op}}\,\zeta_{k,i} \tag{6.59}$$

To optimise, make (6.58) zero. Thus:

$$\left(\frac{\partial \dot{I}_k}{\partial x_i}\right)_{OPT}=-\frac{a^{c}}{t_{op}\,c^{I}_{k,i}}\,\frac{\partial C^{c}_{k}}{\partial x_i} \tag{6.60}$$

Note that the RHSs of (6.53) and (6.58) have the same forms. However, (6.53) is expressed in terms of the quantities which relate to the system and (6.58) in terms related to the element being optimised. This transformation is made possible through the introduction of the concept of the local unit cost of irreversibility, $c^{I}_{k,i}$, which gives an indication of the thermoeconomic non-equivalence of irreversibilities in the different plant components. This thermoeconomic indicator takes into account the thermodynamic non-equivalence of irreversibilities through $\sigma_{k,i}$, while the capital investment (economic) contribution to this indicator comes from $\zeta_{k,i}$.

$\sigma_{k,i}$ has already been discussed in Section 6.1. The second coefficient, $\zeta_{k,i}$, gives a measure of the changes in the capital costs of the elements (other than the k-th element) resulting from the change in the irreversibility rate in the k-th element. For some types of element the contribution to the value of this coefficient may be taken to be zero when:

1. The changes caused in the element due to a change in x_i are such as not to involve any changes in their capital costs, ie the equipment may have to be of modified design but of unaltered capital cost.
2. The component affected by the optimisation of the k-th element is available in a number of standard sizes, and the change produced in it is small enough to be accommodated within one particular size.

In cases when either all elements (except for the k-th element) fall into one of these two categories, or if their contribution to $\zeta_{k,i}$ is small, the expression for the local unit cost reduces to:

$$(c^{I}_{k,i})_{\zeta_{k,i}=0}=c^{\varepsilon}_{IN}\,\sigma_{k,i} \tag{6.61}$$

Optimisation procedure

One of the most important decisions in this type of optimisation method is the correct choice of the variable parameter x_i. Since both $\dot{I}_k$ and C^{c}_{k} must be related to this parameter in the most direct manner, x_i will usually be a local parameter, ie one belonging to the k-th element. Furthermore, the choice of this parameter will be affected by the need for suitable functions of the type $\dot{I}_k(x_i)$ and $C^{c}_{k}(x_i)$ in order to obtain the partial derivatives appearing on the RHS of (6.58). This choice will, clearly, depend on the type of component under consideration.

Heat exchangers

Total heat transfer area, A_{HT}, affects most directly both the performance and the capital cost of heat exchangers. Capital cost for any particular design can usually be expressed

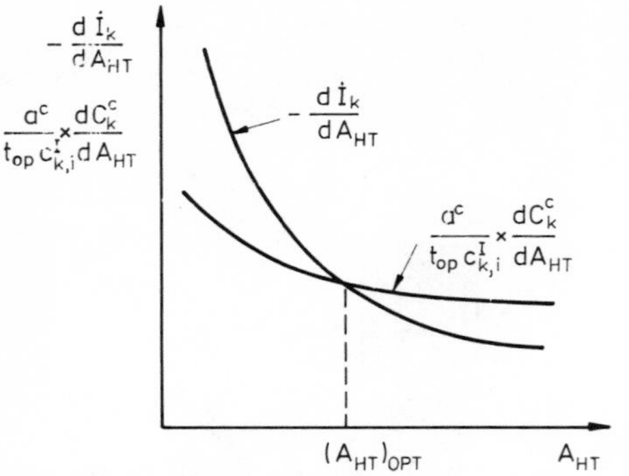

Fig 6.10 Graphical determination of the optimum heat transfer area of a heat exchanger.

in the form of an empirical equation, as a function of its heat transfer area. Hence, from an equation $C^c_k = C^c_k(A_{HT})$ the derivative dC^c_k/dA_{HT} can be obtained. The other required function in the form $\dot{I}_k = \dot{I}_k(A_{HT})$ can be formulated for the given operating conditions of the heat exchanger from its heat transfer characteristics. Where gas streams are involved, and hence the effect of pressure losses on $\dot{I}_k$ must be considered, this function must be determined as part of the process of optimisation of the geometrical parameters of the heat exchanger (Section 6.4). Figure 6.9 shows this relationship. In cases where pressure losses are of minor importance, eg in heat transfer with phase change or in forced convection involving liquid streams, the derivation of the function $\dot{I}_k = \dot{I}_k(A_{HT})$ will, in general, be simpler. The graphical solution of (6.60) is shown qualitatively in Fig 6.10. The intersection of the two curves determines the value of the optimum heat transfer area $(A_{HT})_{OPT}$. From $(A_{HT})_{OPT}$ other quantities such as $(\dot{I}_k)_{OPT}$ and $(C^c_k)_{OPT}$ can be calculated.

Expanders and compressors

The thermodynamic performance of expanders and compressors can be characterised by some form of efficiency, eg isentropic efficiency, η_s. Also, the capital costs, for particular values of pressure ratio and volume flow rate, are functions mainly of their isentropic efficiencies. Thus, the required functions can be put in the form $C^c_k = C^c_k(\eta_s)$ and $\dot{I}_k = \dot{I}_k(\eta_s)$ and the derivatives are obtained from them. The optimum value of η_s is then determined from (6.60).

Once one element of the system is optimised, the procedure can be applied to other elements thought to show promise for reducing the unit cost of the product. In general, because of the mutual interactions or bonds between the elements of the system, this type of optimisation must be carried out iteratively.

The autonomous method of thermoeconomic optimisation

These techniques, based on the version described by El-Sayed and Evans[6.19], will be illustrated on a three-element system with sequential processing of exergy (Fig 6.11(a)). Also, the optimisation will be restricted to one independently adjustable parameter, x, per element. With the output of the system $\dot{E}_{OUT} = \text{const}$, the objective function is the total cost per unit of time and is given by:

$$\dot{C}_T = c^{\varepsilon}_1 \dot{E}_1 + \dot{Z}_1 + \dot{Z}_2 + \dot{Z}_3 \tag{6.62}$$

where c^{ε}_1 is the unit cost of exergy input; $\dot{E}_1$ the exergy input rate; and $\dot{Z}_1$, $\dot{Z}_2$, $\dot{Z}_3$ the

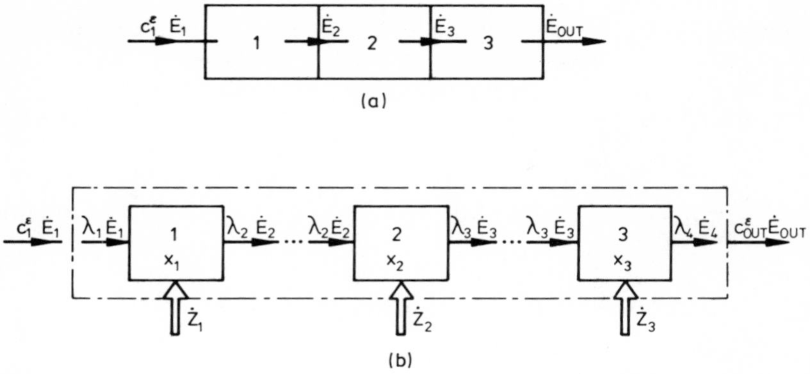

Fig 6.11 A three-element 'chain system' and the unit exergy costs of the elements.

cost of capital investment per unit of time for the three elements 1, 2 and 3 respectively.

Assume that the optimum configuration of each of the three elements, and hence of the whole system, can be obtained through the adjustment of x_1, x_2 and x_3 which are located in the elements 1, 2 and 3 respectively. Thus the minimum of the objective function $\dot{C}_T$ is given by:

$$\frac{\partial \dot{C}_T}{\partial x_1}=\frac{\partial \dot{C}_T}{\partial x_2}=\frac{\partial \dot{C}_T}{\partial x_3}=0 \tag{6.63}$$

With $\dot{E}_1$, $\dot{E}_2$, $\dot{E}_3$ and $\dot{E}_4$ as independent variables, four constraining equations relating these variables are required. As a result a total Lagrangian $\dot{L}_T$ becomes the objective function instead of $\dot{C}_T$. The constraining equations are:

$$\left.\begin{aligned} \dot{E}_1 &= f_1^E(\dot{E}_2, x_1) \\ \dot{E}_2 &= f_2^E(\dot{E}_3, x_2) \\ \dot{E}_3 &= f_3^E(\dot{E}_4, x_3) \\ \dot{E}_4 &= \dot{E}_{OUT} = \text{const} \end{aligned}\right\} \tag{6.64}$$

The capital expenditure rates are subject to the functional relations:

$$\left.\begin{aligned} \dot{Z}_1 &= f_1^Z(\dot{E}_2, x_1) \\ \dot{Z}_2 &= f_2^Z(\dot{E}_3, x_2) \\ \dot{Z}_3 &= f_3^Z(\dot{E}_4, x_3) \end{aligned}\right\} \tag{6.65}$$

Hence, the Lagrangian is:

$$\begin{aligned} \dot{L}_T = {} & c_1^\varepsilon \dot{E}_1 + f_1^Z(\dot{E}_2, x_1) + f_2^Z(\dot{E}_3, x_2) + f_3^Z(\dot{E}_4, x_3) + \lambda_1[f_1^E(\dot{E}_2, x_1) - \dot{E}_1] \\ & + \lambda_2[f_2^E(\dot{E}_3, x_2) - \dot{E}_2] + \lambda_3[f_3^E(\dot{E}_4, x_3) - \dot{E}_3] + \lambda_4(\dot{E}_{OUT} - \dot{E}_4) \end{aligned} \tag{6.66}$$

Minimum $\dot{C}_T$ occurs when the partial derivatives of $\dot{L}_T$ with respect to each of the variables x_1, x_2, x_3, $\dot{E}_1$, $\dot{E}_2$, $\dot{E}_3$ and $\dot{E}_4$ vanish.

$$\frac{\partial \dot{L}_T}{\partial x_1}=\frac{\partial}{\partial x_1}(f_1^Z+\lambda_1 f_1^E)=0 \tag{6.67a}$$

$$\frac{\partial \dot{L}_T}{\partial x_2}=\frac{\partial}{\partial x_2}(f_2^Z+\lambda_2 f_2^E)=0 \tag{6.67b}$$

$$\frac{\partial \dot{L}_T}{\partial x_3}=\frac{\partial}{\partial x_3}(f_3^Z+\lambda_3 f_3^E)=0 \tag{6.67c}$$

$$\frac{\partial \dot{L}_T}{\partial \dot{E}_1}=\frac{\partial}{\partial \dot{E}_1}(c_1^\varepsilon \dot{E}_1-\lambda_1 \dot{E}_1)=0 \qquad \therefore \quad \lambda_1=c_1^\varepsilon \tag{6.67d}$$

$$\frac{\partial \dot{L}_T}{\partial \dot{E}_2}=\frac{\partial}{\partial \dot{E}_2}(f_1^Z+\lambda_1 f_1^E-\lambda_2 \dot{E}_2)=0 \qquad \therefore \quad \lambda_2=\frac{\partial}{\partial \dot{E}_2}(f_1^Z+\lambda_1 f_1^E) \tag{6.67e}$$

$$\frac{\partial \dot{L}_T}{\partial \dot{E}_3}=\frac{\partial}{\partial \dot{E}_3}(f_2^Z+\lambda_2 f_2^E-\lambda_3 \dot{E}_3)=0 \qquad \therefore \quad \lambda_3=\frac{\partial}{\partial \dot{E}_3}(f_2^Z+\lambda_2 f_2^E) \tag{6.67f}$$

$$\frac{\partial \dot{L}_T}{\partial \dot{E}_4}=\frac{\partial}{\partial \dot{E}_4}(f_3^Z+\lambda_3 f_3^E-\lambda_4 \dot{E}_4)=0 \qquad \therefore \quad \lambda_4=\frac{\partial}{\partial \dot{E}_4}(f_3^Z+\lambda_3 f_3^E) \tag{6.67g}$$

An examination of these equations, (6.67a) to (6.67g), shows that the first three represent local objective functions involving trade-off between local capital expenditure rates and local irreversibility rates in which the Lagrange multipliers λ_1, λ_2 and λ_3 take on the role of the local unit costs of exergy. (6.67d) to (6.67f) give the proper values of the local unit costs of input exergy to the three elements while (6.67g) gives this value for the exergy flux $\dot{E}_4$. From the earlier considerations, and in particular if the functional relations (6.64) and (6.65) are satisfied, the system may be decomposed into autonomous sub-systems and optimised for local variables. Figure 6.11(b) shows the system decomposed into its three autonomous sub-systems, with the exergy cost rates and the capital expenditure rates indicated by arrows. The key to the success of this method lies in the determination of the proper values of the local unit costs of exergy. One possible iterative procedure for the optimisation of the system would be to assume the unit exergy costs, taking $\lambda_1=c_1^\varepsilon$, and optimise element 3 with $\dot{E}_4=\dot{E}_{OUT}=\text{const}$, which would yield the value of $\dot{E}_3$. Subsequently, element 2 would be optimised yielding $\dot{E}_2$ and then element 1 yielding $\dot{E}_1$. Then using equations (6.67e) and (6.67f), improved values of λ_2 and λ_3 can be obtained.

An alternative approach, should (6.67a) to (6.67c) prove rather complicated, is to assume values of x_1, x_2 and x_3 from which λ_1, λ_2 and λ_3 can be calculated using equations (6.67d) and (6.67f). Then (6.67a) to (6.67c) can be used to check for zero. The optimisation of the system can be simplified considerably if certain assumptions can be made about the form of the functional relations given by (6.65). Two special cases are considered by El-Sayed and Evans.

Capital expenditure rates relatively insensitive to output exergy rate: Thus the functional relations given by (6.65) reduce to the form $\dot{Z}=f^Z(x)$.

Hence, from (6.67d) to (6.67f):

$$\lambda_1=c_1^\varepsilon \tag{6.68a}$$

$$\lambda_2=\frac{\partial}{\partial \dot{E}_2}(\lambda_1 f_1^E)=c_1^\varepsilon \frac{\partial \dot{E}_1}{\partial \dot{E}_2} \tag{6.68b}$$

where $\dot{E}_1 = f_1^E(\dot{E}_2, x_1)$

$$\lambda_3 = \frac{\partial}{\partial \dot{E}_3}(\lambda_2 f_2^E) = \lambda_2 \frac{\partial \dot{E}_2}{\partial \dot{E}_3} \tag{6.68c}$$

where $\dot{E}_2 = f_2^E(\dot{E}_3, x_2)$.

Furthermore, if x_1, x_2 and x_3 represent rational efficiencies ψ_1, ψ_2 and ψ_3 then:

$$f_1^E = \dot{E}_2/\psi_1 \qquad f_2^E = \dot{E}_3/\psi_2 \tag{6.69}$$

and (6.68a) and (6.68b) reduce to:

$$\lambda_2 = \frac{c_1^{\varepsilon}}{\psi_1} \qquad \lambda_3 = \frac{\lambda_2}{\psi_2} = \frac{c_1^{\varepsilon}}{\psi_1 \psi_2} \tag{6.70}$$

showing that the increase in the unit cost of exergy across an element is due to loss of exergy (irreversibility) in the element under consideration, and is independent of the capital expenditure on the element.

Referring back to (6.30) and (6.61), which apply to the same type of system with sequential processing of exergy, and taking that for the elements $\omega = \psi$, it is clear that the Lagrange multipliers given by (6.70) have an identical form to the local unit cost of irreversibility as determined by the structural method.

Capital expenditure of the elements increases linearly with the exergy output of the elements: Thus the capital expenditure rates for the l-th element can be represented as $\dot{Z}_l = \dot{E}_{(l+1)} f_l^{Z'}(x_l)$.

From expressions (6.67d) to (6.67f):

$$\lambda_1 = c_1^{\varepsilon} \tag{6.71a}$$

$$\lambda_2 = \frac{\partial}{\partial \dot{E}_2}[\dot{E}_2 f_1^{Z'}(x_1) + \lambda_1 f_1^E]$$

$$= f_1^{Z'}(x_1) + \lambda_1 \frac{\partial \dot{E}_1}{\partial \dot{E}_2} \tag{6.71b}$$

where $\dot{E}_1 = f_1^E(\dot{E}_2, x_1)$

$$\lambda_3 = \frac{\partial}{\partial \dot{E}_3}[\dot{E}_3 f_2^{Z'}(x_2) + \lambda_2 f_2^E]$$

$$= f_2^{Z'}(x_2) + \lambda_2 \frac{\partial \dot{E}_2}{\partial \dot{E}_3} \tag{6.71c}$$

where $\dot{E}_2 = f_2^E(\dot{E}_3, x_2)$.

If x_1, x_2 and x_3 represent rational efficiencies, (6.71b) and (6.71c) can be further modified using (6.69).

$$\lambda_2 = f_1^{Z'}(x_1) + \lambda_1 \frac{c_1^{\varepsilon}}{\psi_1} = \frac{\dot{E}_2 f_1^{Z'}(x_1) + c_1 \dot{E}_1}{\dot{E}_2}$$

$$= \frac{\dot{Z}_1 + c_1^{\varepsilon} \dot{E}_1}{\dot{E}_2} \tag{6.72a}$$

$$\lambda_3 = f_2^{Z'}(x_2) + \frac{\lambda_2}{\psi_2} = \frac{\dot{E}_3 f_1^{Z'}(x_1) + \lambda_2 \dot{E}_2}{\dot{E}_3}$$

$$= \frac{\dot{Z}_2 + \lambda_2 \dot{E}_2}{\dot{E}_3} \tag{6.72b}$$

In contrast to the previous case, here the increase in the unit cost of exergy across an element is due to both the capital expenditure rate on the element and the loss of exergy (irreversibility). In this respect, the thermoeconomic balance of the element is the same as that applicable to the system as a whole.

The thermoeconomic model used in the second case has been adopted in the thermoeconomic optimisation of a sea water desalination plant. Details of the optimisation procedure are given in Refs [6.18–6.20].

6.6 Thermoeconomic optimisation of a heat exchanger in a CHP plant—an illustrative example

The structural method will be applied to the CHP plant with an extraction turbine described in Section 6.3, operating without fresh steam heating. The heat exchanger to be optimised is in Sub-region F. Because of the large size of the plant, the total load is shared by three heat exchangers and hence the total heat transfer area A_F is given by $A_F = 3A_{HE}$ where A_{HE} is the heat transfer area of the individual heat exchangers.

Optimum condition can be expressed from (6.60) as:

$$\left(\frac{\partial \dot{I}_F}{\partial A_F}\right)_{OPT} = -\frac{a^c}{t_{op}\, c^I_{F,A}} \frac{\partial C^c_F}{\partial A_F} \tag{a}$$

Note that the variable parameter, x_i, adopted here is A_F rather than P_3, the extraction pressure, which was used in Section 6.3 (see (6.49)).

Capital cost

A shell-and-tube heat exchanger is selected. For a particular size range and method of construction the capital cost per heat exchanger is given by a manufacturer in the form of a power law:

$$C^c_{HE}/[£] = 1\,800(A_{HE}/[m^2])^{0.582}$$

The capital cost of the three heat exchangers in terms of the total heat transfer area is:

$$C^c_F/[£] = 2\,849(A_F/[m^2])^{0.582}$$

Hence, the derivative on the RHS of (a) can be obtained.

$$\frac{\partial C^c_F}{\partial A_F}\bigg/[£/m^2] = 1\,658(A_F/[m^2])^{-0.418} \tag{b}$$

Irreversibility rate

As in Section 6.3, assume that constant average temperature T_u can be adopted without undue loss of accuracy. Thus the total heating rate is:

$$\dot{Q}_u = A_F(U_{ov})_F(T_3 - T_u) \tag{c}$$

where $(U_{ov})_F$ is the overall heat transfer coefficient and T_3 is the saturation temperature at the extraction pressure, P_3. Assuming that:

$$(T_3 - T_u) \ll T_u$$

and using (6.50) and (c) the irreversibility rate for Sub-region F is given by:

$$\dot{I}_F = \frac{\dot{Q}_u^2 T_0}{A_F (U_{ov})_F T_u^2}$$

and its partial derivative with respect to A_F by:

$$\frac{\partial \dot{I}_F}{\partial A_F} = -\frac{\dot{Q}_u^2 T_0}{(U_{ov})_F T_u^2} \frac{1}{A_F^2}$$

Hence, taking the value of the overall heat transfer coefficient as $(U_{ov})_F = 1.5\ \text{kW/m}^2\text{K}$ and making use of the numerical values given in Section 6.3:

$$\frac{\partial \dot{I}_F}{\partial A_F} \Big/ [\text{kW/m}^2] = -0.74 \times 10^3 \left(\frac{1}{A_F/[\text{m}^2]} \right) \tag{d}$$

Financial arrangements

It is assumed that the annual cost of investment is given by the product of the capital cost and the capital-recovery factor a^c, ie capital is borrowed at a fixed interest rate and is repayed in equal annual instalments over a pre-determined number of years. The salvage value of the heat exchanger will be neglected. The capital-recovery factor can be calculated[6.2] from:

$$a^c = \frac{i_R (1 + i_R)^{N_y}}{(1 + i_R)^{N_y} - 1} \tag{e}$$

where N_y is the period (number of years) of repayment and i_R is the interest rate. Taking $i_R = 0.18$ and $N_y = 20$ years:

$$a^c = \frac{0.18(1 + 0.18)^{20}}{(1 + 0.18)^{20} - 1} = \underline{0.1868}$$

Unit cost of input exergy, c^{ε}_{IN}

The fuel cost per volume will be taken as $c^{v}_{IN} = 1.08 \times 10^{-3}\ £/\text{m}^3$ and the net calorific value $(\text{NCV})^v = 38.56\ \text{MJ/m}^3$. Hence the unit cost of *energy* input corresponding to the net calorific value is $c^{en}_{IN} = 2.8 \times 10^{-6}\ £/\text{kJ}$. But it was assumed (point 6 Section 6.3) that the value of φ for the fuel used is unity which gives $\text{NCV} = \varepsilon_{IN}$. Consequently:

$$c^{en}_{IN} = c^{\varepsilon}_{IN} = 2.8 \times 10^{-6}\ £/\text{kJ}$$

Capital cost coefficient

This coefficient is expressed as a sum ((6.56)) of the partial derivatives of the type $(\partial C^c_{i'}/\partial \dot{I}_k)_{x_i = \text{var}}$ for all the elements of the system except for the one which is being optimised. As the effect of minor plant components on the value of $\zeta_{k,i}$ is likely to be negligible, only the turbines, the boiler and the condenser will be considered. Although element E (Fig 6.4) should include the cooling tower, it can be assumed that this component is designed with generous over-capacity sufficient to accommodate any changes in the duty of the condenser which might arise from the optimisation of element F.

The turbine: The effect of a change in the extraction pressure (associated with the change in $\dot{I}_F$) is to shift the division of power generation between the HP turbine and the LP turbine. As the total power output is fixed, it may be assumed that the total cost of the two turbines will be unaffected by the optimisation of element F. Hence, the value of the partial derivative in this case is zero.

The condenser: The condenser is basically a heat exchanger of the same type of construction as that in Sub-region F except that it operates at lower steam pressure. The capital cost of the condenser, C^c_E, will be expressed in terms of the heat transfer area A_E as a linear equation of the type:

$$C^c_E/[£] = a_E + b_E(A_E/[m^2]) \tag{f}$$

For the range of operating conditions under which the condenser works the two coefficients in (f) are:

$$a_E/[£] = 80 \times 10^3$$

$$b_E/[£/m^2] = 40$$

The required derivative for the condenser can be expressed as:

$$\left(\frac{\partial C^c_E}{\partial \dot{I}_F}\right)_{A_F=\text{var}} = \frac{\left(\frac{\partial C^c_E}{\partial A_E}\right)\left(\frac{\partial \dot{I}_T}{\partial \dot{I}_F}\right)_{A_F=\text{var}}}{\left(\frac{\partial \dot{I}_T}{\partial Q_E}\right)\left(\frac{\partial Q_E}{\partial A_E}\right)} \tag{g}$$

The partial derivatives will now be expressed in terms of known quantities.

$$\left(\frac{\partial \dot{I}_T}{\partial \dot{I}_F}\right)_{A_F=\text{var}} = \sigma_{F,A_F} \tag{h}$$

From (6.32), (6.39) and (6.41):

$$\frac{\partial \dot{I}_T}{\partial Q_E} = \frac{\varphi}{\eta_{comb}} \tag{i}$$

From the heat transfer equation:

$$\frac{\partial Q_E}{\partial A_E} = (U_{ov})_E(\Delta T_m)_E \tag{j}$$

Assume for the condenser $(U_{ov})_E = 2$ kW/m^{2}K and $(\Delta T_m)_E = 5$ K. From (f):

$$\frac{\partial C^c_E}{\partial A_E} = b_E \tag{k}$$

Using (g) to (k), the partial derivative for the condenser is:

$$\left(\frac{\partial C^c_E}{\partial \dot{I}_F}\right)_{A_F=\text{var}} = \frac{\sigma_{F,A_F}\eta_{comb}b_E}{\varphi(U_{ov})_E(\Delta T_m)_E} \tag{l}$$

The boiler: The capital cost of the boiler C^c_A can be expressed as a linear function of its heat transfer rate to water substance $\dot{Q}_A$:

$$C^c_A/[£] = a_A/£ + (b_A/[£/kW])(\dot{Q}_A/kW) \tag{m}$$

where, for the type and size of boiler under consideration, the two constants are:

$$a_A/[£] = 2 \times 10^6$$

$$b_A/[£/kW] = 55$$

It should prove useful to express the heat transfer rate $\dot{Q}_A$ as a function of the exergy input.

$$\dot{Q}_A = \dot{E}_{IN} \frac{\eta_{comb}}{\varphi} \tag{n}$$

The required derivative for the boiler can be expressed conveniently in the form:

$$\left(\frac{\partial C_A^c}{\partial \dot{I}_F}\right)_{A_F=\text{var}} = \left(\frac{\partial C_A^c}{\partial \dot{Q}_A}\right)\left(\frac{\partial \dot{Q}_A}{\partial \dot{I}_T}\right)\left(\frac{\partial \dot{I}_T}{\partial \dot{I}_F}\right)_{A_F=\text{var}} \tag{o}$$

The derivatives in (o) will, now, be expressed in terms of known quantities:
From (m):

$$\frac{\partial C_A^c}{\partial \dot{Q}_A} = b_A \tag{p}$$

From (n) with (6.41):

$$\frac{\partial \dot{Q}_A}{\partial \dot{I}_T} = \frac{\partial \dot{Q}_A}{\partial \dot{E}_{IN}} = \frac{\eta_{comb}}{\varphi} \tag{r}$$

Using (h) and (o) to (r):

$$\left(\frac{\partial C_A^c}{\partial \dot{I}_F}\right)_{A_F} = \frac{b_A \eta_{comb} \sigma_{F,A_F}}{\varphi} \tag{s}$$

From (*l*) and (s) the capital cost coefficient is:

$$\zeta_{F,A_F} = \frac{\sigma_{F,A_F}\eta_{comb}}{\varphi}\left(\frac{b_E}{(U_{ov})_E(\Delta T_m)_E} + b_A\right) \tag{t}$$

Local unit cost of irreversibility for element F

Substituting (t) in (6.59):

$$c_{F,A_F}^I = \sigma_{F,A_F}\left[c_{IN}^{\varepsilon} + \frac{a^c \eta_{comb}}{t_{op}\varphi}\left(\frac{b_E}{(U_{ov})_E(\Delta T_m)_E} + b_A\right)\right] \tag{u}$$

The numerical value of c_{F,A_F}^I will now be calculated for the case when $t_{op} = 7\,000$ h. Substituting the data for the plant:

$$\begin{aligned} c_{F,A_F}^I &= 3\left[2.8 \times 10^{-6} + \frac{0.1868 \times 0.85}{7\,000 \times 3\,600 \times 1}\left(\frac{40}{2 \times 5} + 55\right)\right] \\ &= 3[2.8 + 0.0253 + 0.346] \times 10^{-6} \\ &= 9.50 \times 10^{-6}\ \text{£/kJ} \end{aligned}$$

Optimum parameters: The optimum heat transfer area can now be obtained from expressions (a), (b), (d) and the numerical values of a^c, t_{op} and c_{F,A_F}^I. Thus:

$$\begin{aligned} (A_F)_{OPT} &= \left[\frac{0.74 \times 10^6 \times 7\,000 \times 3\,600 \times 9.50}{0.1868 + 1\,658 \times 10^6}\right]^{1/1.582} \\ &= \underline{4\,358\ \text{m}^2} \end{aligned}$$

The optimum mean temperature difference in the heat exchanger, from (c), is:

$$(\Delta T_m)_{F,OPT} = \frac{25\,000}{1.5 \times 4\,358} = 3.82\text{ K}$$

so:

$$(T_3)_{OPT} = T_u + (\Delta T_m)_{F,OPT}$$
$$= 406.15 + 3.82 = 409.97\text{ K}$$

and the corresponding extraction pressure, by interpolation in steam tables, is:

$$\underline{(P_3)_{OPT} = 3.31\text{ bar}}$$

Using this value of $(T_3)_{OPT}$, the optimum irreversibility rate, from (6.50) is:

$$(\dot{I}_F)_{OPT} = 25\,000 \times 283.15\left(\frac{1}{406.15} - \frac{1}{409.97}\right)$$
$$= \underline{168\text{ kW}}$$

The optimum capital cost of the three heat exchangers is:

$$(C_F^c)_{OPT} = 3 \times 1\,800\left(\frac{4\,358}{3}\right)^{0.582}$$
$$= \underline{£373\,916}$$

In the original proposal (Figs 6.5 and 6.6), the steam condenses in the heat exchanger at the saturation temperature (corresponding to $P_3 = 4$ bar) of 143.6°C giving a temperature difference of 10.6 K. The total heat transfer area required would be:

$$A_F' = \frac{25\,000}{1.5 \times 10.6} = 1\,572\text{ m}^2$$

and the corresponding total capital cost of the three heat exchangers:

$$C_F^c = 2\,849(1\,572)^{0.582} = £206\,576$$

This is £167 340 less than the optimum capital cost calculated above. Comparing the irreversibility rates, we see that the optimum configuration gives a saving of $\Delta\dot{I}_F = 459 - 168 = 291$ kW which corresponds to a saving of exergy input $\Delta\dot{E}_{IN} = \sigma_{F,A_F}\Delta\dot{I}_F =$ £873 kW. With unit cost of exergy at 2.8×10^{-6} £/kJ, the annual saving in the cost of fuel would be £61 000. Thus the additional capital cost required for the optimum configuration would pay for itself in less than three years.

Conclusions

This example has shown that the optimisation procedure is relatively uncomplicated and, provided all the necessary data are available, can be carried out quickly using just a pocket calculator.

A large part of the work involved in the optimisation concerns the calculation of the capital cost coefficient. However, as shown by the calculated result, it contributes only about 12% of the value of c_{F,A_F}^I. The corresponding effect on the value of $(A_F)_{OPT}$ is about −8%. An error of this order of magnitude may be quite acceptable in some cases, particularly when making an initial estimate of the value of the parameters corresponding to the optimum configuration. Under these conditions the local unit cost of irreversibility can be calculated from (6.61).

When optimising a more complex type of plant, the most complicated and time absorbing part of the procedure is the calculation of the CSB for the element under consideration. It is in such cases that the use of a computer program based, perhaps on the graph-theoretic approach[6.13,6.14], for calculating irreversibility rates of the elements of the system would be of great assistance.

Having optimised one of the elements, other elements could then be dealt with in turn.

6.7 Exergy costing in multi-product plants

As pointed out in Chapter 2, exergy may be regarded as a measure of the capacity of a given form of energy to cause change, ie do work, heat objects and cause endothermic reactions to take place. It is also exergy which is lost, or consumed, in order to make a particular process proceed at a certain rate and it is therefore reasonable to assess the price of energy on the basis of its exergy 'content'.

In the case of a single product plant, costing of the product can be done on the basis of energy or exergy. For example, in the case of a boiler used for heating, the unit cost of the thermal energy or exergy produced can be obtained from:

$$\dot{C}_{\mathrm{OUT}} = \dot{C}_{\mathrm{EN}} + \dot{Z} \tag{6.73}$$

$$\dot{C}_{\mathrm{OUT}} = \dot{\mathrm{E}}^{\mathrm{Q}}_{\mathrm{OUT}} \times c^{\varepsilon} = \dot{Q}_{\mathrm{OUT}} \times c^{\mathrm{en}} \tag{6.74}$$

where:

$\dot{C}_{\mathrm{OUT}}$—value of the boiler output per unit of time;
$\dot{C}_{\mathrm{EN}}$—cost of the input energy per unit of time;
$\dot{Z}$—capital investment rate;
$\dot{Q}_{\mathrm{OUT}}$, $\dot{\mathrm{E}}^{\mathrm{Q}}_{\mathrm{OUT}}$—the rate of 'heat' output and the corresponding thermal exergy rate;
c^{en}, c^{ε}—unit cost of energy and exergy respectively.

In the case of a multi-product plant, a single cost equation is not sufficient. Additional criteria are required to determine the relationship between the unit costs of the different products. This is where exergy can be of use as a basis for costing the products.

There is a variety of multi-product plants to which exergy costing can be applied including:

1. Combined heating and power plants.
2. Electric power generation with sea water desalination plant (see Ref [6.21]).
3. Combined refrigeration and heating plant, eg a reversed cycle plant in which the evaporator refrigerates ice on an ice-skating rink and the condenser heats water of a swimming pool of a sports complex.
4. Multi-product distillation plant, eg air separation or distillation of petroleum products.

As the exergy method of costing is based on rational thermodynamic considerations, it encourages efficient production and is fair to the consumer. The method is applicable in cases where the producer exercises full control over the price of his products, for instance in the case of generation of heating and power by a municipal corporation or by a company generating heating and power for use within an industrial complex. The prices of the two commodities in the latter case may be required for the purpose of costing products produced in the complex.

This method of costing would not be applicable in cases when the prices of the products in question are governed entirely by commercial factors. For instance, the price of silver produced as a by-product in a metallurgical refining process of a base metal, calculated by this method could be very different from the current world market price. However, even in this or similar cases, the exergy method of costing could be used to establish the commercial feasibility of co-generation of given products by one type of plant or another or separate generation.

As an example of costing of products in a multi-purpose plant we shall consider a simple CHP plant with a back-pressure turbine.

Costing of heating and electric power from a CHP plant

The plant

The plant diagram (Fig 6.12) shows that low-pressure steam exhausted by the turbine is delivered to the point of usage by a pipeline. The feed water is returned at the environmental temperature so that its exergy can be taken as zero. The back-pressure steam can, in general, be supplemented by fresh steam, which is passed first through the reducing and desuperheating station shown by dotted lines in the figure. However, it will be assumed here that all the heating steam comes from the exhaust of the turbine and the Grassmann diagram (Fig 6.13) corresponds to this assumption. The plant operating parameters are:

Electric power output, $\dot{W}_{el} = 6\,000$ kW
Boiler pressure, $P_1 = 40$ bar
Steam superheat temperature, $\theta_1 = 400°C$
Turbine back pressure, $P_2 = 4$ bar
Environmental temperature, $\theta_0 = 20°C$

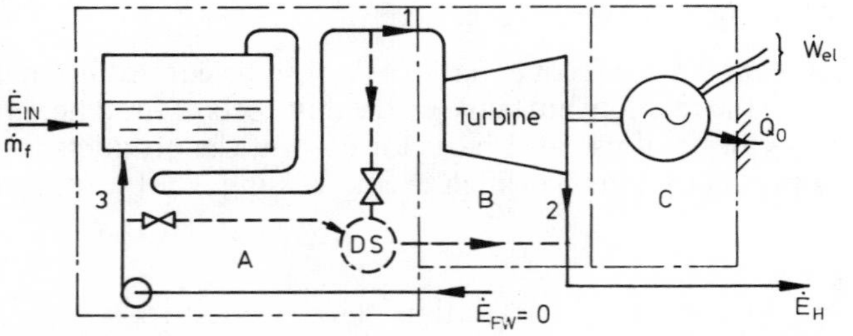

Fig 6.12 Diagram of a simple CHP plant with a back-pressure turbine, DS—desuperheater.

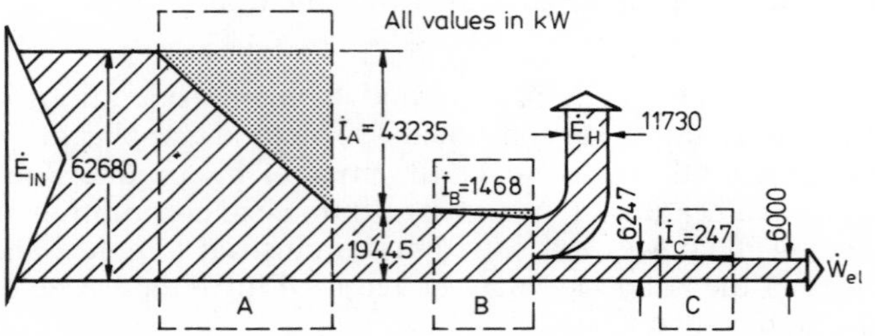

Fig 6.13 Grassmann diagram for the CHP plant shown in Fig 6.12, for the case when no fresh steam is used for heating.

Turbine isentropic efficiency, $\eta_s = 0.75$
Boiler combustion efficiency, $\eta_{comb} = 0.85$
Mechanical efficiency, $\eta_m = 0.98$
Electrical efficiency, $\eta_{el} = 0.98$
Fuel, coal with NCV = 24.44 MJ/kg
Fuel specific exergy, $\varepsilon_F = 26.37$ MJ/kg
Fuel specific cost, $c_F^{sp} = £57.3$/tonne.

Capital costs of principal components:

Boiler, $C_A^c = £4.671 \times 10^6$
Turbine, $C_B^c = £0.286 \times 10^6$
Electric generator, $C_c^c = £0.120 \times 10^6$
Interest rate on the capital, $i_R = 0.18$
Period of repayment, $N_y = 20$ years
Time of operation of the plant per year, $t_{op} = 7\,000$ h

The following simplifying assumptions have been made:

1. All pressure losses are neglected.
2. All heat transfer losses are neglected except those due to dissipation of electrical and mechanical type which are confined to Sub-region C (electric generator).
3. Power input to the feed pump and other auxiliary equipment is neglected.

Using this data and assumptions the following quantities were evaluated:
Turbine shaft power:

$$\dot{W}_{sh} = \frac{\dot{W}_{el}}{\eta_m \eta_{el}} = \frac{6\,000}{0.98 \times 0.98} = 6\,247 \text{ kW}$$

Mass flow rate of steam:

$$\dot{m}_s = \frac{\dot{W}_{sh}}{h_1 - h_2} = \frac{6\,247}{3\,214 - 2\,818} = 15.78 \text{ kg/s}$$

Rate of fuel supply $\dot{m}_F$. From an energy balance of the boiler:

$$\dot{m}_F(\text{NCV})\eta_{comb} = \dot{m}_s(h_1 - h_3)$$

$$\therefore \quad \dot{m}_F = \frac{15.78(3\,214 - 83.9)}{24.44 \times 10^3 \times 0.85} = 2.377 \text{ kg/s}$$

Exergy input rate:

$$\dot{E}_{IN} = \dot{m}_F \varepsilon_F = 2.377 \times 26.37 \times 10^3 = 62\,680 \text{ kW}$$

Steam exergy rates:

$$\dot{E}_1 = 19\,445 \text{ kW}$$

$$\dot{E}_2 = \dot{E}_H = 11\,730 \text{ kW}$$

Rational efficiency of the boiler:

$$\psi_A = \frac{\dot{E}_1}{\dot{E}_{IN}} = \frac{19\,445}{62\,680} = 0.310$$

Two types of rational efficiencies will be formulated for the turbine. If it is considered that both $\dot{W}_{sh}$ and $\dot{E}_H$ constitute the turbine output:

$$\psi_B^{eq} = \frac{\dot{W}_{sh} + \dot{E}_H}{\dot{E}_1} = 0.9245$$

This form of ψ_B will be used in the 'equality method' described below*. When it is considered that the only form of turbine output is $\dot{W}_{sh}$, the rational efficiency takes the form:

$$\psi_B^{ex} = \frac{\dot{W}_{sh}}{\dot{E}_1 - \dot{E}_H} = 0.8017$$

which will be used in the 'extraction method'. Rational efficiency of the electric generator:

$$\psi_c = \frac{\dot{W}_{el}}{\dot{W}_{sh}} = 0.9605$$

Capital-recovery factor:

$$a^c = \frac{i_R(1+i_R)^{N_y}}{(1+i_R)^{N_y} - 1} = 0.1868$$

Capital investment rate for the boiler:

$$\dot{Z}_A = \frac{a^c}{t_{op}} C_A^c = 0.03463 \text{ £/s}$$

Similarly capital investment rates for the turbine and the electric generator are:

$$\dot{Z}_B = 0.00212 \text{ £/s}$$

$$\dot{Z}_C = 0.00089 \text{ £/s}$$

Unit cost of input exergy:

$$c_{IN}^{\varepsilon} = \frac{c_F^{sp}}{\varepsilon_F} = \frac{0.0573}{26.37 \times 10^3} = 2.18 \times 10^{-6} \text{ £/kJ}$$

Exergy costing using the autonomous method

The general cost equation for an autonomous Sub-region X can be expressed as:

$$\sum_{OUT,X} (\dot{E}_i c_i^{\varepsilon}) = \sum_{IN,X} (\dot{E}_i c_i^{\varepsilon}) + \dot{Z}_X \tag{6.75}$$

Applying (6.75) to the boiler (Sub-region A):

$$c_1^{\varepsilon} = \frac{\dot{E}_{IN} c_{IN}^{\varepsilon}}{\dot{E}_1} + \frac{\dot{Z}_A}{\dot{E}_1}$$

$$= \frac{c_{IN}^{\varepsilon}}{\psi_A} + \frac{\dot{Z}_A}{\dot{E}_1} \tag{6.76}$$

Substituting the numerical values obtained earlier, the unit cost of exergy of high

---

* Some of the terms and techniques used in this section are based on Ref [6.23].

pressure steam is:

$$c_1^{\varepsilon} = \frac{2.18 \times 10^{-6}}{0.310} + \frac{0.03463}{19\,445} = 8.81 \times 10^{-6}\ \text{£/kJ}$$

Applying (6.75) to the turbine (Sub-region B):

$$\dot{W}_{sh} c_{sh}^{\varepsilon} + \dot{E}_H c_H^{\varepsilon} = \dot{E}_1 c_1^{\varepsilon} + \dot{Z}_B \tag{6.77}$$

Now there is a problem since (6.77) has two unknowns, c_{sh}^{ε} and c_H^{ε}, but there is only one equation. Thus assumption must be made regarding the allocation of the costs to the two forms of the products, $\dot{W}_{sh}$ and $\dot{E}_H$. Possible options[6.22] include:

In the *equality method* the generation of the two products is considered to have the same priority, so the generation of both $\dot{W}_{sh}$ and $\dot{E}_H$ is charged for the cost of high-pressure steam and the capital cost of the turbine, leading to the relationship:

$$c_{sh}^{\varepsilon} = c_H^{\varepsilon} = \frac{c_1^{\varepsilon}}{\psi_B^{eq}} + \frac{\dot{Z}_B}{\dot{W}_{sh} + \dot{E}_H} \tag{6.78}$$

Substituting the numerical values:

$$c_{sh}^{\varepsilon} = c_H^{\varepsilon} = \frac{8.81 \times 10^{-6}}{0.9245} + \frac{0.002\,12}{6\,247 + 11\,730} = \underline{9.53 \times 10^{-6}\ \text{£/kJ}}$$

Applying (6.75) to the electric generator (Sub-region C):

$$\begin{aligned} c_{el}^{\varepsilon} &= \frac{c_{sh}^{\varepsilon}}{\psi_C} + \frac{\dot{Z}_C}{\dot{W}_{el}} \\ &= \frac{9.53 \times 10^{-6}}{0.9605} + \frac{0.000\,89}{6\,000} \\ &= \underline{10.1 \times 10^{-6}\ \text{£/kJ}} = \underline{3.64\ \text{p/kWh}} \end{aligned}$$

In the *extraction method* it is considered that the sole purpose of the turbine is to generate shaft power and thus the whole cost of the turbine and the irreversibility occurring in it are charged against it. This amounts to assuming the unit cost of exergy of steam entering the turbine to be the same as that of the steam leaving it, ie:

$$c_1^{\varepsilon} = c_H^{\varepsilon} = 8.81 \times 10^{-6}\ \text{£/kJ} \tag{6.79}$$

as calculated above from the cost equation for the boiler.

From (6.77) and (6.79), the unit cost for shaft power is:

$$\begin{aligned} c_{sh}^{\varepsilon} &= \frac{c_1^{\varepsilon}}{\psi_B^{ex}} + \frac{\dot{Z}_B}{\dot{W}_{sh}} = \frac{8.81 \times 10^{-6}}{0.8097} + \frac{0.002\,12}{6\,247} \\ &= \underline{11.22 \times 10^{-6}\ \text{£/kJ}} \end{aligned}$$

Finally, from the cost equation for the electric generator we get:

$$\begin{aligned} c_{el}^{\varepsilon} &= \frac{11.22 \times 10^{-6}}{0.9605} + \frac{0.000\,89}{6\,000} \\ &= \underline{11.83 \times 10^{-6}\ \text{£/kJ}} = \underline{4.26\ \text{p/kWh}} \end{aligned}$$

Thus, as a result of the different assumptions used in the extraction method, the unit cost of steam c_H^{ε} is somewhat less than in the equality method whilst that of electric

power is greater. The justification for the assumption used must be considered carefully in each case before a particular method of calculation is selected. The values of unit costs of exergy calculated above are clearly only approximate, erring on the low side, since a number of different cost items has been omitted from the calculations for the sake of simplicity. Among the omitted costs one could mention the cost of the site, buildings and site facilities, repairs and maintenance, operating labour costs, *etc.*
In addition to the two methods described above, the three following methods may be appropriate in certain circumstances[6.23].

In the *by-product work method* it is assumed that the generation of process steam is essential even if no electric power is to be generated. Therefore the exergy of the process steam is costed as if it were produced alone in a hypothetical low-pressure boiler at the required pressure and temperature. Using this value of $c_{\mathrm{H}}^{\varepsilon}$, the unit cost of shaft power is calculated from the cost balance for the turbine. When this method is used, the unit cost for the exergy of low-pressure steam is substantially higher than those obtained either by the extraction or equality methods. The use of this method may be appropriate when an expansion of steam generation capacity of an industrial complex is being considered. If the cost of electricity calculated by this method proves to be competitive with the purchased electricity then a decision in favour of co-generation should be made. However, once a decision has been made, the actual costing of $\dot{W}_{\mathrm{el}}$ and $\dot{\mathrm{E}}_{\mathrm{H}}$ for internal accounting should be made by the equality or the extraction method.

In the *by-product steam method*, the unit cost of the shaft work is calculated from a hypothetical condensing turbine or estimated on the basis of purchased electric energy, whichever is cheaper. The unit cost of exergy of low-pressure steam can then be obtained from the cost equation for the turbine. This method gives low unit cost of exergy of steam and a high one for electric power.

The four methods described above allocate the costs of the investments and the exergy losses of the various elements of the system to the different plant products according to different criteria. In some types of plants it may be difficult or impossible to find suitable criteria for the allocation of the costs to the different plant products in a component by component manner. For example, in a distillation plant the constituents of the mixture are not separated until the last of a series of processes and hence it would be more appropriate to treat all the products in the same way. In the *global approach* we consider the plant as a whole rather than the individual components. The overall expenditure equation for the plant under consideration can be written as:

$$\dot{\mathrm{E}}_{\mathrm{IN}}\, c_{\mathrm{IN}}^{\varepsilon} + (\dot{Z}_{\mathrm{A}} + \dot{Z}_{\mathrm{B}} + \dot{Z}_{\mathrm{C}}) = \dot{\mathrm{E}}_{\mathrm{H}}\, c_{\mathrm{H}}^{\varepsilon} + \dot{W}_{\mathrm{el}}\, c_{\mathrm{el}}^{\varepsilon} \tag{6.80}$$

Assuming, as with the equality method, that the two products are of equal importance:

$$c_{\mathrm{H}}^{\varepsilon} = c_{\mathrm{el}}^{\varepsilon} = \frac{\dot{\mathrm{E}}_{\mathrm{IN}}\, c_{\mathrm{IN}}^{\varepsilon}}{\dot{\mathrm{E}}_{\mathrm{H}} + \dot{W}_{\mathrm{el}}} + \frac{(\dot{Z}_{\mathrm{A}} + \dot{Z}_{\mathrm{B}} + \dot{Z}_{\mathrm{C}})}{\dot{\mathrm{E}}_{\mathrm{H}} + \dot{W}_{\mathrm{el}}} \tag{6.81}$$

Substituting the numerical values in (6.81):

$c_{\mathrm{H}}^{\varepsilon} = c_{\mathrm{el}}^{\varepsilon} = \underline{9.83 \times 10^{-6}\ \text{£/kJ}} = \underline{3.54\ \text{p/kWh}}$

There is a basic difference in the nature of the unit costs of the two products considered above. The value of the thermal output per unit of time $\dot{C}_{\mathrm{H}}$ can be expressed as:

$$\dot{C}_{\mathrm{H}} = \dot{\mathrm{E}}_{\mathrm{H}}\, c_{\mathrm{H}}^{\varepsilon} = \dot{m}_{\mathrm{s}}\, \varepsilon_{\mathrm{H}}\, c_{\mathrm{H}}^{\varepsilon} \tag{6.82}$$

As the back-pressure of the turbine is reduced, the value of the specific exergy* ε_{H} of the

* Here only the physical exergy of the steam need be considered.

heating steam also goes down, reaching zero when the saturation temperature reaches the environmental temperature. Thus, even though for the unit costs $c_H^\varepsilon = c_{el}^\varepsilon$, $\dot{C}_H$ *is not constant*. This can be seen more directly if specific cost, c_H^{sp}, rather than unit cost is used. In such a case:

$$\dot{C}_H = \dot{m}_s c_H^{sp} \tag{6.83}$$

From (6.82) and (6.83):

$$c_H^{sp} = \varepsilon_H c_H^\varepsilon$$

which shows that the commercial value of 1 kg of steam is proportional to its specific exergy. Since the quality of electrical energy is invariable, the cost of the electrical output is not subject to such variations.

6.8 Other thermoeconomic applications of exergy

There are many applications of exergy in the area of thermoeconomics which, for lack of space, cannot be reported here in full. They will only be briefly reviewed.

Approximate thermoeconomic modelling of thermal plants

The exergy method can be used to generalise the results of thermoeconomic analysis of thermal plants because of the following exergy concepts and their particular characteristics:

1. The rational efficiency of a plant is a general criterion of thermodynamic perfection. Its range of values lies between 0 and 1, and it tends to unity as the behaviour of the plant approaches full reversibility.
2. Exergy is well suited for use as a generalised measure of plant output. For example, it would be wrong to compare the refrigeration duties of two refrigeration plants operating at different temperatures, but they can be compared on a rational basis if we express the refrigeration duties in terms of their thermal exergy outputs.

Thermoeconomic modelling depends on the general principle that for plants of the same type and the same structure, an improvement in the plant performance can be usually obtained at the expense of additional capital expenditure. Szargut[6.24,6.25] models this behaviour using a general expression of the type:

$$C^c = C_0^c + \dot{E}_{OUT} k \left(\frac{\psi}{1-\psi} \right)^m \tag{6.84}$$

where:

C^c—capital cost of the plant;
C_0^c—component of the capital cost which does not affect plant efficiency;
$\dot{E}_{OUT}$—nominal plant exergy output;
ψ—rational efficiency of the plant;
k, m—empirical constants which characterise a particular type of plant.

In addition to (6.84), an expression is used for the running cost of the form:

$$C^R = \dot{E}_{OUT} t_{op} \left[\frac{c_{IN}^\varepsilon}{\psi} + c_M \right] + C_0^R \tag{6.85}$$

where c_M is the cost of maintenance, repairs, *etc* per unit of exergy output; and C_0^R the fixed component of the annual running cost.

From (6.84) and (6.85) the objective function is:

$$\dot{C}_T = \frac{a^c C^c + C^R}{\dot{E}_{OUT}\, t_{op}} \rightarrow \min \tag{6.86}$$

The minimum of the objective function yields the optimum rational efficiency:

$$\psi_{OPT} = \frac{1}{1 + (mL)^{1/m+1}} \tag{6.87}$$

where:

$$L = \frac{a^c k}{c^{\varepsilon}_{IN}\, t_{op}} \tag{6.88}$$

L has been called *dimensionless number of thermoeconomic similarity*.

Substituting (6.87) and (6.88) in (6.84), an expression for the optimum capital cost of the plant is obtained. Once the constants k and m are determined from a known optimum solution, these expressions can be used to extend its applicability to other similar plants. One can also use these expressions to study the effect of various thermoeconomic parameters on the optimum capital cost of the plant.

Using an earlier version of these expressions (with $m = 1$), Szargut determined[6.24] the optimum parameters for an electric power station, a refrigerator for an air-conditioning plant and for a blast-furnace.

Minimisation of the consumption of natural resources

Non-renewable natural resources (NRNR) are used not only in the production of manufactured products but also in the production of food (artificial fertilisers, farm machinery, *etc*). There are no economic incentives to conserve NRNR, since an optimum economic solution does not in general coincide with the minimum use of NRNR. As an aid towards more sparing use of NRNR Szargut introduces[6.25] the concept of the ecological cost coefficient, which determines the cumulative consumption of NRNR in the various stages of production. Szargut considers that exergy should be used as a measure of the ecological value of the various NRNR. This he justifies by the fact that exergy is reckoned in relation to the dead state, which could be looked upon as the state of equilibrium on the Earth when there are no more natural resources left. Consequently the ecological cost could be expressed as the index of the cumulative consumption of the *exergy* of NRNR.

In addition to the exergy of NRNR, the ecological cost coefficient should take into account the harmful effect of production processes on the environment. In general the greater is the exergy of the pollutants, the greater is the disturbance of the equilibrium of the environment. Thus the effect of environmental pollution can be taken into account by adding to the ecological cost the exergy of the waste products rejected into the environment at various stages of production. The index of cumulative consumption of exergy is defined in a similar way to the index of cumulative consumption of energy, which has been dealt with extensively in a number of publications[6.27].

Considering the growth in concern regarding the uncontrolled depletion of NRNR, the concepts and techniques proposed by Szargut could play an important part in limiting this process in the future.

Application of exergy in maintenance decisions

This type of application of exergy is well illustrated by an example given by Fehring and

Gaggioli[6.28] concerning a steam power plant in which one of the seven feed heaters used, the fifth in the feed train denoted FH5, has deteriorated through film build-up on the heat transfer surface and by a reduction in the number of operational tubes through elimination (by plugging) of leaking tubes. The necessary decision concerning the faulty feed heater was arrived at by considering the following three cases of operating conditions of the plant.

Case A corresponds to the design operating conditions, ie to the state of the plant before any deterioration in the feed heater FH5 set in.

Case B is the situation when, after a number of years in service, FH5 has a proportion of the tubes plugged and has suffered some reduction in the overall heat transfer coefficient due to fouling of the heat transfer surfaces. The two factors lead to an increase in the temperature difference at the water outlet from the feed heater (terminal temperature difference). If the next feed heater, FH6, is in good condition it will take over most of the duty which FH5 failed to carry, but it will do this in a less efficient way.

At some stage in the life of the plant the deterioration in the performance of FH5 may become so great that the heater should be taken temporarily out of service in order to either re-tube it or replace it with a new heater. The former alternative involves in general a greater time out of service (down-time) than heater replacement.

Case C is when FH5 is out of service. This leads to a more serious upset in the operation of the feed train than that represented by Case B, in particular the maximum temperature of the feed water is lower, requiring additional fuel input to the boiler in order to maintain normal plant output.

To carry out the economic analysis on which to base the maintenance decision, Fehring and Gaggioli obtained first the relevant values of exergy rates and the corresponding costs of exergy. The particular costs calculated are those for raising the feed-water temperature from the level at the inlet to FH4 to the design temperature at entry to the boiler (or exit from FH7), and under conditions of Case C, the cost of the additional fuel required when FH5 is out of service. The cost rate (eg the hourly cost), of feed water heating corresponding to Case A and Case B was obtained from:

$$\dot{C}_{FH} = \sum_{i=1}^{7} c^{\varepsilon}_{B,i}\dot{E}_{B,i} \tag{6.89}$$

where $c^{\varepsilon}_{B,i}$ is the unit cost of bled steam to the i-th feed heater and $\dot{E}_{B,i}$ is the exergy rate of the bled steam supplied to the i-th feed heater. For Case C, the cost rate was evaluated from:

$$\dot{C}_{FH} = \sum_{i=4}^{7} c^{\varepsilon}_{B,i}\dot{E}_{B,i} + c^{\varepsilon}_{F}\Delta\dot{E}_{IN} \tag{6.90}$$

where c^{ε}_{F} is the unit cost of the boiler fuel (coal) and $\Delta\dot{E}_{IN}$ is the additional exergy input to the boiler required because of the lowering of the feed water temperature.

The unit costs of the exergy of the bled steam supplied to the feed heaters was calculated by the autonomous method in a similar manner to the calculations shown in Section 6.7. In this case, however, since the plant was already operational, the cost equations did not include the capital expenditure terms.

The additional annual operating cost ΔC^{an}, incurred due to plant operation with the deteriorated heater was evaluated from data obtained for Case A and Case B. The cost component due to loss of efficiency is:

$$\Delta C^{an}_{ef} = t_{op}[\dot{C}^{B}_{FH} - \dot{C}^{A}_{FH}] \tag{6.91}$$

where t_{op} is the time of operation of the plant per year. Taking a typical annual time out of service t_{os}, required for plugging leaking tubes, the additional cost component associated with operation of the plant without FH5 was evaluated from Case A and C as:

$$\Delta C_{os}^{an} = t_{os}[\dot{C}_{FH}^{C} - \dot{C}_{FH}^{A}] \tag{6.92}$$

If the annual labour cost associated with the maintenance of FH5 is denoted as ΔC_{lab}^{an}, the total additional cost is:

$$\Delta C_{T}^{an} = \Delta C_{ef}^{an} + \Delta C_{os}^{an} + \Delta C_{lab}^{an} \tag{6.93}$$

To decide if the replacement of the deteriorated heater was economically justified, the authors employed discounted cash flow analysis. This showed that for the given replacement cost, projected life, assumed salvage value, inflation rate and interest rate on the capital, the net cumulative discounted cash flow had a positive value and hence the replacement of the heater was justified.

As an alternative to replacement, re-tubing was considered. This, however, would have involved a longer period of the heater out of service although the cost of re-tubing was less than that of replacement. The authors calculated the maximum out of service period for re-tubing which would have made re-tubing economically justified.

This account shows how use of the Exergy Method allowed the required answer to be obtained with the aid of local exergy-related parameters, without the need for laborious calculations involving the whole plant. Such methods can be applied both to improvements in existing plant and at the design stage to appraise options such as the introduction of a more efficient heat exchanger, improvement of pipe insulation or determination of an optimum pipe diameter[6.29].

Optimum division of capital cost in a two-element system

In the special case when the plant can be divided into two functionally identifiable parts and the total capital cost available for the plant is fixed, the distribution of the total capital cost between the two parts can be expressed in a particularly simple form[6.30]. Denoting quantities belonging to the two components with subscripts A and B the exergy balance for the plant is:

$$\dot{E}_{IN} = \dot{E}_{OUT} + \dot{I}_A + \dot{I}_B \tag{6.94}$$

and the total capital cost:

$$C_T^c = C_A^c + C_B^c \tag{6.95}$$

When all the plant components possess optimum internal configurations their efficiencies will generally increase with their capital costs. However, since the total capital cost available is fixed, the plant efficiency will depend on the way the total sum is distributed between the two parts of the plant. Hence, with $\dot{E}_{OUT} = \text{const}$ and $C_T^c = \text{const}$:

$$\left.\begin{aligned} \dot{E}_{IN} &= f_1^T(C_A^c) = f_2^T(C_B^c) \\ \dot{I}_A &= f_1^A(C_A^c) = f_2^A(C_B^c) \\ \dot{I}_B &= f_1^B(C_A^c) = f_2^B(C_B^c) \end{aligned}\right\} \tag{6.96}$$

The condition for maximum plant efficiency is:

$$\frac{\partial \dot{E}_{IN}}{\partial C_A^c} = \frac{\partial \dot{I}_A}{\partial C_A^c} + \frac{\partial \dot{I}_B}{\partial C_A^c} = 0 \tag{6.97}$$

But from (6.95):

$$\delta C_A^c = -\delta C_B^c \tag{6.98}$$

hence, from (6.97) and (6.98):

$$\frac{\partial \dot{I}_A}{\partial C_A^c} = \frac{\partial \dot{I}_B}{\partial C_B^c} \tag{6.99}$$

showing that, in a two element system, the division of the capital cost is an optimum when the partial derivative of the irreversibility rate with respect to the capital cost is the same for both elements.

Appendix A Chemical exergy and enthalpy of devaluation

This appendix extends the contents of the first three chapters of the book and is intended as a deeper and more specialist treatment of the concepts relating to chemical exergy. Also the concept of enthalpy of devaluation is introduced. Both these concepts are of particular relevance in chemical and process engineering and power generation.

General expression for chemical exergy

The concept of chemical exergy has been introduced in Chapter 2 via some simple numerical examples. This treatment can be generalised so that the expressions obtained are applicable to any homogeneous substance of a single chemical species. The expression for chemical exergy will be derived with reference to an ideal device consisting of two connected modules, Module Y and Module Z (Fig A.1). These modules are generalised models of the ideal devices shown in Fig 2.12.

Module Y: Within this module the substance reacts reversibly with one or more suitable reference substances, called here *co-reactants*, to form *products* which are also reference substances. If, for example, the substance under consideration is methane, the reaction is:

$$CH_4 + \underbrace{2O_2}_{\text{co-reactant}} \rightarrow \underbrace{CO_2 + 2H_2O}_{\text{products}}$$

This type of reaction in which, besides the substance under consideration, all the other substances are reference substances is called a *reference reaction.* All the substances delivered to the module and removed from it are at the environmental pressure P_0 and temperature T_0. The function of this module can be fulfilled by such idealised devices as the van't Hoff equilibrium box (Section 1.12) or a reversible fuel cell with the required reversible isothermal compressors and expanders. From (1.69) and (1.77), the reversible work delivered by the module is equal to the difference between the chemical potentials of the reactants and products. Thus per mole of the substance:

$$[(\tilde{W}_x)_{REV}]_{0R}^{0P} = \mu_0 - \left(\sum_k n_k\mu_{0k} - \sum_j n_j\mu_{0j}\right) = -\Delta G_0 \tag{A.1}$$

where ΔG_0 is the Gibbs function of the reaction, n_k and n_j are the numbers of moles and μ_{0k} and μ_{0j} are the chemical potentials of the co-reactants and products necessary for

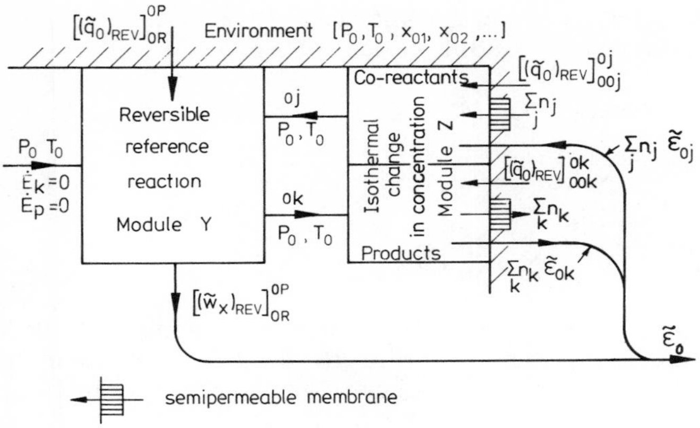

Fig A.1 An ideal device for determining the chemical exergy of a homogenous substance of a single chemical species.

the stoichiometric reaction with 1 mole of the substance under consideration with chemical potential μ_0.

Module Z: This module brings about reversibly an isothermal change of concentration of the reference substances which are either delivered from the environment to Module Y or vice versa. Module Z consists of a number of cells equal to the number of reference substances involved in the reaction; only two cells, one for a co-reactant and one for a product, are shown in Fig A.1. Each cell is equipped with semipermeable membranes to enable the required substance to be separated from the environmental mixture. The calculation of the reversible work associated with the changes of concentration of the substances passing through the cells involves in general complex calculations, particularly in the case of non-gaseous reference substances. The results of such calculations[A.1–A.3] are available in the form of tables of standard chemical exergies of chemical elements and compounds (see Tables A.2, A.3 and A.4). In general, as shown by (1.69), reversible isothermal work is given by the change in the Gibbs function of the substance, so the total work required by the cell handling the co-reactants is:

$$[(\tilde{w}_x)_{REV}]^{0j}_{00j} = -\sum_j n_j \tilde{\varepsilon}_{0j} = -\sum_j n_j(\mu_{0j} - \mu_{00j}) \tag{A.2}$$

Similarly, the work delivered by the cells handling the products is:

$$[(\tilde{w}_x)_{REV}]^{0k}_{00k} = \sum_k n_k \tilde{\varepsilon}_{0k} = \sum_k n_k(\mu_{0k} - \mu_{00k}) \tag{A.3}$$

In the case of components of the atmosphere, which can be treated as perfect gases, the reversible isothermal work terms given above in terms of changes in chemical potential can be put in the form:

$$[(\tilde{w}_x)_{REV}]^{0j}_{00j} = -\tilde{R}T_0 \sum_j n_j \ln \frac{P_0}{P_{00j}} \tag{A.4}$$

$$[(\tilde{w}_x)_{REV}]^{0k}_{00k} = \tilde{R}T_0 \sum_k n_k \ln \frac{P_0}{P_{00k}} \tag{A.5}$$

The net work delivered by modules Y and Z combined is equal, according to the definition, to the chemical exergy of the substance, ie:

$$\tilde{\varepsilon}_0 = -\Delta G_0 - \sum_j n_j(\mu_{0j} - \mu_{00j}) + \sum_k n_k(\mu_{0k} - \mu_{00k})$$

or:

$$\tilde{\varepsilon}_0 = -\Delta G_0 - \sum_j n_j \tilde{\varepsilon}_{0j} + \sum_k n_k \tilde{\varepsilon}_{0k} \tag{A.7}$$

Note that when the substance under consideration is a reference substance, Module Y becomes redundant and the chemical exergy of the reference substance $(\tilde{\varepsilon}_0)_{RS}$ will be given by the reversible isothermal work of a product-cell of Module Z. From (A.3):

$$(\tilde{\varepsilon}_0)_{RS} = (\mu_0 - \mu_{00})_{RS} \tag{A.8}$$

and if the reference substance is an ideal gas:

$$(\tilde{\varepsilon}_0)_{RS} = \tilde{R} T_0 \ln \frac{P_0}{(P_{00})_{RS}} \tag{A.9}$$

Standard chemical exergy

For general application to chemical processes and to processes involving chemical interaction with the environment, a system of reference substances originally devised by Szargut and his co-workers[A.1,A.2] is most convenient. Values of standard chemical exergy for various chemical elements, based on (A.6) have been computed and tabulated.

The *standard state* for which the values of chemical exergy have been computed is the standard state used for quoting thermo-chemical data and defined by the pressure $P^0 =$ 1.01325 bar and temperature $T^0 = 298.15$ K. P^0 and T^0 may be looked upon as standardised environmental parameters. Standard values are usually quoted in tables for a phase which exists in a stable form under P^0 and T^0. Occasionally thermo-chemical data are quoted for phases in which the substances under consideration do not exist in a stable form in the standard state. This applies, in particular, to water substances which, in the standard state, exists as compressed liquid, but values of different energy functions, including chemical exergy, are quoted for it in both gaseous and liquid phases (see Table A.3). Here, the gaseous state is a fictitious state in which the substance is assumed to behave as an ideal gas at the standard pressure and temperature.

The reference substances selected for chemical elements come from different parts of the environment. The choice being based on the assumption that these parts of the environment are in mutual equilibrium. The reference substances selected fall into four groups:

(1) Gaseous constituents of the atmosphere.
(2) Solid reference substances from the Earth's crust.
(3) Ionic reference substances from the seas.
(4) Reference substances in molecular, non-ionized form, from the seas.

Detailed descriptions of the method of calculation of standard chemical exergies for the various chemical elements are given in the original publications[A.1,A.2].

Table A.1 Chemical elements with gaseous reference substances. Reproduced from Ref [3.3], by permission. ($T^0 = 298.15$ K, $P^0 = 1.013\ 25$ bar.)

	Reference substances				
Chemical element	Chemical symbol	Mole fraction in dry air	Standard partial pressure in the environment P_i^{00}/bar	$-\tilde{R}T^0 \ln P_i^{00}/P^0$ kJ/kmol	Reference reaction
Ar	Ar	0.009 33	0.009 07	11 690	$Ar \rightarrow Ar$
C	CO_2	0.000 3	0.000 294	20 170	$C + O_2 \rightarrow CO_2$
D	D_2O g	—	0.000 001 37	33 500	$D + \frac{1}{4}O_2 \rightarrow \frac{1}{2}D_2O$
H	H_2O g	—	0.008 8	11 760	$H + \frac{1}{4}O_2 \rightarrow \frac{1}{2}H_2O$
He	He	0.000 005	0.000 004 9	30 360	$He \rightarrow He$
Kr	Kr	0.000 001	0.000 000 98	34 320	$Kr \rightarrow Kr$
N	N_2	0.780 3	0.758 3	720	$N \rightarrow \frac{1}{2}N_2$
Ne	Ne	0.000 018	0.000 017 7	27 150	$Ne \rightarrow Ne$
O	O_2	0.209 9	0.204 0	3 970	$O \rightarrow \frac{1}{2}O_2$
Xe	Xe	0.000 000 09	0.000 000 088	40 300	$Xe \rightarrow Xe$

In the case of chemical elements with gaseous (atmospheric) reference substances, the calculation of chemical exergy is quite simple and Table A.1 shows the scheme used in the calculations. As the composition of dry air is known quite precisely (column 3), it was only necessary to assume an average annual value of the partial pressure of water vapour to obtain the partial pressures of all the gaseous reference substances (column 4). In column 5 are the values of standard molar chemical exergy of the gaseous reference substances. All the gaseous substances, including water vapour in its fictitious gaseous state, as explained above, are treated as ideal gases. Consequently their values of standard chemical exergy are given by the reversible isothermal work of expansion at $T^0 = \text{const}$ from the standard partial pressure P_i^{00} of the reference substance under consideration in the atmosphere. In cases when the chemical element listed in column 1 serves as its own reference substance, the isothermal work in column 5 is equal to the standard molar chemical exergy of the element. This is the case for all the elements listed in Table A.1 except for carbon, hydrogen and deuterium. The values of standard molar chemical exergy and the reference substances are given for most of the common chemical elements in Table A.2. Values of standard chemical exergy are given for a wide selection of chemical compounds in Tables A.3 and A.4.

Chemical exergy of chemical compounds

For chemical compounds not listed in Tables A.3 and A.4, molar chemical exergy can be calculated from the appropriate value of Gibbs function of formation and the values of chemical exergy of the constituent chemical elements. Considering the reversible reaction of formation of the compound, in which exergy is conserved, the standard molar chemical exergy of the compound can be written as:

$$\tilde{\varepsilon}^0 = \Delta\tilde{g}_f^0 + \sum E_{el}^0 \tag{A.10}$$

where $\Delta\tilde{g}_f^0$ is the standard molar Gibbs function of formation and E_{el}^0 the standard chemical exergy of the constituent elements, per mole of the chemical compound.

Table A.2 Standard enthalpy of devaluation and standard chemical exergy of chemical elements. Reproduced from Ref [6.31], by permission. ($T^0 = 298.15$ K, $P^0 = 1.013\ 25$ bar.)

		Enthalpy of devaluation		Chemical exergy		
Chemical element	Standard state of the element	Reference substance	Standard enthalpy of devaluation, h_d^0/[kJ/kmol]	Reference substance	Concentration of reference substance in standard environment	Standard chemical exergy, $\tilde{\varepsilon}^0$/[kJ/kmol]
1	2	3	4	5	6	7
Ag	s	AgCl, s	46 260	$AgCl_2^-$, i	2.7×10^{-9} (w)	73 700
Al	s	Al_2SiO_5, s sillimanite	927 800	Al_2SiO_5, s sillimanite	2×10^{-3} (c)	887 890
Ar	g	Ar, g	0	Ar, g	0.907 (a)	11 690
As	s	—	—	$HAsO_4^{2-}$, i	1.5×10^{-8} (w)	477 040
Au	s	Au, s	0	$AuCl_2^-$, i	5.8×10^{-11} (w)	18 900
B	s	—	—	$H_2BO_3^-$, i	4.6×10^{-6} (w)	615 920
Ba	s, II	$BaSO_4$, s, barite	741 640	Ba^{2+}, i	5×10^{-8} (w)	760 050
Bi	s	Bi_2O_3, s	288 680	Bi_2O_3, s	7×10^{-10} (c)	271 370
Br_2	l	—	—	Br^-, i	6.5×10^{-5} (w)	91 770
C	s, graphite	CO_2, g	393 780	CO_2, g	0.03 (a)	410 820
Ca	s, II	$CaCO_3$, s, calcite	813 910	Ca^{2+}, i	4×10^{-4} (w)	717 400
Cd	s, α	$CdCO_3$, s	354 410	Cd^{2+}, i	5×10^{-11} (w)	290 920
Cl_2	g	NaCl, s	161 710	Cl^-, i	19×10^{-3} (w)	117 520
Co	s, III	Co_3O_4, s	293 080	Co^{2+}, i	9×10^{-11} (w)	260 520
Cr	s	Cr_2O_3, s	564 590	Cr_2O_3, s	4×10^{-7} (c)	538 610
Cs	s	CsCl, s	352 480	Cs^+, i	2×10^{-9} (w)	408 530
Cu	s	$CuCO_3$, s	201 590	Cu^{2+}, i	5×10^{-9} (w)	134 400
D_2	g	D_2O, g	249 370	D_2O, g	0.00014 (a)	266 220
F_2	g	CaF_2, s	401 500	F^-, i	1.4×10^{-6} (w)	448 820
Fe	s	Fe_2O_3, s, hematite	411 350	Fe_2O_3, s	2.7×10^{-4} (c)	377 740

H_2	g	H_2O, g	242 000	H_2O, g	0.9	(a)	238 490
He	g	He, g	0	He, g	0.0005	(a)	30 290
Hg	l	$HgCl_2$, s	68 560	$HgCl_4^-$, i	3.4×10^{-10}	(w)	122 700
J_2	s	—	—	JO_3^-, i	5×10^{-8}	(w)	184 190
K	s	KCl, s	355 330	K^+, i	3.8×10^{-4}	(w)	371 520
Kr	g	Kr, g	0	Kr, g	0.0001	(a)	34 280
Li	s	LiCl, s	328 200	Li^+, i	1×10^{-7}	(w)	396 170
Mg	s	$MgCO_3$, s	719 920	Mg^{2+}, i	12.7×10^{-4}	(w)	626 710
Mn	s, α, IV	MnO_2, s, I	521 260	MnO_2, s	2×10^{-4}	(c)	483 240
Mo	s	MoO_3, s	755 010	MoO_3, s	2×10^{-8}	(c)	715 540
N_2	g	N_2, g	0	N_2, g	75.83	(a)	720
Na	s	Na_2SO_4, s, II	330 410	Na^+, i	10.56×10^{-3}	(w)	343 830
Ne	g	Ne, g	0	Ne, g	0.0018	(a)	27 120
Ni	s	NiO, s	244 510	Ni^{2+}, i	9×10^{-11}	(w)	252 800
O_2	g	O_2, g	0	O_2, g	20.40	(a)	3 970
P	s, white	$Ca_3(PO_4)_2$, s, α	843 640	HPO_4^{2-}, i	5×10^{-8}	(w)	859 600
Pb	s	$PbCO_3$, s	306 680	Pb^{2+}, i	4×10^{-9}	(w)	226 940
Rb	s	RbCl, s	349 970	Rb^{2+}, i	2×10^{-7}	(w)	398 800
S	s, rhombic	$CaSO_4\cdot2H_2O$, s, gypsum	724 580	SO_4^{2-}, i	8.84×10^{-4}	(w)	598 850
Sb	s, III	Sb_2O_5, s	490 690	Sb_2O_5, s	7×10^{-10}	(c)	359 190
Se	s	—	—	SeO_4^{2-}, i	4×10^{-9}	(w)	326 960
Si	s	SiO_2, s, α, quartz	859 970	SiO_2, s	4.72×10^{-1}	(c)	803 010
Sn	s, white	SnO_2, s	581 130	SnO_2, s	2×10^{-5}	(c)	542 660
Sr	s	$SrCO_3$, s, III, strontianite	825 430	Sr^{2+}, i	1.3×10^{-5}	(w)	737 650
Ti	s, II	TiO_2, s, III, rutile	912 720	TiO_2, s, III	9×10^{-5}	(c)	876 000
U	s, III	UO_3, s	1 264 400	UO_3, s	2×10^{-8}	(c)	1 224 180
V	s	V_2O_5, s	780 840	V_2O_5, s	2×10^{-6}	(c)	725 880
W	s	WO_3, s, yellow	840 880	WO_3, s	4×10^{-8}	(c)	799 680
Xe	g	Xe, g	0	Xe, g	0.000 009	(a)	40 250
Zn	s	$ZnCO_3$, s	419 310	Zn^{2-}, i	5×10^{-10}	(w)	353 160

Key: s—solid, l—liquid, g—gaseous, i—ionic, aq—in aqueous solution, (a) standard partial pressure in air, kPa, (c) standard mass fraction in the Earth's crust, (w) standard mass fraction in sea water.

Table A.3 Standard enthalpy of devaluation and standard chemical exergy of inorganic substances. Reproduced from Ref [6.31], by permission. ($T^0 = 298.15$ K, $P^0 = 1.013\ 25$ bar.)

Substance	State*	Relative molecular mass	Standard enthalpy of devaluation, $\tilde{h}_d^0$/[kJ/kmol]	Standard chemical exergy, $\tilde{\varepsilon}^0$/[kJ/kmol]
1	2	3	4	5
Ag	s	107.870	46 260	73 730
Ag_2CO_3	s	275.749	−20 180	121 940
AgCl	s	143.323	0	23 420
AgF	s	126.868	43 950	120 810
$AgNO_3$	s	169.875	−76 960	46 580
Ag_2O	s	231.739	61 910	64 540
Ag_2O_2	s	247.739	67 620	179 036
Ag_2S	s, α	247.804	785 280	714 060
Ag_2SO_4	s	311.802	103 250	144 220
Al	s	26.981 5	927 800	887 890
Al_4C_3	s	143.959	4 763 080	4 661 880
$AlCl_3$	s	133.340 5	474 510	426 940
Al_2O_3	s, corundum	101.961 2	184 690	204 270
$Al_2O_3 \cdot H_2O$	s	119.976 5	125 520	199 450
$Al_2O_3 \cdot 3H_2O$	s, gibbsite	156.007 2	12 160	209 210
Al_2S_3	s	150.155	3 520 230	3 079 540
$Al_2(SO_4)_3$	s	342.148	592 060	502 100
Al_2SiO_5	s, andalusite	162.046	28 030	45 940
Al_2SiO_5	s, cyanite	162.046	25 940	49 200
Al_2SiO_5	s, sillimanite	162.046	0	15 400
Ar	g	39.948	0	11 690
Au	s	196.967	0	18 920
AuCl	s	232.42	46 100	63 410
$AuCl_3$	s	303.326	124 070	152 110
AuF_3	s	253.962	253 490	437 260
Au_2O_3	s	441,932	−80 810	121 550
Ba	s, II	137.34	741 640	760 050
$BaCO_3$	s, II, witherite	197.35	−84 190	37 170
$BaCl_2$	s	208.25	42 710	63 610
BaF_2	s	175.34	−58 050	61 470
BaO	s	153.34	183 120	244 880
BaO_2	s	169.34	111 530	175 770
$Ba(OH)_2$	s	171.36	36 590	145 520
BaS	s	169.40	1 022 420	896 670
$BaSO_3$	s	217.40	282 750	257 450
$BaSO_4$	s, barite	233.40	0	13 470
Bi	s	208.980	288 680	271 370
BiO	s	224.979	81 240	92 440
Bi_2O_3	s	465.958	0	52 260
Bi_2S_3	s	514.152	2 567 720	2 165 480
C	s, graphite	12.011 15	393 780	410 820
C	s, diamond	12.011 15	395 680	413 690
CCl_4	l	153.823	584 270	583 180
CO	g	28.010 5	283 150	275 430
CO_2	g	44.009 5	0	20 140

CS_2	l	76.139	1 932 410	1 673 670
Ca	s, II	40.08	813 930	717 400
CaC_2	s	64.10	1 538 670	1 471 210
$CaCO_3$	s, calcite	100.09	0	5 050
$CaCO_3$	s, aragonite	100.09	−170	6 100
$CaCl_2$	s	110.99	180 130	84 910
$CaCl_2 \cdot H_2O$	s	129.00	107 700	63 610
$CaFe_2O_4$	s	215.77	104 600	56 480
$Ca_2Fe_2O_5$	s	271.85	324 680	213 620
$Ca(NO_3)_2$	s	164.089 8	−123 290	−11 970
$Ca(NO_3)_2 \cdot 2H_2O$	s	200.120 5	−241 790	−17 020
$Ca(NO_3)_2 \cdot H_2O$	s	218.135 8	−289 290	−8 740
$Ca(NO_3)_2 \cdot 4H_2O$	s	236.151 2	−349 290	−8 870
CaO	s	56.08	177 940	119 620
$CaO \cdot Al_2O_3$	s	158.04	388 690	326 780
$2CaO \cdot Al_2O_3$	s	214.12	105 760	339 540
$4CaO \cdot Al_2O_3$	s	326.28	—	485 420
$12CaO \cdot 7Al_2O_3$	s	1 386.68	3 362 640	2 211 220
$Ca(OH)_2$	s	74.09	68 660	62 500
$Ca_3(PO_4)_2$	s, α	310.18	0	31 910
CaS	s	72.14	1 055 750	838 880
$CaSO_4$	s, anhydrite	136.14	104 950	4 300
$CaSO_4 \cdot \frac{1}{2}H_2O$	s, α	145.15	83 260	4 350
$CaSO_4 \cdot \frac{1}{2}H_2O$	s, β	145.15	85 350	5 310
$CaSO_4 \cdot 2H_2O$	s, gypsum	172.17	0	2 760
$CaSiO_3$	s	116.16	93 780	27 590
Ca_2SiO_4	s	172.24	235 290	116 680
Ca_3SiO_5	s	228.32	419 240	250 770
Cd	s, α	112.40	354 410	290 920
$CdCO_3$	s	172.41	0	36 960
$CdCl_2$	s	183.31	126 750	65 810
CdO	s	128.40	99 600	65 980
$Cd(OH)_2$	s	146.41	38 480	61 370
CdS	s	144.46	934 550	749 050
$CdSO_4$	s	208.46	152 200	81 290
$CdSO_4 \cdot H_2O$	s	226.48	88 520	63 070
Cl_2	g	70.906	161 710	117 520
Cl	g	35.453	201 850	163 940
Co	s, III	58.933 2	293 080	260 520
Co_3C	s	188.810 75	1 312 790	1 222 110
$CoCO_3$	s	118.942 6	−36 200	29 630
$CoCl_2$	s	129.839	129 060	103 800
CoO	s	74.932 6	53 600	51 070
Co_3O_4	s	240.797 2	0	37 560
$Co(OH)_2$	s	92.947 9	−14 230	47 120
$Co(OH)_3$	s	109.955 3	−83 310	27 240
CoS	s	90.997	933 090	765 580
Co_2S_3	s	214.058	2 590 640	2 108 250
$CoSO_4$	s	154.995	148 900	103 220
Cr	s	51.996	564 590	539 260
Cr_3C_2	s	180.010	2 393 410	2 348 290
Cr_4C	s	219.995	2 583 480	2 494 900
Cr_7C_3	s	400.005	4 955 530	4 819 010
$CrCl_2$	s	122.902	330 400	306 000
$CrCl_3$	s	158.355	243 610	221 120

Table A.3—*continued*

Substance	State*	Relative molecular mass	Standard enthalpy of devaluation, $\tilde{h}_d^0$/[kJ/kmol]	Standard chemical exergy, $\tilde{\varepsilon}^0$/[kJ/kmol]
1	2	3	4	5
CrO_3	s	99.994	−14 850	42 910
Cr_2O_3	s	151.990	0	36 510
Cs	s	132.905	352 480	408 530
CsCl	s	168.358	0	62 850
$CsNO_3$	s	194.910	−141 980	22 310
Cs_2O	s	281.809	387 180	544 370
CsOH	s	149.912	66 520	164 090
Cs_2S	s	297.874	1 089 990	1 091 020
Cs_2SO_4	s	361.872	8 590	123 020
Cu	s	63.54	201 590	134 400
$CuCO_3$	s	123.55	0	33 210
CuCl	s	98.99	146 370	77 050
$CuCl_2$	s	134.45	157 310	120 450
$CuFe_2O_4$	s	239.23	36 580	22 790
CuO	s	79.54	46 260	6 590
Cu_2O	s	143.08	236 380	123 500
$Cu(OH)_2$	s	97.55	−5 230	19 770
CuS	s	95.60	877 600	668 020
Cu_2S	s, II	159.14	1 048 210	773 000
$CuSO_4$	s	159.60	155 800	80 940
Cu_2SO_4	s	223.14	377 490	222 450
D_2	g	4.029 46	249 370	266 220
D_2O	g	20.028 86	0	33 450
D_2O	l	20.028 86	−45 430	24 520
F_2	g	37.996 8	401 500	448 820
Fe	s	55.847	411 350	377 740
Fe_3C	s, cementite	179.552	1 648 760	1 558 570
$FeCO_3$	s, siderite	115.856	56 950	120 410
$FeCl_2$	s	126.753	231 840	193 130
$FeCl_3$	s	162.206	248 630	225 600
$FeCr_2O_4$	s	223.837	108 990	133 490
FeO	s	71.846	140 800	133 750
Fe_2O_3	s, hematite	159.692	0	20 370
Fe_3O_4	s, magnetite	231.539	116 170	126 960
$Fe(OH)_3$	s	106.869	−50 450	46 530
FeS	s, α	87.911	1 040 810	879 000
FeS	s, β	87.911	1 046 540	890 180
FeS_2	s, pyrite	119.975	1 682 480	1 447 410
$FeSO_4$	s	151.909	212 740	170 240
FeSi	s	83.933	1 190 140	1 099 080
$FeSiO_3$	s	131,931	115 690	111 260
Fe_2SiO_4	s	203.778	243 670	228 310
$FeTiO_3$	s	151.75	86 870	99 530
H_2	g	2.015 94	242 000	238 490
H	g	1.007 97	360 860	322 410
HCl	g	36.461	109 490	85 950

HF	g	20.006 4	52 960	71 840
HNO_3	l	63.012 9	−52 330	45 650
H_2O	l	18.015 34	−44 030	3 120
H_2O	g	18.015 34	0	11 710
H_3PO_4	s	98.001 3	−75 360	98 850
H_2S	g	34.080	946 420	804 770
H_2SO_4	l	98.077	154 720	161 010
He	g	4.002 6	0	30 290
Hg	l	200.59	68 560	122 700
$HgCO_3$	s	260.60	−89 390	193 550
$HgCl_2$	s	271.50	0	63 100
Hg_2Cl_2	s	472.09	33 720	153 740
HgO	s, red	216.59	−22 210	64 240
HgS	s, red	232.65	734 940	679 430
$HgSO_4$	s	296.65	88 500	150 580
Hg_2SO_4	s	497.24	119 210	234 880
K	s	39.102	355 330	371 520
K_2CO_3	s	138.213	−42 450	90 110
KCl	s	74,555	0	21 390
$KClO_4$	s	138.553	2 430	125 900
KF	s	58.100	−6 880	59 760
$KMnO_4$	s	158.038	62 680	148 840
KNO_3	s	101.106 9	−137 380	−15 290
K_2O	s	94.203	348 820	425 540
KOH	s	56.109	50 200	101 320
K_2S	s	110.268	1 016 560	928 230
K_2SO_3	s	158.266	317 780	384 910
K_2SO_4	s, II	174.266	590	38 470
K_2SiO_3	s	154.288	1 308 010	1 565 320
Kr	g	83.80	0	34 280
Li	s	6.939	328 200	396 170
Li_2CO_3	s	73.887	−166 250	77 180
LiCl	s	42.392	0	70 160
$LiNO_3$	s	68.934	−154 450	12 700
Li_2O	s	29.877	60 200	233 460
LiOH	s	23.946	−38 350	75 620
Li_2SO_4	s	109.940	−54 380	76 520
Mg	s	24.312	719 920	626 710
$MgCO_3$	s	84.321	0	13 700
$MgCl_2$	s	95.218	239 370	151 860
$MgFe_2O_4$	s	200.004	76 640	38 230
MgO	s	40.311	117 690	59 170
$Mg(OH)_2$	s	58.327	36 610	33 830
$Mg(NO_3)_2$	s	148.321 8	−69 680	50 950
$Mg(NO_3)_2 \cdot 6H_2O$	s	256.413 8	−440 400	29 400
MgS	s	56.376	1 097 000	875 550
$MgSO_3$	s	104.374	438 730	305 400
$MgSO_4$	s	120.374	165 430	67 480
$MgSiO_3$	s	100.396	81 430	25 060
Mg_2SiO_4	s	140.708	255 810	140 770
Mg_2TiO_4	s	160.52	185 980	92 200
Mn	s, α, IV	54.938 1	521 260	483 240
Mn_3C	s, II	176.825 45	1 953 370	1 856 770
$MnCO_3$	s	114.947 5	19 480	35 360
$MnCl_2$	s	125.844	200 230	159 340

Table A.3—*continued*

Substance	State*	Relative molecular mass	Standard enthalpy of devaluation, $\tilde{h}_d^0$/[kJ/kmol]	Standard chemical exergy, $\tilde{\varepsilon}^0$/[kJ/kmol]
1	2	3	4	5
$MnFe_2O_4$	s	230.630	118 430	131 730
MnO	s, I	70.937 5	136 070	122 390
MnO_2	s, I	86.936 9	0	21 110
Mn_2O_3	s, II	157.874 4	70 760	79 110
Mn_3O_4	s, I	228.811 9	176 270	177 600
$Mn(OH)_2$	s	88.952 8	78 620	110 810
$Mn(OH)_3$	s, green	105.960 2	−41 630	89 130
MnS	s, I	87.002	1 041 520	873 540
$MnSO_4$	s	151.000	181 390	137 540
$MnSO_4 \cdot H_2O$	s	169.015	127 400	115 130
$MnSiO_3$	s	131.022	114 720	106 800
Mo	s	95.94	755 010	715 540
Mo_2C	s	203.89	1 921 800	1 853 700
MoO_2	s	127.94	210 730	228 310
MoO_3	s	143.94	0	43 940
Mo_2S_3	s	288.07	3 256 710	2 811 400
N_2	g	28.013 4	0	720
N_2 (air)	g	28.953 4	0	690
NH_3	g	17.030 5	316 780	341 250
NH_4Cl	s	53.491	249 260	331 150
NH_4NO_3	s	80.043 48	118 320	300 020
$(NH_4)_2SO_4$	s	132.138	513 240	660 470
NO	g	30.006 1	90 430	89 040
NO_2	g	46.005 5	33 870	56 220
N_2O	g	44.012 8	81 600	106 220
N_2O_4	g	92.011 0	9 670	107 040
Na	s	22.989 8	330 410	343 830
$NaAlO_2$	s	81.970 1	114 050	120 280
$NaAlSi_2O_6 \cdot H_2O$	s	220 055	5 730	95 250
Na_2CO_3	s	105.989 1	−77 090	53 130
$Na_2CO_3 \cdot H_2O$	s	124.004 3	−138 710	52 790
$Na_2CO_3 \cdot 7H_2O$	s	232.096 3	−457 330	69 640
$NaCl$	s	58.443	0	22 200
$NaHCO_3$	s	84.007 1	−103 120	224 370
$NaNO_3$	s	84.994 7	−94 430	−15 745
Na_2O	s	61.979 0	244 650	313 260
$NaOH$	s, II	39.997 2	23 100	84 490
Na_2S	s	78.044	1 011 940	930 340
Na_2SO_3	s	126.042	294 320	82 840
Na_2SO_4	s, II	142.041	0	35 000
$Na_2SO_4 \cdot 10H_2O$	s	322.195	−521 570	54 380
Na_2SiO_3	s	122.064	980	71 660
$Na_2Si_2O_5$	s	182.149	−8 620	61 190
Na_4SiO_4	s	184.043	175 640	312 120
Ne	g	20.183	0	27 120
Ni	s, II	58.71	244 510	252 800

Ni_3C	s	188.14	1 173 370	1 209 430
$NiCO_3$	s	118.72	−42 890	42 980
$NiCl_2$	s	129.62	90 120	111 150
$Ni(NO_3)_2$	s	182.72	−183 380	25 110
NiO	s	74.71	0	38 460
$Ni(OH)_2$	s	92.72	−51 910	49 160
$Ni(OH)_3$	s	109.73	−71 170	74 370
NiS	s, I	90.77	895 820	775 450
Ni_3S_2	s	240.26	1 999 700	1 774 810
$NiSO_4$	s	154.77	77 300	76 080
$NiSO_4 \cdot 6H_2O$	s, green	262.86	−279 310	72 830
O_2	g	31.998 8	0	3 970
O	g	15.999 4	247 520	232 110
O_3	g	47.998 2	142 350	169 100
P	s, III, white	30.973 8	843 640	859 600
PCl_3	l	137.333	764 650	744 060
P_2O_5	s	141.944 6	574 430	319 540
Pb	s	207.19	306 680	226 940
$PbCO_3$	s	267.20	0	17 090
$PbCl_2$	s	278.10	108 950	30 410
PbO	s, I, yellow	223.19	88 670	40 020
PbO	s, II, red	223.19	87 290	39 400
PbO_2	s	239.19	29 850	17 380
Pb_3O_4	s	685.57	184 840	71 510
$Pb(OH)_2$	s	241.20	33 700	41 560
PbS	s	239.25	936 890	724 390
$PbSO_4$	s, II	303.25	112 260	21 660
$PbSiO_3$	s	283.27	83 110	36 120
Pb_2SiO_4	s	506.46	164 120	69 530
Rb	s	85.47	349 370	398 800
Rb_2CO_3	s	230.95	−36 240	170 630
RbCl	s	120.92	0	51 440
$RbNO_3$	s	147.47	−140 650	14 490
Rb_2O	s	186.94	368 400	608 590
RbOH	s	102.48	56 300	147 840
Rb_2S	s	203.00	1 074 980	1 052 290
Rb_2SO_4	s	267.00	−2 290	92 110
S	s, II, rhombic	32.064	724 580	598 850
S	s, I, monoclinic	32.064	724 880	598 950
SO_2	g	64.062 8	427 480	303 500
SO_3	g	80.062 2	329 140	225 070
Sb	s, III	121.75	490 690	359 190
Sb_2O_3	s	291.50	270 040	269 730
Sb_2O_4	s	307.50	85 400	104 530
Sb_2O_5	s	323.60	0	52 250
Si	s	28.086	859 970	803 010
SiC	s	40.097	1 141 960	1 104 220
$SiCl_4$	l	169.898	542 810	465 010
SiO_2	s, II, quartz	60.085	0	1 860
SiO_2	s, II	60.085	1 680	3 280
SiS_2	s, white	92.214	2 102 470	996 740
Sn	s, II, white	118.69	581 130	542 660
Sn	s, III, grey	118.69	583 640	547 180
$SnCl_2$	s	189.60	392 810	355 630
SnO	s	134.69	294 380	287 550

Table A.3—*continued*

Substance	State*	Relative molecular mass	Standard enthalpy of devaluation, $\tilde{h}_d^0$/[kJ/kmol]	Standard chemical exergy, $\tilde{\varepsilon}^0$/[kJ/kmol]
1	2	3	4	5
SnO_2	s	150.69	0	26 820
SnS	s	150.75	1 227 840	1 059 060
SnS_2	s	182.82	1 862 820	1 581 260
Sr	s	87.62	825 430	737 650
$SrCO_3$	s, III, strontianite	147.63	0	16 570
$SrCl_2$	s	158.53	158 150	74 480
$SrCl_2 \cdot 6H_2O$	s	266.62	– 186 400	66 330
SrO	s	103.62	234 670	165 750
SrO_2	s	119.62	182 340	153 800
$Sr(OH)_2$	s	121.68	107 400	109 310
SrS	s	119.68	1 097 420	988 710
$SrSO_4$	s	183.68	104 310	9 170
Ti	s, II	47.90	912 720	876 000
TiC	s	59.91	1 080 410	1 064 120
TiO	s, II	63,90	393 710	388 840
TiO_2	s, III, rutile	79.90	0	23 090
Ti_2O_3	s, II	143.80	306 050	325 100
Ti_3O_5	s, II	223.70	281 760	322 200
TiS_2	s	112.03	2 026 940	1 743 640
U	s, III	238.03	1 264 400	1 224 180
UCl_3	s	344.39	615 170	534 850
UCl_4	s	379.84	536 100	483 790
UCl_5	s	415.30	571 310	508 710
UO_2	s	270.03	179 180	151 450
UO_3	s	286.03	0	43 940
V	s	50.942	780 840	725 880
VC	s	62.953	1 121 520	1 086 170
VO	s	66.941	355 550	328 880
VO_2	s	82.940	164 850	68 990
V_2O_3	s	149.882	330 760	324 280
V_2O_5	s	181.881	0	32 530
W	s	183.85	840 880	799 680
WC	s	195.86	1 196 600	1 173 270
WO_2	s	215.85	270 220	284 540
WO_3	s, yellow	231.85	0	42 220
WS_2	s	249.98	2 096 190	1 803 920
Xe	g	131.30	0	40 250
Zn	s	65.37	419 310	353 160
$ZnCO_3$	s	125.38	0	38 080
$ZnCl_2$	s	136.28	164 850	101 340
$ZnFe_2O_4$	s	241.06	40 830	27 630
ZnO	s	81.37	71 090	37 080
$Zn(OH)_2$	s	99.38	18 640	34 260
ZnS	s, II, sphalerite	97.43	940 830	764 860
$ZnSO_4$	s	161.43	164 680	93 280
$ZnSiO_3$	s	141.45	45 810	12 590
Zn_2SiO_4	s	222.82	187 860	121 280

* s—solid, l—liquid, g—gaseous.

Table A.4 Standard enthalpy of devaluation and standard chemical exergy of organic substances. Reproduced from Ref [3.3], by permission. ($T^0 = 298.15$ K, $P^0 = 1.01325$ bar.)

No.	Substance		State	Relative molecular mass	Standard enthalpy of devaluation, $\tilde{h}_d^0$ kJ/kmol	Standard chemical exergy, $\tilde{\varepsilon}^0$ kJ/kmol
1	2		3	4	5	6
1	CH_4	methane	g	16.042	802 320	836 510
2	C_2H_6	ethane	g	30.068	1 428 780	1 504 360
3	C_3H_8	propane	g	44.094	2 045 380	2 163 190
4	C_4H_{10}	butane	g	58.120	2 658 830	2 818 930
5	C_5H_{12}	pentane	g	72.146	3 274 290	3 477 050
			l	72.146	3 247 240	3 475 590
6	C_6H_{14}	hexane	g	86.172	3 889 280	4 134 590
			l	86.172	3 857 630	4 130 570
7	C_7H_{16}	heptane	l	100.198	4 467 820	4 786 300
8	C_8H_{18}	octane	l	114.224	5 078 000	5 440 030
9	C_9H_{20}	nonane	l	128.250	5 688 230	6 093 550
10	$C_{10}H_{22}$	decane	l	142.276	6 298 410	6 749 750
11	$C_{11}H_{24}$	undecane	l	156.302	6 908 600	7 404 520
12	$C_{12}H_{26}$	dodecane	l	170.328	7 518 820	8 059 340
13	$C_{13}H_{28}$	tridecane	l	184.354	8 129 010	8 714 200
14	$C_{14}H_{30}$	tetradecane	l	198.380	8 739 190	9 368 970
15	$C_{15}H_{32}$	pentadecane	l	212.406	9 349 370	10 023 870
16	$C_{16}H_{34}$	hexadecane	l	226.432	9 959 640	10 678 810
17	C_3H_6	cyclopropane	g	42.078	1 960 640	2 052 490
18	C_4H_8	cyclobutane	g	56.104	2 569 770	2 707 730
19	C_6H_{12}	cyclohexane	g	84.156	3 691 380	3 928 100
			l	84.156	3 658 260	3 922 990
20	C_7H_{14}	methylcyclo-hexane	g	98.182	4 295 490	4 573 030
			l	98.182	4 260 070	4 566 080
21	C_8H_{16}	ethylcyclo-hexane	g	112.208	4 914 260	5 246 900
			l	112.208	4 873 770	5 224 290
22	C_9H_{18}	propylcyclo-hexane	l	126.234	5 483 330	5 878 430
23	$C_{10}H_{20}$	butylcyclo-hexane	l	140.260	6 094 260	6 534 510
24	C_2H_4	ethylene	g	28.052	1 323 870	1 366 610
25	C_3H_6	propene	g	42.078	1 927 730	2 010 840
26	C_4H_8	1-butene	g	56.104	2 542 940	2 668 920
27	C_6H_{12}	1-hexene	g	84.156	3 772 890	3 984 330
			l	84.156	3 742 240	3 981 650
28	C_7H_{14}	1-heptene	g	98.182	4 387 970	4 641 570
			l	98.182	4 338 490	4 620 680
29	C_2H_2	acetylene	g	26.036	1 256 460	1 269 310
30	C_3H_4	propyne	g	40.062	1 850 860	1 904 070
31	C_4H_6	1-butyne	g	54.088	2 466 360	2 561 190
32	C_6H_{10}	1-hexyne	g	82.140	3 696 320	3 876 600
33	C_7H_{12}	1-heptyne	g	96.166	4 311 440	4 534 300
34	C_6H_6	benzene	g	78.108	3 171 630	3 310 540
			l	78.108	3 137 670	3 305 350

Table A.4—*continued*

No.	Substance		State	Relative molecular mass	Standard enthalpy of devaluation, $\tilde{h}_d^0$ kJ/kmol	Standard chemical exergy, $\tilde{\varepsilon}^0$ kJ/kmol
1	2		3	4	5	6
35	C_7H_8	toluene	g	92.134	3 774 440	3 952 550
			l	92.134	3 736 420	3 940 240
36	C_8H_{10}	ethylbenzene	g	106.160	4 389 980	4 610 250
			l	106.160	4 347 700	4 599 370
37	C_9H_{12}	propylbenzene	l	120.186	4 957 510	5 262 930
38	$C_{10}H_{14}$	butylbenzene	l	134.212	5 567 730	5 908 120
39	$C_{16}H_{26}$	decylbenzene	l	218.368	9 198 310	9 730 670
40	$C_{10}H_8$	naphthalene	s	128.164	4 984 220	5 264 190
41	$C_{10}H_{14}$	1,2,4,5-tetra-methylbenzene	s	134.212	5 532 980	5 896 060
42	$C_{11}H_{10}$	2-methylnaphthalene	s	142.190	5 574 930	5 892 920
43	$C_{11}H_{16}$	pentamethylbenzene	s	148.238	6 131 610	6 534 420
44	$C_{12}H_{18}$	hexamethylbenzene	s	162.264	6 739 110	7 191 670
45	$C_{14}H_{10}$	anthracene	s	178.220	6 850 940	7 229 600
46	$C_{14}H_{10}$	phenanthrene	s	178.220	6 835 870	7 213 270
47	$C_{14}H_{14}$	1,1-diphenylethane	s	182.252	7 250 910	7 682 020
48	$C_{18}H_{38}$	octadecane	s	254.484	11 116 710	11 981 110
49	$C_{19}H_{16}$	triphenylmethane	s	244.318	9 579 730	10 127 620
50	$C_{24}H_{18}$	1,3,5-triphenylbenzene	s	246.384	11 850 110	12 510 990
51	$C_{25}H_{20}$	tetraphenylmethane	s	320.410	12 544 110	13 254 570
52	CH_2O	formaldehyde	g	30.026	519 870	541 650
53	CH_2O_2	formic acid	g	46.026	259 080	303 580
			l	46.026	212 980	294 040
54	C_2H_6O	ethanol	g	46.068	1 278 230	1 370 800
			l	46.068	1 235 940	1 364 560
55	C_2H_6O	dimethylether	g	46.068	1 328 140	1 426 440
56	C_2H_4O	acetaldehyde	g	44.052	1 105 520	1 167 860
57	C_2H_4O	ethylene oxide	g	44.052	1 220 530	1 288 990
58	$C_2H_6O_2$	ethylene glycol	l	62.068	1 058 630	1 214 210
59	$C_2H_4O_2$	acetic acid	g	60.052	834 140	923 570
			l	60.052	786 610	912 640
60	C_3H_8O	propan-2-ol	l	60.094	1 830 590	2 007 820
61	C_3H_6O	acetone	g	58.078	1 690 880	1 798 440
			l	58.078	1 659 600	1 795 380
62	C_4H_8O	butanol	l	72.104	2 296 460	2 472 470
63	C_4H_8O	butan-2-one	l	72.104	2 264 050	2 441 780
64	C_4H_4O	furan	g	68.072	2 024 360	2 123 420
			l	68.072	1 996 730	2 122 750
65	$C_4H_8O_2$	butyric acid	l	88.104	2 018 750	2 224 950
66	$C_4H_8O_2$	ethyl acetate	l	88.104	2 073 550	2 278 750
67	$C_5H_{12}O$	pentan-1-ol	l	88.146	3 060 720	3 325 530
68	$C_5H_{12}O$	2-methylbutan-2-ol	l	88.146	3 017 220	3 289 530
69	$C_5H_{10}O$	cyclopentanol	l	86.130	2 878 680	3 121 220
70	$C_5H_6O_2$	furfuryl alcohol	l	98.098	2 418 550	2 694 580
71	$C_6H_{14}O$	hexan-1-ol	l	102.172	3 668 890	3 977 170

72	$C_6H_{12}O$	cyclohexanol	l	100.156	3 465 410	3 764 560
73	$C_7H_{16}O$	heptan-1-ol	l	116.198	4 285 570	4 637 550
74	C_7H_8O	benzyl alcohol	l	108.134	3 563 430	3 804 960
75	$C_4H_{10}O_4$	erithritol	s	122.124	1 874 680	2 204 470
76	$C_4H_6O_4$	succinic acid	s	118.092	1 356 860	1 616 310
77	$C_4H_4O_4$	maleic acid	s	116.076	1 271 280	1 500 300
78	$C_4H_4O_4$	fumaric acid	s	116.076	1 249 090	1 476 100
79	C_6H_6O	phenol	s	94.114	2 925 860	3 135 370
80	$C_6H_{14}O_6$	galactitol	s	182.178	2 729 580	3 212 360
81	$C_6H_{14}O_6$	mannitol	s	182.178	2 739 590	3 220 860
82	$C_6H_{12}O_6$	α-D-galactose	s	180.162	2 529 580	2 942 570
83	$C_6H_{12}O_6$	L-sorbose	s	180.162	2 544 610	2 952 820
84	$C_7H_6O_2$	benzoic acid	s	138.125	3 097 230	3 350 440
85	$C_7H_6O_3$	o-hydroxybenzoic acid	s	154.125	2 888 100	3 158 140
86	$C_8H_4O_3$	phthalic anhydride	s	148.120	3 173 800	3 439 410
87	$C_8H_6O_4$	phthalic acid	s	166.136	3 094 300	3 419 530
88	$C_{12}H_{10}O$	diphenyl ether	s	170.212	5 903 140	6 293 850
89	$C_{12}H_{22}O_{11}$	β-lactose	s	342.308	5 154 200	6 013 420
90	$C_{12}H_{22}O_{11}$	sucrose	s	342.308	5 166 180	6 033 090
91	$C_{12}H_{24}O_{12}$	α-lactose monohydrate	s	360.324	5 152 190	6 070 860
92	$C_{12}H_{24}O_{12}$	β-maltose monohydrate	s	360.324	5 173 210	6 091 000
93	$C_{16}H_{34}O$	hexadecan-1-ol	s	242.448	9 731 250	10 532 980
94	$C_{16}H_{32}O_2$	palmitic acid	s	256.432	9 290 260	10 089 060
95	$C_2H_2O_4$	oxalic acid	s	90.038	202 720	370 950
96	$C_2H_4N_4$	cyanoguanidine	s	84.084	1 296 480	1 481 960
97	$C_3H_6N_6$	melamine	s	126.124	1 835 620	2 127 400
98	$C_5H_5N_5$	adenine	s	135.130	2 664 860	2 946 840
99	$C_6H_4N_2$	2-cyanopyridine	s	104.108	3 106 060	3 251 510
100	$C_{12}H_{11}N$	diphenyl amine	s	169.216	6 188 050	6 553 430
101	CH_4ON_2	urea	s	60.058	544 660	693 580
102	$CH_6O_2N_2$	ammonium carbonate	s	78.074	474 780	673 570
103	$C_2H_5O_2N$	gylcine	s	75.068	867 630	1 055 280
104	$C_3H_7O_2N$	(±)alanine	s	89.094	1 462 410	1 697 540
105	$C_4H_7O_4N$	L-aspartic acid	s	133.104	1 445 160	1 751 920
106	$C_4H_8O_3N_2$	L-asparagine	s	132.120	1 749 870	2 070 500
107	$C_4H_2O_4N_2$	alloxan	s	142.072	813 120	1 056 120
108	$C_4H_7ON_3$	creatinine	s	113.120	2 179 650	2 448 730
109	$C_4H_9O_2N_3$	creatine	s	131.136	2 123 080	2 453 170
110	$C_4H_6O_3N_4$	allantoin	s	158.120	1 579 970	1 916 760
111	$C_5H_9O_4N$	D-glutamic acid	s	147.130	2 047 850	2 403 600
112	$C_5H_4ON_4$	hypoxanthine	s	136.114	2 337 280	2 606 950
113	$C_5H_4O_2N_4$	xanthine	s	152.114	2 068 780	2 366 340
114	$C_5H_5O_3N_4$	uric acid	s	169.122	1 950 590	2 294 870
115	$C_5H_5ON_5$	guanine	s	151.130	2 385 180	2 697 010
116	$C_9H_9O_3N$	hippuric acid	s	179.170	4 014 340	4 398 530
117	$C_{12}H_5O_{12}N_7$	bis(1,3,5-trinitrophenyl)-amine	s	439.216	5 397 200	6 173 640
118	C_2H_6S	ethanethiol	l	62.136	2 163 610	2 139 620
119	C_2H_6S	dimethyl sulphide	l	62.136	2 171 860	2 150 970
120	C_3H_8S	propane-1-thiol	l	76.163	2 782 970	2 802 640
121	C_3H_8S	methyl ethyl sulphide	l	76.163	2 782 250	2 802 940
122	$C_4H_{10}S$	butane-1-thiol	l	90.190	3 384 940	3 448 920
123	$C_4H_{10}S$	2-methyl propane-1-thiol	l	90.190	3 377 620	3 444 480
124	$C_4H_{10}S$	diethyl sulphide	l	90.190	3 391 560	3 456 540

Table A.4—*continued*

No.	Substance		State	Relative molecular mass	Standard enthalpy of devaluation, $\tilde{h}_d^0$ kJ/kmol	Standard chemical exergy, $\tilde{\varepsilon}^0$ kJ/kmol
1	2		3	4	5	6
125	$C_4H_{10}S$	methyl propyl sulphide	l	90.190	3 384 940	3 452 810
126	C_4H_4S	thiophene	l	84.142	2 864 570	2 850 290
127	$C_4H_{10}S_2$	ethyl butyl disulphide	l	122.256	4 114 540	4 063 330
128	$C_5H_{12}S$	pentane-1-thiol	l	104.217	3 993 510	4 101 510
129	C_5H_6S	2-methyl thiopene	l	98.169	3 373 220	3 402 110
130	C_5H_6S	3-methyl thiopene	l	98.169	3 374 770	3 403 780
131	C_6H_6S	thiophenol	l	110.180	3 875 260	3 921 690
132	$C_6H_{14}S_2$	dipropyl disulphide	l	150.310	5 333 860	5 370 870
133	$C_6H_{10}O_4S_2$	3,3′-dithiodipropionic acid	s	210.278	4 050 480	4 176 460
134	C_3H_7OSN	cysteine	s	105.163	2 218 000	2 292 610
135	$C_6H_{12}O_4S_2N_2$	cystine	s	240.310	4 210 160	4 425 660

Key: s—solid, l—liquid, g—gas.

Example A.1

Calculate the value of standard molar chemical exergy of methane vapour. Its Gibbs function of formation is[A.4]

$$(\Delta\tilde{g}_f^0)_{CH_4(g)} = -50\,810 \text{ kJ/kmol}$$

Solution: The reaction of formation is:

$$C + 2H_2 \rightarrow CH_4$$

Hence (A.10) becomes:

$$\tilde{\varepsilon}^0_{CH_4(g)} = (\Delta\tilde{g}_f^0)_{CH_4(g)} + \tilde{\varepsilon}^0_{c(s)} + 2\tilde{\varepsilon}^0_{H_2(g)}$$

From Table A.2:

$$\tilde{\varepsilon}_{c(s)} = 410\,530 \text{ kJ/kmol}$$

$$\tilde{\varepsilon}_{H_2(g)} = 238\,350 \text{ kJ/kmol}$$

Substituting

$$\tilde{\varepsilon}^0_{CH_4(g)} = -50\,810 + 410\,530 + 2 \times 238\,350 = \underline{836\,420 \text{ kJ/kmol}}$$

Correcting $\tilde{\varepsilon}^0$ for environmental temperature

If the environmental temperature T_0 differs appreciably from the standard temperature T^0 a correction for this effect on the value of chemical exergy may be required. The main error arises[A.5] in such a case because modules Y and Z interact thermally with the actual environment over a finite temperature difference. If the heat transfers to the two modules are denoted by Q_Y and Q_Z, the correction is equal to the reversible work associated with these heat transfers and can be expressed as:

$$\tilde{\varepsilon}_0 = \tilde{\varepsilon}^0 - (Q_Y + Q_Z)\frac{T^0 - T_0}{T^0} \tag{A.11}$$

By considering the mode of operation of Module Y and Z:

$$Q_Y = T^0 \Delta S^0 = -\Delta G^0 + \Delta H^0 \tag{A.12}$$

and:

$$Q_Z = \sum_k n_k \tilde{\varepsilon}_k^0 - \sum_j n_j \tilde{\varepsilon}_j^0 \tag{A.13}$$

The assumption that, in a reversible isothermal process, the work done by the system is equal to the heating done on the system, implicit in (A.13), is only exactly true for an ideal gas, but holds reasonably well for real gases. Substituting (A.12) and (A.13) in (A.11) and rearranging:

$$\tilde{\varepsilon}_0 = \tilde{\varepsilon}^0 \frac{T^0}{T_0} - \Delta H^0 \frac{T^0 - T_0}{T^0} \tag{A.14}$$

Example A.2

Calculate the molar chemical exergy of methane vapour when the environmental temperature is 283.15 K. Standard enthalpy of combustion of methane vapour, when all the products are in gaseous phase, is:

$$\Delta H^0 = -802\ 300 \text{ kJ/kmol}$$

Solution

Using the standard molar chemical exergy of methane vapour calculated in Example A.1:

$$(\tilde{\varepsilon}_0)_{CH_4(g)} = 836\ 420 \frac{283.15}{298.15} + 802\ 300 \frac{15}{298.15} = 834\ 703 \text{ kJ/kmol}$$

As will be seen, the difference between $\tilde{\varepsilon}^0$ and $\tilde{\varepsilon}_0$ (for $T_0 = 283.15$ K) is quite small, about 0.2%.

Standard environment as a source and sink of reference substances and zero-grade thermal energy

Before considering reference datum states for exergy, consider an open system undergoing a prescribed flow process while it interacts thermally and chemically solely with the environment. The purpose of this exercise is to demonstrate the relationships between exergy of material streams, exergy of heat transfers, work transfer, irreversibility rate for the process, and the balance of reference substances participating in the secondary reactions inside the injection and extraction modules associated with the prescribed process. In this arrangement the process under consideration may be any real, steady-flow process of arbitrary complexity involving any substances and any chemical reaction, but excluding nuclear reactions. As shown in Fig A.2, *the actual environment has been replaced* by injection and extraction modules and heat pumps interacting with the standard environment which becomes the sole source and sink of both matter and thermal energy as required by the prescribed process. Each stream of

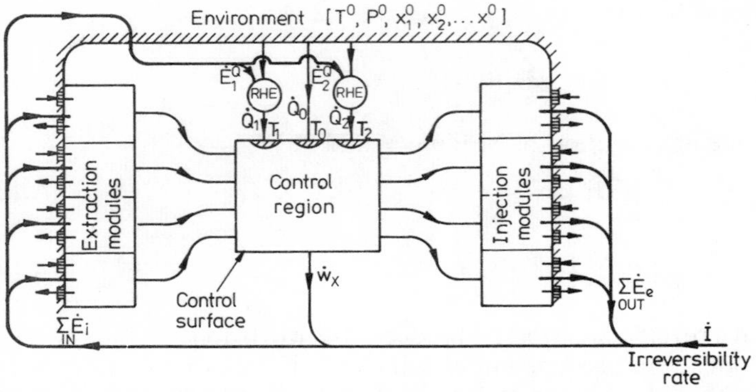

Fig A.2 An open system undergoing a steady flow process whilst interacting solely with the standard environment.

matter entering the control region is synthesised beforehand in reversible extraction modules from reference substances withdrawn from the standard environment.

The streams of matter leaving the control region are passed through similar modules operating in a reverse mode where matter is decomposed reversibly into reference substances which are then returned to the standard environment. The net reversible work delivered or absorbed by a module represents the exergy of the particular stream. The heat transfers at the temperatures required at the system boundary are provided in this arrangement by heat pumps or heat engines operating reversibly between the system boundary and the standard environment. The work input or output is equal to the thermal exergy, E^Q. When all the reversible work quantities associated with the supply and removal of matter and with the provision of the necessary heat transfers are summed and added to the net shaft work of the system, the net work which has to be supplied from outside (say, from an MER) is equal to the irreversibility of the control region.

A comparison of reference datum states

Absolute zero

Figure A.3 shows Module XYZ which combines the roles of the three separate modules shown in Figs 2.2 and A.1. From this ideal model, the exergy of the stream may be written as the difference in the exergy function $(\tilde{h} - T^0\tilde{s})$ for the substances entering and leaving the module. In the case of reference substances, the co-reactants and the products, this function will have the form:

$$\left.\begin{aligned}\tilde{h}_j^{00} - T^0\tilde{s}_j^{00} = \mu_j^{00}\\ \tilde{h}_k^{00} - T^0\tilde{s}_k^{00} = \mu_k^{00}\end{aligned}\right\} \tag{A.15}$$

Thus, neglecting kinetic and potential exergy terms:

$$\tilde{\varepsilon}_1 = (\tilde{h}_1 - T^0\tilde{s}_1) + \sum_j n_j\mu_j^{00} - \sum_k n_k\mu_k^{00} \tag{A.16}$$

Clearly, as there is a number of different substances involved, (A.16) will have no validity unless the properties of all these substances are evaluated with respect to a

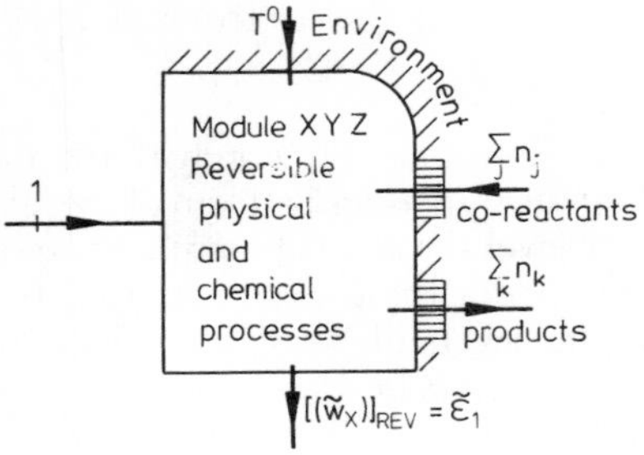

Fig A.3 A single-module ideal device for determining the exergy of a stream of substance.

common reference level. If we consider that at the absolute zero both the enthalpy as well as entropy of substances are zero, then absolute zero is a suitable common reference level for all the substances. Hence, (A.16) will be valid if all the properties of the substances are absolute.

Referring now to the ideal model shown in Fig A.2, it should be realised that under steady operating conditions the quantities of the different reference substances processed by the extraction modules will be the same as those processed by the injection modules. Hence it follows that, in an exergy balance, the sum of the terms $(\sum_j n_j\mu_j^{00} - \sum_k n_k\mu_k^{00})$ for all the streams entering the control region will cancel with the sum of all such terms for the streams leaving the control region. This leads to the conclusion that the exergy function, $(\tilde{h} - T^0\tilde{s})$, formed from the absolute values h and s, may be used in an exergy balance instead of exergy. As the absolute values of enthalpy and entropy of substances are not easily available, this conclusion is of little practical value to the analysis of real processes. However, it is relevant in the derivation of the exergy balance in Appendix B.

Environmental state P_0, T_0

The expression for molar exergy based on the three separate modules X, Y and Z (see Figs 2.3 and A.1) can be written (neglecting kinetic and potential exergy) in the form:

$$\tilde{\varepsilon}_1 = [(\tilde{h}_1 - T_0\tilde{s}_1) - (\tilde{h}_0 - T_0\tilde{s}_0)] - \Delta G_0 + \left[\sum_k n_k\tilde{\varepsilon}_{0k} - \sum_j n_j\tilde{\varepsilon}_{0j}\right] \tag{A.17}$$

The RHS of (A.17) can be considered in three parts as:

(i) The physical exergy of the substance under consideration. This term represents reversible work delivered by Module X. The enthalpies and entropies for the substance can be evaluated with respect to *any convenient reference base.*
(ii) The Gibbs function of the reference reaction involving 1 mole of the substance under consideration and stoichiometric quantities of reference substances, co-reactants and products. This term represents the maximum work of the chemical reaction delivered by Module Y.
(iii) The net chemical exergy of the reference substances, ie products and co-reactants which is delivered by Module Z. Note that the magnitude of this term depends on the molar quantities of the reference substances involved in the reference reaction of the substance under consideration and their molar chemical exergies. When applying an exergy balance to a control region which operates under steady conditions (Fig A.2) the

sum of all the terms $[\sum_k n_k\tilde{\varepsilon}_{0k} - \sum_j n_j\tilde{\varepsilon}_{0j}]$ for the streams entering the control region *will cancel out* with those for the streams leaving the control region.

As this shows, the calculation of irreversibilities, within a control region with the aid of an exergy balance (see Chapter 3) may be carried out using a '*relative*' form of exergy. This is equivalent to using as an alternative reference level the environmental state of the reference substances rather than the dead state. Under these circumstances, the total exergy of a stream of substance is equal to the reversible work delivered by Modules X and Y, which means that *the relative form of chemical exergy is equal to the negative value of the Gibbs function of the reference reaction.* Adopting the notation $\tilde{\varepsilon}^{\mathrm{r}}$ for the relative values of exergy, from the above:

$$\tilde{\varepsilon}_1^{\mathrm{r}} = \tilde{\varepsilon}_{\mathrm{ph}1} + \tilde{\varepsilon}_0^{\mathrm{r}} \tag{A.18}$$

where:

$$\tilde{\varepsilon}_0^{\mathrm{r}} = -\Delta G_0 \tag{A.19}$$

Thus $\tilde{\varepsilon}_0^{\mathrm{r}}$ can be obtained for any pure substance from tables of thermo-chemical data. In the case of an ideal-gas mixture the relative chemical exergy can be calculated from (2.21). Clearly, in contrast to chemical exergy, no knowledge of the chemical composition of the environment is required in the calculation of the relative values of chemical exergy. This may offer some advantages in a process analysis in which the states of the streams at the control surface are all specified. However, when the process involves exchange of substances between the system and the environment, knowledge of the chemical composition of the environment is indispensable*.

Selection of reference datum

Clearly, the reference datum for exergy evaluation is different in principle from the reference base used for property tabulation in, for example, steam tables. In general, the reference datum is established from the composite system consisting of the system under analysis and the relevant reference substances. In most cases one can use a standard, universal environment such as the one used in the compilation of Tables A.1 to A.4, but in some processes, such as sea-water distillation, the reference system consists of the local environment. This may be a particular sea or a lake with brackish water with the particular salinity.

An interesting example of a local environment has been provided by Beyer[A.7]. To carry out an exergy analysis of an ethanol distillation process, Beyer plotted an exergy-concentration chart for the ethanol–water mixture. The reference substance selected for the mixture was the waste product, a weak ethanol–water mixture containing 0.1% of ethanol by mass (the maximum permitted concentration for the waste), with $T_0 = 285$ K and $P_0 = 1$ bar.

When choosing a reference environment one must make sure that its state remains steady during the process under consideration. Should this not be the case, these changes in the environment must be taken into account in the analysis. For example, ambient air is a time-dependent environment when analysing an air-conditioning system because the variations in the atmospheric pressure, temperature and relative humidity are important. Such changes may be irrelevant when considering, for example, a chemical plant.

When considering H_2O as a reference substance for itself or for H_2 (see Table A.1) it

* As shown in Ref [A.6], it is also possible to use $\tilde{\varepsilon}_0^{\mathrm{r}} = -\Delta\tilde{g}_{\mathrm{f}}$.

will be found that the choice of the reference state is not quite straightforward. As is well known, water vapour in atmospheric air never reaches the saturation state, although this is its equilibrium state in the presence of the liquid phase (ie surface water). However, such an unstable environment is maintained by the effect on the atmosphere of a very much larger system, namely the sun. Thus, the unsaturated water vapour in the atmosphere is in a relatively steady state despite the fact that it is not in a state of equilibrium. Thus Szargut and his co-workers have adopted for water vapour an average annual value of its partial pressure $P^{00} = 0.0088$ bar with $T^0 = 298.15$ K as the standard reference state. Other workers[A.8] have favoured the use of liquid water or saturated vapour for this purpose.

The different reference states for water substance can be conveniently represented on a $T-S$ diagram (Fig A.4). The environmental state, defined by P_0 and T_0 (or by P^0 and T^0 when the standard state is considered), will be found to lie in the compressed liquid

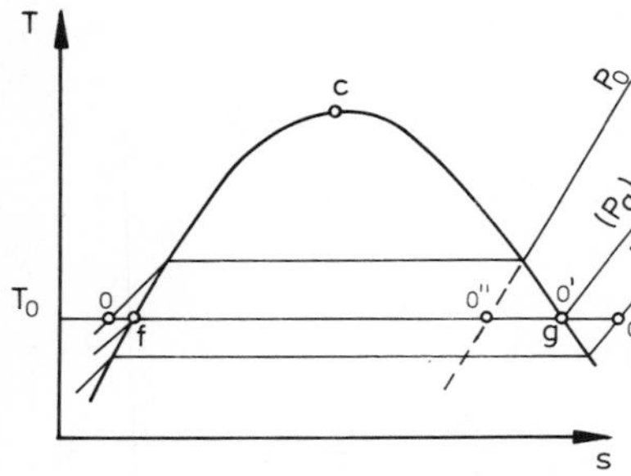

Fig A.4 Reference states for water substance.

region, marked by 0 on the diagram. The value of the Gibbs function at this state, $g_0 = h_0 - T_0 s_0$, is not very different from that calculated for the saturated liquid state at T_0. Further, the latter state, indicated by f in Fig A.4, has the same value of Gibbs function as the saturated vapour state, g, and any state at T_0 between state f and state g. The numerical values of the Gibbs functions calculated for $T_0 = 298.15$ K and $P_0 = 1$ bar are[A.9]:

$$g_0 = -0.25 \text{ kJ/kg}$$

$$g_f = g_g = -4.7 \text{ kJ/kg}$$

While the difference between g_0 and g_f (or g_g) is large in comparison with their absolute values, the effect of this difference on practical calculations of exergy is small since the value of g_0 is usually a small fraction of the total. The saturated vapour state may be used as a convenient alternative environmental state and is therefore marked as 0′ in Fig A.4. Recall that for tabulation of different thermo-chemical data, including chemical exergy, a fictitious standard state, T^0, P^0, for H_2O is adopted at which it is assumed to behave as an ideal gas. This state is marked as 0″ on the diagram. With this assumption, the standard chemical exergy of H_2O as a reference substance is given as the isothermal work at $T^0 = \text{const}$ when the vapour is expanded from P^0 to P^{00}, which is (see Table A.1) 11 760 kJ/kmol. Using the saturated vapour state as the alternative environmental state, 0′, the value of chemical exergy of water substance is:

$$(\tilde{\varepsilon}_0)_{H_2O} = \tilde{R}T_0 \ln(1/\phi_0) \tag{A.20}$$

where:

$$\phi_0 = P_{00}/(P_g)_{T_0}$$

With $P_{00} = 0.0088$ bar (from Table A.1) and $(P_g)_{T_0} = 0.0317$ bar (at 25°C)[A.9]:

$$(\tilde{\varepsilon}_0)_{H_2O} = 8.3144 \times 298.15 \ln \frac{0.00317}{0.0088}$$

$$= 3\ 177 \text{ kJ/kmol}$$

As is pointed out above, the value of the Gibbs function at T_0 between states f and g is constant and not very different from that at the environmental state, 0. The Gibbs function at T_0 is, of course, identical with exergy function, ie $g_0 = \beta_0$. Utilising this and (2.11) simplifies the construction for determining ε_{ph} of H_2O from an *h–s* diagram (Fig 2.11) as shown in Fig A.5. As this diagram shows, the physical exergy corresponding to

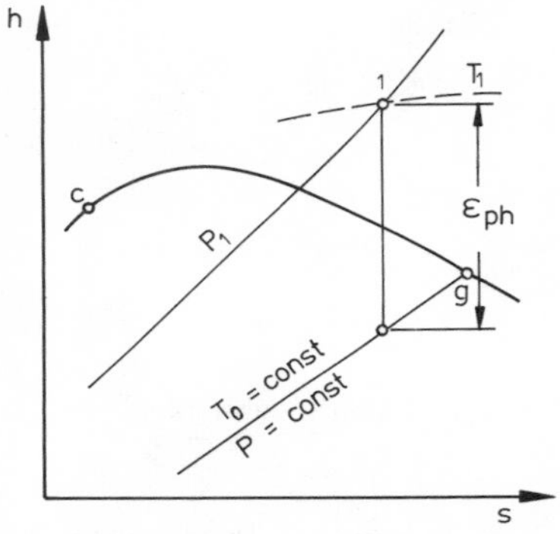

Fig A.5 Determination of an approximate value of physical exergy of water substance from an *h–s* diagram.

state 1 is equal to the isentropic enthalpy drop from this state to the isotherm $T_0 =$ const. The value of ε_{ph} obtained from this construction when added to the value of chemical exergy calculated from (A.13) will yield an approximate value of the total exergy of H_2O corresponding to state 1.

Enthalpy of devaluation

When analysing a process which involves no chemical reactions, ie in a purely physical process, the enthalpy of each substance participating in the process may be evaluated relative to an arbitrary reference state. For a process which includes chemical reactions it is convenient to take account of chemical energy by including it in the enthalpy of each substance involved. If this scheme is adopted, the number of reference states must be fewer than the number of participating chemical compounds. Since in a chemical reaction the same quantities of chemical elements are present in the reactants and in the products, the number of reference states should be the same as the number of chemical elements participating in the process.

Just as in the case of exergy, the enthalpy of a substance may be divided into two components, *physical enthalpy* and *chemical enthalpy*. The latter of the two components of enthalpy is the enthalpy of the substance in the environmental state (T_0, P_0), and thus it is analogous to chemical exergy. Thus for specific enthalpy:

$$h = h_{ph} + h_0 = \Delta h\Big|_{T_0,P_0}^{T,P} + h(T_0, P_0) \qquad \text{(A.21)}$$

When the substance can exist in the environmental state in more than one phase, say, liquid and vapour phase, it is clear that at T_0 and P_0 the value of h_{ph} should be zero in the phase selected for h_0.

The procedure to be followed when evaluating the chemical enthalpy is similar to that used with chemical exergy. First, for each chemical element a reference substance containing the element in question must be selected. Next, an equation of the reference reaction involving the element in question, its reference substance as well as some reference substances of other chemical elements, must be formulated. Although other alternatives are possible, it is convenient to choose as reference substances common environmental substances of low chemical potential from the atmosphere and the Earth's crust. Because of the chemical inactivity of these substances, they are of no use as natural energy resources and consequently it is appropriate to allocate to them zero chemical enthalpy. A reference reaction based on such a system of reference substances is known as a *reaction of devaluation*[A.10] and the corresponding chemical enthalpy of a chemical element or a chemical compound is known as *the enthalpy of devaluation* h_d. Hence, for this system of reference substances (A.21) takes the form:

$$h = h_{ph} + h_d = \Delta h\Big|_{T_0,P_0}^{T,P} + h_d(T_0, P_0) \qquad \text{(A.22)}$$

Take calcium as an example of a reaction of devaluation of a chemical element. The reference substance selected for Ca is calcite, $CaCO_3$, and the reaction of devaluation is:

$$Ca + \underbrace{\tfrac{1}{2}O_2 + CO_2}_{\text{co-reactants}} \rightarrow \underbrace{CaCO_3}_{\text{product}} \qquad \text{(A.23)}$$

Similarly for hydrogen, H_2, the reference substance is water vapour and the reaction of devaluation has the form:

$$H_2 + \underbrace{\tfrac{1}{2}O_2}_{\text{co-reactant}} \rightarrow \underbrace{(H_2O)_g}_{\text{product}} \qquad \text{(A.24)}$$

As an illustration of a reaction of devaluation of a chemical compound, for calcium hydroxide, $Ca(OH)_2$:

$$Ca(OH)_2 + \underbrace{CO_2}_{\text{co-reactant}} \rightarrow \underbrace{CaCO_3 + H_2O}_{\text{products}} \qquad \text{(A.25)}$$

The equation for the reaction of devaluation of a chemical compound is obtained by adding side by side equations for the relefant chemical elements. (A.25) is a sum of (A.23) and (A.24).

Note that all the co-reactants and products in (A.23) to (A.25) are common environmental substanceswhich exist in a state of equilibrium. In some instances these

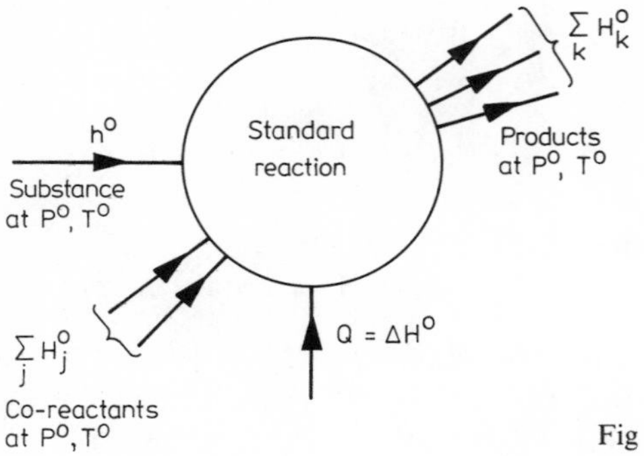

Fig A.6 A model for a standard reaction.

are the same reference substances which have been used in the compilation of the tables of standard chemical exergy of chemical elements, although in that case the use of substances present in sea water has been found to give better results for some elements.

As in the case of other standard thermo-chemical properties, the use of standard values of enthalpy of devaluation, h_d^0, can be most convenient. The model shown in Fig A.6 may be used to formulate the energy balance for a standard reaction which can be written in a general form as:

$$h^0 = -\Delta H^0 - \sum_j H_j^0 + \sum_k H_k^0 \tag{A.26}$$

where ΔH^0 is standard enthalpy of the reaction and $\sum_j H_j^0$ and $\sum_k H_k^0$ the joint enthalpy of the co-reactants and the products respectively.

In the special case when the standard reaction is a reaction of devaluation, however, all the co-reactants and all the products are reference substances selected from among the common environmental substances and for each of them the enthalpy of devaluation is allocated the value zero. Thus for a reaction of devaluation:

$$\sum_j \dot{H}_j = \sum_j \dot{H}_{d,j} = 0$$

$$\sum_k \dot{H}_k = \sum_k \dot{H}_{d,k} = 0 \tag{A.27a}$$

$$h^0 = h_d^0 \tag{A.27b}$$

and hence, from (A.26):

$$h_d^0 = -\Delta H^0 \tag{A.27c}$$

The usual differences between the environmental pressure and temperature and the standard values ($P^0 = 0.101\ 325$ MPa and $T^0 = 298.15$ K) are usually small enough to have a negligible effect on the value of the enthalpy of reaction. Hence, for most practical applications:

$$h_d \simeq h_d^0 \tag{A.28}$$

It follows from this discussion that the enthalpy of devaluation is a generalisation of the concept of the calorific value. The calorific value may be regarded as the chemical enthalpy of the fuel evaluated for the case when the enthalpy of the components of air and that of the products of complete combustion is taken to be zero. When the only elements involved in the chemical process are C, H, O and N, the reference substances for the calorific value are the same as those for the enthalpy of devaluation. Hence:

$$\text{for} \quad \text{C, H, O, N} \qquad h_d^0 \equiv (CV)^0 \tag{A.29}$$

However, the calorific value of sulphur is not equal to its enthalpy of devaluation since SO_2 is not a common environmental substance. Hence, in the case of fuels which contain sulphur (A.29) is not applicable.

In Table A.2 values of enthalpy of devaluation and the corresponding reference substances are given for the more important chemical elements. Further, in Tables A.3 and A.4, values of enthalpy of devaluation are given for numerous chemical compounds. For compounds not listed in Tables A.3 and A.4, the enthalpy of devaluation may be calculated from the enthalpy of formation ΔH_f^0 and the enthalpies of devaluation of the chemical elements, $(h_d^0)_{el}$, contained in the compound. For a reaction of formation:

$$h_d^0 = \Delta H_f^0 + \sum_{el} [m_{el}(h_d^0)_{el}] \tag{A.30}$$

where m_{el} is the mass of the element in the compound per unit mass of the compound.

The enthalpy of devaluation serves a similar purpose in energy balances to that of the enthalpy of formation, the advantage of the former being that it usually has a positive value and a more easily understandable physical significance.

Application of enthalpy of devaluation

Standard enthalpies of devaluation ease the computation of energy balances in processes involving chemical reactions. Neglecting changes in kinetic energy and potential energy, the SFEE may be written:

$$\dot{Q} - \dot{W}_x = \sum_k \dot{n}_k \tilde{h}_k - \sum_j \dot{n}_j \tilde{h}_j \tag{A.31}$$

where the subscripts j and k refer to the reactants and the products respectively. Using (A.22), (A.31) can be modified as:

$$\dot{Q} - \dot{W}_x = \left[\sum_k \dot{n}_k \tilde{h}_{d,k}^0 - \sum_j \dot{n}_j \tilde{h}_{d,j}^0\right] + \left[\sum_k \dot{n}_k \tilde{h}_{ph,k} - \sum_j \dot{n}_j \tilde{h}_{ph,j}\right] \tag{A.32}$$

The first term (in the square brackets) on the RHS of (A.32) represents the enthalpy changes due to chemical effects and is equal to the rate of the enthalpy of reaction $\Delta \dot{H}^0$. Thus:

$$\Delta \dot{H}^0 = \sum_k \dot{n}_k \tilde{h}_{d,k}^0 - \sum_j \dot{n}_j \tilde{h}_{d,j}^0 \tag{A.33}$$

This may be conveniently expressed per mole of one of the reactants, eg per mole of the fuel, as:

$$\Delta H^0 = \sum_k \nu_k \tilde{h}_{d,k}^0 - \sum_j \nu_j \tilde{h}_{d,j}^0 \tag{A.34}$$

where v_j and v_k are the numbers of moles of the reactants and the products respectively per mole of one of the reactants (the fuel). Using values of the enthalpy of devaluation from Tables A.3 or A.4, the enthalpy of the reaction can be evaluated easily.

The second term on the RHS of (A.32) represents the physical enthalpies of the reactants and the products involved in the reaction. These can be evaluated using the specific heat capacities which are available in the form of empirical equations or in a tabular form (see Table D.1).

Example A.3

Calculate the enthalpy of reaction per mole of Ca for the following reaction

$$Ca + 2H_2O_{(1)} \rightarrow Ca(OH)_2 + H_2$$

Solution: From Table A.3:

$$(h_d^0)_{Ca} = 813\,910 \text{ kJ/kmol}$$
$$(h_d^0)_{H_2O.1} = -44\,030 \text{ kJ/kmol}$$
$$(h_d^0)_{Ca(OH)_2} = 68\,660 \text{ kJ/kmol}$$
$$(h_d^0)_{Ca(OH)_2} = 242\,000 \text{ kJ/kmol}$$

Using (A.34):

$$\Delta H^0 = 68\,660 + 242\,000 - 813\,910 - 2(-44\,030)$$
$$= \underline{-591\,310 \text{ kJ}}$$

As the negative sign indicates, the reaction is exothermic.

Example A.4

Calculate the enthalpy of devaluation for barium cyanide $Ba(CN)_2$. The enthalpy of formation for $Ba(CN)_2$ is -2.01×10^{-5} kJ/kmol[A.11].

Solution: The reaction of formation for $Ba(CN)_2$ is:

$$Ba + 2C + N_2 \rightarrow Ba(CN)_2$$

From Table A.2 the values of enthalpy of devaluation of the elements are:

$$(h_d^0)_{Ba} = 741\,640 \text{ kJ/kmol}$$
$$(h_d^0)_C = 393\,780 \text{ kJ/kmol}$$
$$(h_d^0)_{N_2} = 0$$

Using (A.30), for $Ba(CN)_2$:

$$\Delta h_d^0 = -2.01 \times 10^5 + (741\,640 + 2 \times 393\,780)$$
$$= \underline{1.328 \times 10^6 \text{ kJ/kmol}}$$

*Appendix B Derivation of the exergy balance for a control region**

Consider a control region fixed in relation to the environment. For the sake of generality the control region is assumed (Fig B.1) to have several streams of matter entering and leaving it. Also, several constant temperature reservoirs of thermal energy, including the environment, interact with the system. Two types of work are considered, shaft work, W_x, and work due to the changes in the volume of the control region W_{CS}. Work done by the system and heat transfer to the system are taken as positive. The processes taking place may be non-steady and may include chemical reactions. The differential form of the energy equation (1.28b) can be written for such a control region as:

$$\sum_r \dot{Q}_r + \sum_{IN} \dot{m}_i \left(h_i + \frac{C_i^2}{2} + g_E Z_i \right) = \frac{d\mathbf{E}_{CR}}{dt} + \sum_{OUT} \dot{m}_e \left(h_e + \frac{C_e^2}{2} + g_E Z_e \right) + \dot{W}_x + \dot{W}_{CS} \quad \text{(B.1)}$$

In the above equation and in the remainder of this appendix all the thermodynamic properties used are assumed to have absolute values. Also, velocities are reckoned relative to the quiescent environment and the altitude, Z, is measured from the surface of the oceans.

The term $d\mathbf{E}_{CR}/dt$ represents the rate of increase of energy inside the control region. The work due to change in the volume of the control region can be expressed in terms of the net useful work and the work necessary to displace the atmosphere at P_0. Thus:

$$\dot{W}_{CSN} = \dot{W}_{CS} - P_0 \frac{dV_{CR}}{dt} \quad \text{(B.2)}$$

Similarly, the rate of entropy production within the control region can be shown (see (1.58)) to be:

$$\dot{\Pi} = \sum_{OUT} (s_e \dot{m}_e) - \sum_{IN} s_i \dot{m}_i + \frac{dS_{CR}}{dt} - \sum_r \frac{\dot{Q}_r}{T_r} \quad \text{(B.3)}$$

The first two terms on the RHS of (B.3) give the transport of entropy associated with the flow of matter in and out of the control region. The third term represents the rate of change of entropy inside the control region. This can be expressed for a continuous

* Reproduced from Ref [3.5], by permission.

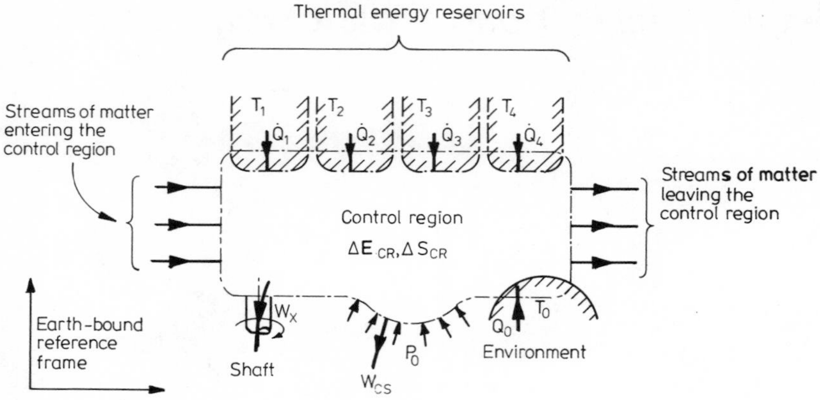

Fig B.1 Control region analysis.

substance as a volume integral:

$$\frac{\mathrm{d}S_{CR}}{\mathrm{d}t}=\frac{\mathrm{d}}{\mathrm{d}t}\int_V s\rho\,\mathrm{d}V \tag{B.4}$$

Since the heat transfers at the system boundary (Fig B.1) take place at the temperatures of the TERS interacting with the system, the heat transfers between thermal reservoirs and the system take place reversibly. The sum of the rates of entropy fluxes due to the reversible heat transfers is given by the fourth term on the RHS of (B.3).

A more useful form of the entropy equation (B.3) is obtained by first multiplying by T_0 and then, using the identity:

$$T_0\sum_r \frac{\dot{Q}_r}{T_r}\equiv -\sum_r\left(\dot{Q}_r\frac{T_r-T_0}{T_r}\right)+\sum_r \dot{Q}_r \tag{B.5}$$

and with some rearrangement of the terms:

$$\sum_r \dot{Q}_r=T_0\frac{\mathrm{d}S_{CR}}{\mathrm{d}t}+\sum_{OUT}(T_0 s_e \dot{m}_e)-\sum_{IN}(T_0 s_i \dot{m}_i)+\sum_r\left(\dot{Q}_r\frac{T_r-T_0}{T_r}\right)-T_0\dot{\Pi} \tag{B.6}$$

The energy and the entropy equations are now combined by eliminating the term $\sum_r \dot{Q}_r$ between them. Also, making use of expression (B.2), gives:

$$\dot{W}_x+\dot{W}_{CSN}=-\frac{\mathrm{d}}{\mathrm{d}t}(\mathbf{E}_{CR}+P_0V_{CR}-T_0S_{CR})+\sum_{IN}[\dot{m}_i(h_{Ti}-T_0s_i)]$$
$$-\sum_{OUT}[\dot{m}_e(h_{Te}-T_0s_e)]+\sum_r\left[\dot{Q}_r\frac{T_r-T_0}{T_r}\right]-T_0\dot{\Pi} \tag{B.7}$$

where

$$h_T=h+\frac{C_0^2}{2}+g_E Z_0 \tag{B.8}$$

By the entropy postulate, entropy production inside an isolated system comprising the control region, the TERS and the environment is zero if all the processes taking place within it are reversible. Furthermore, the work done by the system under these circumstances is a maximum. Hence:

$$(\dot{W}_x + \dot{W}_{CSN})_{\dot{\Pi}=0} = (\dot{W}_x + \dot{W}_{CSN})_{MAX} \tag{B.9}$$

Applying (B.9) to expression (B.7):

$$\begin{aligned}(\dot{W}_x + \dot{W}_{CSN})_{MAX} = &-\frac{d}{dt}(\mathbf{E}_{CR} + P_0 V_{CR} - T_0 S_{CR})\\ &+\sum_{IN}[\dot{m}_i(h_{Ti} - T_0 s_i)] - \sum_{OUT}[\dot{m}_e(h_{Te} - T_0 s_e)]\\ &+\sum_r\left[\dot{Q}_r\frac{T_r - T_0}{T_r}\right]\end{aligned} \tag{B.10}$$

Now, the rate at which work is lost through irreversibility, $\dot{I}$, inside the control region is obtained by subtracting (B.7) from (B.10):

$$\begin{aligned}\dot{I} &= (\dot{W}_x + \dot{W}_{CSN})_{MAX} - (\dot{W}_x + \dot{W}_{CSN})\\ &= T_0\dot{\Pi}\end{aligned} \tag{B.11}$$

(B.7) and (B.11) give the differential form of the exergy balance:

$$\begin{aligned}\dot{W}_x + \dot{W}_{CSN} = &-\left(\frac{d\Xi^*}{dt}\right)_{CR} + \sum_{IN}(\varepsilon_i^*\dot{m}_i)\\ &-\sum_{OUT}(\varepsilon_e^*\dot{m}_e) + \sum_r\left(\dot{Q}_r\frac{T_r - T_0}{T_r}\right) - \dot{I}\end{aligned} \tag{B.12}$$

where:

$$\Xi^* = m\zeta = m(\mathbf{e} + P_0 v - T_0 s) \tag{B.13a}$$

$$\varepsilon^* = E^*/\dot{m} = (h_T - T_0 s) \tag{B.13b}$$

(B.13a) and (B.13b) give the absolute non-flow exergy function and the absolute exergy function. The use of absolute functions (relative to the absolute zero) is limited to this derivation since in the rest of this book the environment is used as the reference datum. However, it has also been demonstrated in Appendix A that the balance of exergy and non-flow exergy is unaffected by the use of either of the two reference states, since the differences arising from the two types of reference datum levels cancel out in an exergy balance. It follows, therefore, that in an exergy balance either Ξ^* and E^* evaluated relative to the absolute zero or Ξ and E reckoned in relation to the environment may be used. Therefore, for convenience, the superscripts * will be dropped in all subsequent expressions. Note that the absolute functions are of little practical value since the absolute values of h and u are not obtainable.

The expression for the irreversibility rate for the control region is obtained from (B.3) and (B.11):

$$\dot{I} = T_0\dot{\Pi} = T_0\left[\sum_{OUT}(s_e\dot{m}_e) - \sum_{IN}(s_i\dot{m}_i) + \frac{dS_{CR}}{dt} - \sum_r\frac{\dot{Q}_r}{T_r}\right] \tag{B.14}$$

Given the necessary information, (B.12) and (B.14) can be integrated over a time interval. In many cases of interest the system under investigation can be considered either as a closed system or an open steady flow system.

Closed system—control mass analysis

Here:

$$\dot{m}_i = \dot{m}_e = 0 \tag{B.15}$$

Thus, (B.12) becomes:

$$\dot{W}_x + \dot{W}_{CSN} = -\left(\frac{d\Xi}{dt}\right)_{CR} + \sum_r \left(\dot{Q}_r \frac{T_r - T_0}{T_0}\right) - \dot{I} \tag{B.16}$$

Integrating with respect to time between the limits t_1 and t_2 and assuming $T_r = \text{const}$:

$$W_x + W_{CSN} = \Xi_1 - \Xi_2 + \sum_r \left(Q_r \frac{T_r - T_0}{T_r}\right) - I \tag{B.17}$$

Similarly, with (B.15), (B.14) becomes:

$$\dot{I} = T_0 \dot{\Pi} = T_0 \left[\frac{dS_{CR}}{dt} - \sum_r \frac{\dot{Q}_r}{T_r}\right] \tag{B.18}$$

Integrating, as before, the expression for irreversibility during the time interval from t_1 to t_2 is:

$$I = T_0 \Pi = T_0 \left[(S_2 - S_1)_{CR} - \sum_r \frac{Q_r}{T_r}\right] \tag{B.19}$$

Open system, steady flow case—control region analysis

Here:

$$\begin{gathered}
\sum_{IN} \dot{m}_i = \sum_{OUT} \dot{m}_e \\
\frac{dS_{CR}}{dt} = 0 \\
\left(\frac{d\Xi}{dt}\right)_{CR} = 0 \\
\dot{W}_{CSN} = 0
\end{gathered} \tag{B.20}$$

Hence, the steady-flow exergy balance is:

$$\dot{W}_x = \sum_{IN} \dot{m}_i \varepsilon_i - \sum_{OUT} \dot{m}_e \varepsilon_e + \sum_r \left(\dot{Q}_r \frac{T_r - T_0}{T_r}\right) - \dot{I} \tag{B.21}$$

and the irreversibility rate is:

$$\dot{I} = T_0 \dot{\Pi} = T_0 \left[\sum_{OUT} \dot{m}_e s_e - \sum_{IN} \dot{m}_i s_i - \sum_r \frac{\dot{Q}_r}{T_r}\right] \tag{B.22}$$

Since most thermal plants operate under steady or quasi-steady conditions, (B.21) and (B.22) are particularly useful.

Appendix C Chemical exergy of industrial fuels

Provided the necessary data are available, the chemical exergy of a combustible substance can be obtained from:

$$\tilde{\varepsilon}^0 = -\Delta\tilde{h}^0 + T^0\Delta\tilde{s}^0 + \tilde{R}T^0\left(x_{O_2}\ln\frac{P^{00}_{O_2}}{P^0} - \sum_k x_k \ln\frac{P^{00}_k}{P^0}\right) \tag{C.1}$$

where the subscript k refers to the components of the products of combustion. This expression may be applied to the calculation of chemical exergy of gaseous fuels for which the chemical composition can be determined and the thermochemical data for the components can be obtained. Solid and liquid industrial fuels are solutions of numerous chemical compounds of, usually, unknown nature. This makes it difficult to determine the entropy of the reaction, Δs^0, of these fuels with a reasonable degree of accuracy. Szargut and Styrylska[C.1] assumed that the ratio of chemical exergy, ε^0, to the net calorific value $(\mathrm{NCV})^0$ for solid and liquid industrial fuels is the same as for pure chemical substances having the same ratios of constituent chemicals. This ratio is denoted by φ, ie:

$$\varphi = \frac{\varepsilon^0}{(\mathrm{NCV})^0} \tag{C.2}$$

After computing values of φ for numerous pure organic substances, containing C, H, O, N and S, correlations expressing the dependence of φ on the atomic ratios H/C, O/C, N/C, and in some cases S/C were derived. It was assumed that the applicability of the expressions obtained can be extended to cover industrial fossil fuels.

Solid fuels

For *dry* organic substances contained in *solid* fossil fuels consisting of C, H, O and N with a mass ratio of oxygen to carbon less than 0.667, the following expression was obtained in terms of mass ratios.

$$\varphi_{\mathrm{dry}} = 1.0437 + 0.1882\frac{h}{c} + 0.0610\frac{o}{c} + 0.0404\frac{n}{c} \tag{C.3}$$

where c, h, o and n are the mass fractions of C, H, O and N, respectively. Within the restriction regarding the upper limit of o/c, (C.3) is applicable to a wide range of industrial solid fuels but not to wood. The accuracy of the expression is estimated to be better than $\pm 1\%$.

For fossil fuels with the mass ratio $2.67 > o/c > 0.667$, which, in particular, includes *wood*:

$$\varphi_{\mathrm{dry}} = \frac{1.0438 + 0.1882\dfrac{h}{c} - 0.2509\left(1 + 0.7256\dfrac{h}{c}\right) + 0.0383\dfrac{n}{c}}{1 - 0.3035\dfrac{o}{c}} \tag{C.4}$$

This expression is estimated to be accurate to within $\pm 1\%$.

As (C.3) and (C.4) have been obtained from data applicable to dry substances, it is necessary when calculating ε^0 for a moist fuel using (C.2) to use $(\mathrm{NCV})^0$ of that dry part of the fuel which corresponds to a unit (unit mass or mole) of the moist fuel. This value is given by the sum $[(\mathrm{NCV})^0 + wh_{\mathrm{fg}}^0]$ where w is mass fraction of moisture in the fuel and h_{fg}^0 is the enthalpy of evaporation of H_2O at the standard temperature T^0. Because of lack of sufficient data, the effect of sulphur has not been taken into account in (C.3) and (C.4). It was decided, therefore, to neglect the effect of the energy of the chemical bonds of sulphur and to treat it as a free element, introducing an appropriate correction for its effect on ε^0 of the fuel. Also, the exergy of the moisture and the mineral matter (ash) contained in the fuel was neglected. This leads to the expression for the chemical exergy of solid industrial fossil fuels:

$$\varepsilon^0 = [(\mathrm{NCV})^0 + wh_{\mathrm{fg}}]\varphi_{\mathrm{dry}} + [\varepsilon_s^0 - (\mathrm{NCV})_s^0]s \tag{C.5}$$

where s denotes the mass fraction of sulphur in the fuel. Where s is used as a subscript it indicates sulphur quantities. For rhombic sulphur we have $\varepsilon^0 = 18\,676$ kJ/kg and $(\mathrm{NCV})^0 = 9\,259$ kJ/kg. The difference in the two quantities for use in (C.5) is $[\varepsilon_s^0 - (\mathrm{NCV})_s^0] = 9\,417$ kJ/kg. Also for water substance at $T^0 = 298.15$ K, $h_{\mathrm{fg}} = 2\,442$ kJ/kg. Hence (C.5) can be written for convenience in the form:

$$\varepsilon^0/[\mathrm{kJ/kg}] = [(\mathrm{NCV})^0/[\mathrm{kJ/kg}] + 2\,442w]\varphi_{\mathrm{dry}} + 9\,417s \tag{C.5a}$$

Liquid fuels

For liquid fuels the effect of sulphur was included in the correlation giving the expression:

$$\varphi = 1.0401 + 0.1728\frac{h}{c} + 0.0432\frac{o}{c} + 0.2169\frac{s}{c}\left(1 - 2.0628\frac{h}{c}\right) \tag{C.6}$$

The accuracy of this expression is estimated to be $\pm 0.38\%$.

Gaseous fuels

The chemical exergy of a fuel consisting of a known mixture of gaseous components can be calculated from (C.1). However, as the composition of common gaseous fuels varies within relatively narrow limits, once the value of φ has been calculated for a fuel of a typical composition it can be used with reasonable accuracy in other cases. Table C.1 gives typical values for three common gaseous industrial fuels as well as for a number of other combustible substances.

As will be seen from the table, φ can have values smaller or greater than unity, and for most industrial fuels, except for peat and wood, this value lies within a few per-cent of unity.

Table C.1 Typical values of φ for some industrial fuels and other combustible substances

Fuel	$\varphi = \varepsilon^0/(\text{NCV})^0$
Coke	1.05
Different types of coal	1.06–1.10
Peat	1.16
Wood	1.15–1.30
Different fuel oils and petrol	1.04–1.08
Natural gas	1.04 ± 0.5%
Coal gas	1.00 ± 1%
Blast furnace gas	0.98 ± 1%
Hydrogen	0.985
Carbon monoxide	0.973
Sulphur (rhombic)	2.017

Example C.1

Calculate the standard chemical exergy and the ratio φ of anthracite with the following composition by mass: $c=0.782$, $h=0.024$, $n=0.009$, $s=0.010$, $o=0.015$, $w=0.080$, ash $=0.080$.

The net calorific value of the moist fuel is $(\text{NCV})^0 = 28\,940$ kJ/kg.

Solution: From (C.3):

$$\varphi_{\text{dry}} = 1.0437 + \frac{0.1882 \times 0.024 + 0.0610 \times 0.015 + 0.0404 \times 0.009}{0.782}$$

$$= 1.0511$$

Using (C.5a):

$$\varepsilon^0 = [28\,940 + 0.080 \times 2\,442]1.0511 + 9\,417 \times 0.010$$

$$= 30\,718 \text{ kJ/kg}$$

The ratio $\varphi = \varepsilon^0/(\text{NCV})^0 = \dfrac{30\,718}{28\,940} = \underline{1.061}$

Example C.2

Calculate the standard chemical exergy of a gas oil of the following chemical composition by mass: $c=0.861$, $h=0.132$, $s=0.007$, $o=0$.

The net calorific value of the fuel is $(\text{NCV})^0 = 42\,800$ kJ/kg.

Solution: From (C.6):

$$\varphi = 1.0401 + 0.1728\frac{0.132}{0.861} + 0.2169\frac{0.007}{0.861}\left(1 - 2.0628\frac{0.132}{0.861}\right)$$

$$= \underline{1.068\,36}$$

As this fuel contains no moisture and the effect of sulphur is included in the correlation:

$$\varepsilon^0 = \varphi(\text{NCV})^0$$

$$= 1.068\,36 \times 42\,800$$

$$= \underline{45\,726 \text{ kJ/kg}}$$

Appendix D Mean heat capacity and exergy capacity of ideal gases

To simplify calculation of changes in the physical components of enthalpy, entropy and exergy of ideal gases at constant pressure, the mean molar isobaric values of heat and exergy capacities have been listed in Tables D.1, D.2 and D.3. These mean values have been evaluated in each case for a temperature interval lying between the standard temperature, T^0, and the gas temperature, T. They are defined as:

Mean molar isobaric heat capacity for evaluating enthalpy changes

$$\tilde{c}_P^h = \left[\frac{\tilde{h} - \tilde{h}^0}{T - T^0}\right]_P = \frac{1}{T - T^0}\int_{T^0}^{T} \tilde{c}_P \, dT \qquad (D.1)$$

Mean molar isobaric capacity for evaluating entropy changes

$$\tilde{c}_P^s = \left[\frac{\tilde{s} - \tilde{s}^0}{\ln(T/T^0)}\right]_P = \frac{1}{\ln(T/T^0)}\int_{T^0}^{T} \frac{\tilde{c}_P \, dT}{T} \qquad (D.2)$$

Mean molar isobaric exergy capacity for evaluating changes in physical exergy

$$\tilde{c}_P^\varepsilon = \frac{\tilde{\varepsilon}^{\Delta T}}{T - T^0} = \frac{1}{T - T^0}\left[\int_{T^0}^{T} \tilde{c}_P \, dT - T^0 \int_{T^0}^{T} \frac{\tilde{c}_P \, dT}{T}\right] \qquad (D.3)$$

The tabulated values have been calculated from data on $\tilde{c}_P$ given in polynomial form in Refs [A.11] and [5.3]. A fuller discussion of these quantities and examples of application are given in Section 4.5.

Table D.1 Mean isobaric heat capacity *for enthalpy* of some gases, $\tilde{c}_P^{\mathrm{h}}$. $T^0 = 298.15$ K, $P^0 = 0.101\,325$ MPa, $\theta = T - 273.15$ K.

T	θ	$\tilde{c}_P^{\mathrm{h}}$, kJ/kmol K									
K		N_2	O_2	Air	H_2	CO	CO_2	CH_4	SO_2	SO_3	H_2O
273.15	0.00	29.04	25.31	29.01	28.69	29.07	36.29	35.02	39.42	49.67	32.77
298.15	25.00	29.11	26.07	29.07	28.73	29.13	37.10	35.71	39.89	50.77	32.92
323.15	50.00	29.18	26.83	29.12	28.77	29.19	37.90	36.39	40.35	51.86	33.06
349.15	75.00	29.25	27.43	29.18	28.81	29.26	38.61	37.08	40.81	52.91	33.21
373.15	100.00	29.32	27.95	29.23	28.85	29.32	39.27	37.76	41.24	53.92	33.35
398.15	125.00	29.39	28.41	29.29	28.90	29.38	39.88	38.45	41.67	54.90	33.50
423.15	150.00	29.46	28.81	29.35	28.94	29.44	40.46	39.14	42.08	55.85	33.64
448.15	175.00	29.53	29.17	29.41	28.98	29.51	41.00	39.83	42.49	56.77	33.79
473.15	200.00	29.59	29.50	29.47	29.02	29.57	41.52	40.52	42.88	57.65	33.93
498.15	225.00	29.66	29.79	29.53	29.07	29.63	42.00	41.20	43.26	58.51	34.07
523.15	250.00	29.73	30.06	29.60	29.11	29.70	42.46	41.89	43.63	59.34	34.22
548.15	275.00	29.80	30.30	29.66	29.15	29.76	42.89	42.58	43.98	60.15	34.36
573.15	300.00	29.86	30.52	29.73	29.19	29.82	43.31	43.26	44.33	60.92	34.50
598.15	325.00	29.93	30.73	29.79	29.24	29.88	43.70	43.94	44.67	61.67	34.64
623.15	350.00	30.00	30.92	29.86	29.28	29.95	44.08	44.62	44.99	62.39	34.78
648.15	375.00	30.06	31.09	29.92	29.32	30.01	44.44	45.30	45.31	63.09	34.92
673.15	400.00	30.13	31.26	29.99	29.36	30.07	44.79	45.98	45.62	63.77	35.06
698.15	425.00	30.19	31.41	30.06	29.41	30.14	45.12	46.65	45.91	64.42	35.19
723.15	450.00	30.26	31.56	30.13	29.45	30.20	45.44	47.32	46.20	65.05	35.33
748.15	475.00	30.32	31.69	30.20	29.49	30.26	45.75	47.99	46.48	65.66	35.47
773.15	500.00	30.38	31.82	30.27	29.53	30.32	46.04	48.65	46.75	66.25	35.60
798.15	525.00	30.45	31.94	30.34	29.58	30.39	46.33	49.31	47.01	66.81	35.74
823.15	550.00	30.51	32.05	30.41	29.62	30.45	46.61	49.97	47.26	67.36	35.87
848.15	575.00	30.57	32.16	30.48	29.66	30.51	46.87	50.62	47.50	67.89	36.01
873.15	600.00	30.63	32.26	30.55	29.70	30.58	47.13	51.26	47.74	68.40	36.14
898.15	625.00	30.69	32.36	30.62	29.74	30.64	47.38	51.91	47.97	68.89	36.28
923.15	650.00	30.75	32.45	30.69	29.79	30.70	47.62	52.54	48.19	69.37	36.41
948.15	675.00	30.82	32.54	30.76	29.83	30.76	47.85	53.18	48.40	69.83	36.54

Table D.1—*continued*

T	θ	$\tilde{c}_P^{h}$, kJ/kmol K									
K		N_2	O_2	Air	H_2	CO	CO_2	CH_4	SO_2	SO_3	H_2O
973.15	700.00	30.88	32.63	30.83	29.87	30.83	48.08	53.80	48.60	70.28	36.67
998.15	725.00	30.94	32.71	30.90	29.91	30.89	48.30	54.42	48.80	70.71	36.80
1 023.15	750.00	31.00	32.79	30.97	29.96	30.95	48.51	55.04	48.99	71.13	36.93
1 048.15	775.00	31.05	32.86	31.05	30.00	31.01	48.72	55.65	49.17	71.54	37.06
1 073.15	800.00	31.11	32.93	31.12	30.04	31.08	48.92	56.25	49.35	71.93	37.19
1 098.15	825.00	31.17	33.00	31.19	30.08	31.14	49.12	56.84	49.52	72.32	37.32
1 123.15	850.00	31.23	33.07	31.26	30.13	31.20	49.31	57.43	49.68	72.69	37.45
1 148.15	875.00	31.29	33.14	31.33	30.17	31.27	49.50	58.01	49.84	73.06	37.57
1 173.15	900.00	31.34	33.20	31.40	30.21	31.33	49.68	58.58	49.99	73.41	37.70
1 198.15	925.00	31.40	33.26	31.47	30.25	31.39	49.86	59.15	50.14	73.76	27.83
1 223.15	950.00	31.46	33.32	31.54	30.30	31.45	50.03	59.71	50.28	74.10	37.95
1 248.15	975.00	31.51	33.37	31.61	30.34	31.52	50.20	60.25	50.41	74.43	38.08
1 273.15	1 000.00	31.57	33.43	31.68	30.38	31.58	50.37	60.79	50.55	74.76	38.20
1 298.15	1 025.00	31.62	33.48	31.74	30.42	31.64	50.53	61.33	50.67	75.09	38.33
1 323.15	1 050.00	31.68	33.53	31.81	30.47	31.71	50.68	61.85	50.79		38.45
1 348.15	1 075.00	31.73	33.58	31.88	30.51	31.77	50.84	62.36	50.91		38.57
1 373.15	1 100.00	31.79	33.63	31.94	30.55	31.83	50.99	62.86	51.02		38.69
1 398.15	1 125.00	31.84	33.68	32.01	30.59	31.89	51.14	63.36	51.13		38.81
1 423.15	1 150.00	31.89	33.73	32.07	30.63	31.96	51.28	63.84	51.24		38.93
1 448.15	1 175.00	31.95	33.77	32.14	30.68	32.02	51.42	64.31	51.34		39.05
1 473.15	1 200.00	32.00	33.82	32.20	30.72	32.08	51.56	64.77	51.44		39.17
1 498.15	1 225.00	32.05	33.86	32.26	30.76	32.15	51.70	65.22	51.53		39.29
1 523.15	1 250.00	32.10	33.90	32.32	30.80	32.21	51.83		51.62		39.41
1 548.15	1 275.00	32.15	33.94	32.38	30.85	32.27	51.96		51.71		39.53
1 573.15	1 300.00	32.20	33.99	32.44	30.89	32.33	52.09		51.80		39.64
1 598.15	1 325.00	32.25	34.02	32.50	30.93	32.40	52.21		51.88		39.76
1 623.15	1 350.00	32.30	34.06	32.56	30.97	32.46	52.33		51.96		39.88
1 648.15	1 375.00	32.35	34.10	32.61	31.02	32.52	52.45		52.04		39.99

1 673.15	1 400.00	32.40	34.14	32.67	31.06	32.58	52.57	52.12	40.11
1 698.15	1 425.00	32.45	34.18	32.72	31.10	32.65	52.69	52.20	40.22
1 723.15	1 450.00	32.50	34.21	32.77	31.14	32.71	52.80	52.28	40.33
1 748.15	1 475.00	32.55	34.25	32.82	31.19	32.77	52.92	52.35	40.45
1 773.15	1 500.00	32.60	34.28	32.87	31.23	32.84	53.03	52.42	40.56
1 798.15	1 525.00	32.64	34.32	32.92	31.27	32.90	53.13	52.49	40.67
1 823.15	1 550.00	32.69	34.35		31.31	32.96	53.24	52.57	40.78
1 848.15	1 575.00	32.74	34.38		31.35	33.02	53.34	52.64	40.89
1 873.15	1 600.00	32.78	34.42		31.40	33.09	53.45	52.71	41.00
1 898.15	1 625.00	32.83	34.45		31.44	33.15	53.55	52.78	41.11
1 923.15	1 650.00	32.87	34.48		31.48	33.21	53.65	52.85	41.22
1 948.15	1 675.00	32.92	34.51		31.52	33.28	53.75	52.92	41.33
1 973.15	1 700.00	32.96	34.54		31.57	33.34	53.84	53.00	41.43
1 998.15	1 725.00	33.01	34.57		31.61	33.40	53.94	53.07	41.54
2 023.15	1 750.00	33.05	34.60		31.65	33.46	54.03	53.15	41.65
2 048.15	1 775.00	33.09	34.63		31.69	33.53	54.12	53.22	41.75
2 073.15	1 800.00	33.14	34.66		31.74	33.59	54.21	53.30	41.86
2 098.15	1 825.00	33.18	34.69		31.78	33.65	54.30	53.38	41.96
2 123.15	1 850.00	33.22	34.72		31.82	33.72	54.39	53.46	42.06
2 148.15	1 875.00	33.26	34.74		31.86	33.78	54.48	53.54	42.17
2 173.15	1 900.00	33.30	34.77		31.91	33.84	54.56	53.63	42.27
2 198.15	1 925.00	33.34	34.80		31.95	33.90	54.65	53.71	42.37
2 223.15	1 950.00	33.39	34.83		31.99	33.97	54.73	53.80	42.47
2 248.15	1 975.00	33.43	34.85		32.03	34.03	54.81	53.90	42.57
2 273.15	2 000.00	33.46	34.88		32.08	34.09	54.89	53.99	42.67
2 298.15	2 025.00	33.50	34.91		32.12	34.15	54.97	54.09	42.77

Table D.2 Mean, isobaric heat capacity *for entropy* of some gases, $\tilde{c}_P^s$. $T^0 = 298.15$ K, $P^0 = 0.101\ 325$ MPa, $\theta = T - 273.15$ K.

T	θ	$\tilde{c}_P^s$, kJ/kmol K									
K		N_2	O_2	Air	H_2	CO	CO_2	CH_4	SO_2	SO_3	H_2O
273.15	0.00	29.04	25.30	29.01	28.68	29.07	36.28	35.01	39.41	49.65	32.81
298.15	25.00	29.11	26.06	29.06	28.73	29.13	37.08	35.70	39.88	50.75	32.97
323.15	50.00	29.18	26.82	29.12	28.77	29.19	37.88	36.38	40.35	51.85	33.12
348.15	75.00	29.25	27.40	29.17	28.81	29.25	38.57	37.04	40.78	52.85	33.26
373.15	100.00	29.31	27.89	29.23	28.85	29.31	39.19	37.69	41.19	53.80	33.41
398.15	125.00	29.38	28.31	29.28	28.89	29.37	39.76	38.32	41.59	54.71	33.54
423.15	150.00	29.44	28.68	29.33	28.93	29.43	40.28	38.94	41.96	55.56	33.68
448.15	175.00	29.50	29.00	29.39	28.96	29.48	40.77	39.55	42.32	56.38	33.81
473.15	200.00	29.56	29.29	29.44	29.00	29.54	41.22	40.15	42.66	57.15	33.94
498.15	225.00	29.62	29.54	29.49	29.04	29.59	41.63	40.74	42.98	57.89	34.07
523.15	250.00	29.67	29.77	29.55	29.07	29.64	42.02	41.31	43.30	58.60	34.20
548.15	275.00	29.73	29.98	29.60	29.11	29.70	42.39	41.88	43.60	59.27	34.32
573.15	300.00	29.78	30.17	29.65	29.14	29.75	42.74	42.44	43.88	59.91	34.44
598.15	325.00	29.84	30.34	29.70	29.18	29.80	43.06	42.99	44.16	60.52	34.56
623.15	350.00	29.89	30.50	29.76	29.21	29.85	43.37	43.54	44.42	61.11	34.68
648.15	375.00	29.94	30.65	29.81	29.24	29.90	43.66	44.07	44.68	61.67	34.80
673.15	400.00	29.99	30.78	29.86	29.28	29.95	43.94	44.60	44.92	62.21	34.91
698.15	425.00	30.04	30.91	29.91	29.31	29.99	44.20	45.12	45.15	62.72	35.02
723.15	450.00	30.09	31.03	29.97	29.34	30.04	44.46	45.64	45.38	63.21	35.14
748.15	475.00	30.14	31.14	30.02	29.38	30.09	44.70	46.14	45.59	63.69	35.25
773.15	500.00	30.19	31.24	30.07	29.41	30.14	44.93	46.64	45.80	64.14	35.36
798.15	525.00	30.23	31.34	30.12	29.44	30.18	45.15	47.13	46.00	64.57	35.46
823.15	550.00	30.28	31.43	30.17	29.47	30.23	45.36	47.61	46.19	64.99	35.57
848.15	575.00	30.33	31.51	30.22	39.50	30.28	45.56	48.09	46.37	65.39	35.68
873.5	600.00	30.37	31.59	30.27	29.53	30.32	45.76	43.56	46.55	65.78	35.78
898.15	625.00	30.42	31.67	30.32	29.56	30.37	45.95	49.03	46.72	66.15	35.88
923.15	650.00	30.46	31.74	30.37	29.59	30.41	46.13	49.48	46.88	66.50	35.99
948.15	675.00	30.50	31.81	30.42	29.62	30.46	46.30	49.93	47.04	66.85	36.09

973.15	700.00	30.55	31.88	30.47	29.65	30.50	46.47	50.38	47.19	67.18	36.19
998.15	725.00	30.59	31.94	30.52	29.68	30.54	46.63	50.81	47.34	67.50	36.29
1 023.15	750.00	30.63	32.00	30.57	29.71	30.59	46.79	51.24	47.48	67.81	36.39
1 048.13	775.00	30.67	32.06	30.62	29.74	30.63	46.95	51.67	47.61	68.10	36.49
1 073.15	600.00	30.71	32.12	30.67	29.77	30.67	47.09	52.08	47.74	68.39	36.58
1 098.15	825.00	30.75	32.17	30.72	29.80	30.72	47.24	52.49	47.86	68.67	36.68
1 123.15	850.00	30.79	32.22	30.77	29.83	30.76	47.38	52.90	47.98	68.94	36.77
1 148.15	875.00	30.83	32.27	30.81	29.85	30.80	47.51	53.29	48.10	69.21	36.87
1 173.15	900.00	30.87	32.32	30.86	29.88	30.84	47.64	53.68	48.21	69.46	36.96
1 198.15	925.00	30.91	32.37	30.91	29.91	30.88	47.77	54.06	48.32	69.71	37.06
1 223.15	950.00	30.94	32.41	30.95	29.94	30.93	47.89	54.44	48.42	69.96	37.15
1 248.15	975.00	30.98	32.46	31.00	29.97	30.97	48.02	54.81	48.52	70.19	37.24
1 273.15	1 000.00	31.02	32.50	31.04	29.99	31.01	48.13	55.17	48.62	70.43	37.33
1 298.15	1 025.00	31.06	32.54	31.09	30.02	31.05	48.25	55.53	48.71	70.66	37.42
1 323.15	1 050.00	31.09	32.58	31.13	30.05	31.09	48.36	55.87	48.80		37.51
1 348.15	1 075.00	31.13	32.62	31.17	30.08	31.13	48.47	56.22	48.88		37.60
1 375.15	1 100.00	31.16	32.65	31.22	30.10	31.17	48.58	56.55	48.97		37.69
1 398.15	1 125.00	31.20	32.69	31.26	30.13	31.21	48.68	56.88	49.04		37.78
1 423.15	1 150.00	31.23	32.72	31.30	30.16	31.25	48.78	57.20	49.12		37.86
1 448.15	1 175.00	31.27	32.76	31.34	30.18	31.29	48.88	57.51	49.20		37.95
1 473.15	1 200.00	31.30	32.79	31.38	30.21	31.33	48.98	57.81	49.27		38.04
1 498.15	1 225.00	31.33	32.82	31.42	30.24	31.37	49.07	58.11	49.34		38.12
1 523.15	1 250.00	31.37	32.85	31.46	30.26	31.41	49.16		49.41		38.21
1 548.15	1 275.00	31.40	32.89	31.50	30.29	31.45	49.25		49.47		38.29
1 573.15	1 300.00	31.43	32.92	31.53	30.32	31.48	49.34		49.54		38.38
1 598.15	1 325.00	31.46	32.95	31.57	30.34	31.52	49.43		49.60		38.46
1 623.15	1 350.00	31.49	32.97	31.61	30.37	31.56	49.52		49.66		38.54
1 648.15	1 375.00	31.52	33.00	31.64	30.39	31.60	49.60		49.72		38.63
1 673.15	1 400.00	31.56	33.03	31.68	30.42	31.64	49.68		49.78		38.71
1 698.15	1 425.00	31.59	33.06	31.71	30.45	31.68	49.76		49.83		38.79
1 723.15	1 450.00	31.62	33.08	31.74	30.47	31.71	49.84		49.89		38.87
1 748.15	1 475.00	31.65	33.11	31.78	30.50	31.75	49.92		49.94		38.95
1 773.15	1 500.00	31.68	33.13	31.81	30.52	31.79	49.99		50.00		39.03
1 798.15	1 525.00	31.70	33.16	31.84	30.55	31.83	50.07		50.05		39.11
1 823.15	1 550.00	31.73	33.18		30.57	31.86	50.14		50.10		39.19
1 848.15	1 575.00	31.76	33.21		30.60	31.90	50.21		50.15		39.27

Table D.2—*continued*

T	θ	$\tilde{c}_P^s$, kJ/kmol K									
K		N_2	O_2	Air	H_2	CO	CO_2	CH_4	SO_2	SO_3	H_2O
1 873.15	1 600.00	31.79	33.23		30.62	31.94	50.28		50.20		39.34
1 898.15	1 625.00	31.82	33.25		30.65	31.98	50.35		50.26		39.42
1 923.15	1 650.00	31.85	33.28		30.67	32.01	50.42		50.31		39.50
1 948.15	1 675.00	31.87	33.30		30.70	32.05	50.48		50.36		39.58
1 973.15	1 700.00	31.90	33.32		30.72	32.09	50.55		50.41		39.65
1 998.15	1 725.00	31.93	33.34		30.75	32.12	50.61		50.46		39.73
2 023.15	1 750.00	31.95	33.36		30.77	32.16	50.68		50.51		39.80
2 048.15	1 775.00	31.98	33.39		30.80	32.20	50.74		50.56		39.88
2 073.15	1 800.00	32.01	33.41		30.82	32.23	50.80		50.51		39.95
2 098.15	1 825.00	32.03	33.43		30.84	32.27	50.86		50.66		40.03
2 123.15	1 850.00	32.06	33.45		30.87	32.30	50.92		50.71		40.10
2 148.15	1 875.00	32.08	33.47		30.89	32.34	50.98		50.76		40.18
2 173.15	1 900.00	32.11	33.49		30.92	32.38	51.04		50.82		40.25
2 198.15	1 925.00	32.13	33.51		30.94	32.41	51.09		50.38		40.32
2 223.15	1 950.00	32.16	33.52		30.97	32.45	51.15		50.93		40.40
2 248.15	1 975.00	32.18	33.54		30.99	32.48	51.21		50.98		40.47
2 273.15	2 000.00	32.20	33.56		31.01	32.52	51.26		51.04		40.54
2 298.15	2 025.00	32.23	33.58		31.04	32.55	51.31		51.10		40.61

Table D.3 Mean isobaric exergy capacity of some gases, $\tilde{c}_P^{\varepsilon}$. $T^0 = 298.15$ K, $P^0 = 0.101\,325$ MPa, $\theta = T - 273.15$ K.

T (K)	θ (K)	N_2	O_2	Air	H_2	CO	CO_2	CH_4	SO_2	SO_3	H_2O
		$\tilde{c}_P^{\varepsilon}$, kJ/kmol K									
273.15	0.00	−1.29	−1.11	−1.29	−1.27	−1.29	−1.60	−1.55	−1.74	−2.19	−1.51
298.15	25.00	0.00	0.00	0.00	0.00	0.00	0.00	0.00	0.00	0.00	0.00
323.15	50.00	1.16	1.07	1.16	1.14	1.16	1.52	1.45	1.61	2.07	1.26
348.15	75.00	2.21	2.10	2.21	2.18	2.21	2.95	2.83	3.10	4.05	2.45
373.15	100.00	3.17	3.07	3.16	3.12	3.17	4.31	4.15	4.50	5.93	3.55
398.15	125.00	4.06	3.99	4.04	3.98	4.06	5.60	5.41	5.81	7.72	4.57
423.15	150.00	4.87	4.86	4.85	4.78	4.87	6.82	6.62	7.04	9.45	5.52
448.15	175.00	5.63	5.68	5.61	5.52	5.63	7.98	7.79	8.21	11.10	6.40
473.15	200.00	6.34	6.45	6.31	6.21	6.33	9.09	8.93	9.31	12.69	7.22
498.15	225.00	7.00	7.18	6.97	6.85	6.99	10.14	10.03	10.36	14.21	8.00
523.15	250.00	7.62	7.87	7.58	7.45	7.61	11.15	11.11	11.37	15.68	8.74
548.15	275.00	8.21	8.52	8.17	8.01	8.19	12.11	12.16	12.32	17.10	9.43
573.15	300.00	8.76	9.14	8.72	8.54	8.74	13.03	13.19	13.24	18.47	10.09
598.15	325.00	9.28	9.73	9.24	9.05	9.27	13.91	14.19	14.11	19.79	10.72
623.15	350.00	9.78	10.29	9.73	9.52	9.76	14.75	15.18	14.95	21.07	11.32
648.15	375.00	10.26	10.82	10.21	9.98	10.23	15.56	16.15	15.76	22.30	11.90
673.15	400.00	10.71	11.33	10.66	10.41	10.68	16.34	17.10	16.53	23.49	12.45
698.15	425.00	11.14	11.81	11.09	10.82	11.11	17.09	18.03	17.28	24.64	12.98
723.15	450.00	11.55	12.27	11.50	11.21	11.52	17.81	18.95	18.00	25.76	13.49
748.15	475.00	11.95	12.71	11.90	11.58	11.92	18.50	19.86	18.69	26.84	13.98
773.15	500.00	12.33	13.13	12.28	11.94	12.30	19.17	20.75	19.36	27.88	14.46
798.15	525.00	12.69	13.54	12.65	12.29	12.66	19.82	21.64	20.00	28.90	14.92
823.15	550.00	13.04	13.93	13.00	12.62	13.02	20.45	22.50	20.62	29.88	15.36
848.15	575.00	13.38	14.30	13.35	12.94	13.35	21.05	23.36	21.22	30.83	15.79
873.15	600.00	13.71	14.66	13.68	13.25	13.68	21.63	24.21	21.80	31.75	16.21
898.15	625.00	14.03	15.01	14.00	13.55	14.00	22.20	25.04	22.37	32.65	16.61
923.15	650.00	14.33	15.34	14.31	13.83	14.30	22.75	25.86	22.91	33.52	17.01
948.15	675.00	14.63	15.66	14.61	14.11	14.60	23.28	26.68	23.43	34.36	17.39

Table D.3—*continued*

T	*θ*	$\tilde{c}_P^{\varepsilon}$, kJ/kmol K									
K		N_2	O_2	Air	H_2	CO	CO_2	CH_4	SO_2	SO_3	H_2O
973.15	700.00	14.92	15.97	14.91	14.38	14.89	23.80	27.48	23.94	35.13	17.76
998.15	725.00	15.19	16.27	15.19	14.64	15.17	24.30	28.27	24.44	35.97	18.13
1 023.15	750.00	15.46	16.56	15.47	14.89	15.44	24.79	29.05	24.91	36.75	18.48
1 048.15	775.00	15.73	16.84	15.74	15.14	15.71	25.26	29.82	25.38	37.50	18.83
1 073.15	800.00	15.98	17.11	16.00	15.37	15.96	25.72	30.59	25.82	38.24	19.17
1 098.15	825.00	16.23	17.37	16.26	15.61	16.22	26.17	31.34	26.26	38.95	19.50
1 123.15	850.00	16.47	17.63	16.51	15.83	16.46	26.60	32.08	26.68	39.65	19.82
1 148.15	875.00	16.71	17.87	16.76	16.05	16.70	27.03	32.81	27.09	40.33	20.14
1 173.15	900.00	16.94	18.11	16.99	16.26	16.93	27.44	33.53	27.49	40.99	20.45
1 198.15	925.00	17.16	18.34	17.23	16.47	17.16	27.85	34.24	27.87	41.64	20.75
1 223.15	950.00	17.38	18.57	17.46	16.67	17.38	28.24	34.94	28.25	42.27	21.05
1 248.15	975.00	17.59	18.79	17.68	16.87	17.60	28.62	35.63	28.61	42.89	21.34
1 273.15	1 000.00	17.80	19.00	17.90	17.07	17.82	29.00	36.30	28.96	43.50	21.63
1 298.15	1 025.00	18.00	19.21	18.11	17.26	18.02	29.37	36.97	29.31	44.10	21.91
1 323.15	1 050.00	18.20	19.41	18.32	17.44	18.23	29.72	37.63	29.64		22.19
1 348.15	1 075.00	18.40	19.61	18.52	17.62	18.43	30.07	38.27	29.97		22.46
1 373.15	1 100.00	18.59	19.80	18.72	17.80	18.63	30.41	38.91	30.28		22.73
1 398.15	1 125.00	18.77	19.99	18.92	17.97	18.82	30.75	39.53	30.59		22.99
1 423.15	1 150.00	18.96	20.17	19.11	18.14	19.01	31.07	40.15	30.89		23.25
1 448.15	1 175.00	19.14	20.35	19.30	18.31	19.20	31.39	40.75	31.18		23.50
1 473.15	1 200.00	19.31	20.52	19.48	18.47	19.38	31.71	41.34	31.46		23.75
1 498.15	1 225.00	19.48	20.69	19.66	18.63	19.56	32.01	41.91	31.74		24.00
1 523.15	1 250.00	19.65	20.86	19.84	18.79	19.74	32.31		32.01		24.24
1 548.15	1 275.00	19.82	21.02	20.01	18.95	19.92	32.61		32.27		24.48
1 573.15	1 300.00	19.98	21.18	20.18	19.10	20.09	32.89		32.53		24.72
1 598.15	1 325.00	20.14	21.34	20.34	19.25	20.26	33.18		32.78		24.95
1 632.15	1 350.00	20.30	21.49	20.51	19.39	20.42	33.45		33.03		25.18
1 648.15	1 375.00	20.45	21.64	20.66	19.54	20.59	33.73		33.27		25.41

1 673.15	1 400.00	20.60	21.79	20.82	19.68	20.75	33.99	33.51	25.63
1 698.15	1 425.00	20.75	21.93	20.97	19.82	20.91	34.25	33.74	25.85
1 723.15	1 450.00	20.90	22.07	21.12	19.96	21.07	34.51	33.96	26.07
1 748.15	1 475.00	21.04	22.21	21.27	20.09	21.23	34.76	34.19	26.28
1 773.15	1 500.00	21.18	22.34	21.41	20.23	21.38	35.01	34.40	26.49
1 798.15	1 525.00	21.32	22.47	21.55	20.36	21.53	35.25	34.62	26.70
1 823.15	1 550.00	21.46	22.60		20.49	21.68	35.49	34.83	26.91
1 848.15	1 575.00	21.59	22.73		20.62	21.83	35.72	35.04	27.11
1 873.15	1 600.00	21.72	22.85		20.74	21.98	35.95	35.24	27.31
1 898.15	1 625.00	21.85	22.98		20.87	22.12	36.18	35.45	27.51
1 923.15	1 650.00	21.98	23.10		20.99	22.26	36.40	35.65	27.71
1 948.15	1 675.00	22.11	23.22		21.11	22.41	36.62	35.85	27.90
1 973.15	1 700.00	22.23	23.33		21.23	22.55	36.84	36.04	28.09
1 998.15	1 725.00	22.35	23.45		21.35	22.68	37.05	36.24	28.28
2 023.15	1 750.00	22.47	23.56		21.47	22.82	36.26	36.43	28.47
2 048.15	1 775.00	22.59	23.67		21.58	22.96	37.46	36.62	28.66
2 073.15	1 800.00	22.71	23.78		21.70	23.09	37.66	36.81	28.84
2 098.15	1 825.00	22.83	23.88		21.81	23.22	37.86	37.00	29.02
2 123.15	1 850.00	22.94	23.99		21.92	23.36	38.06	37.20	29.20
2 148.15	1 875.00	23.05	24.09		22.03	23.49	38.25	37.39	29.38
2 173.15	1 900.00	23.16	24.19		22.14	23.62	38.44	37.58	29.56
2 198.15	1 925.00	23.27	24.30		22.25	23.74	38.63	37.77	29.73
2 223.15	1 950.00	23.38	24.39		22.36	23.87	38.81	37.96	29.90
2 248.15	1 975.00	23.48	24.49		22.46	24.00	38.99	38.15	30.07
2 273.15	2 000.00	23.59	24.59		22.57	24.12	39.17	38.34	30.24
2 298.15	2 025.00	23.69	24.68		22.67	24.24	39.35	38.54	30.41

Appendix E Charts of thermodynamic properties

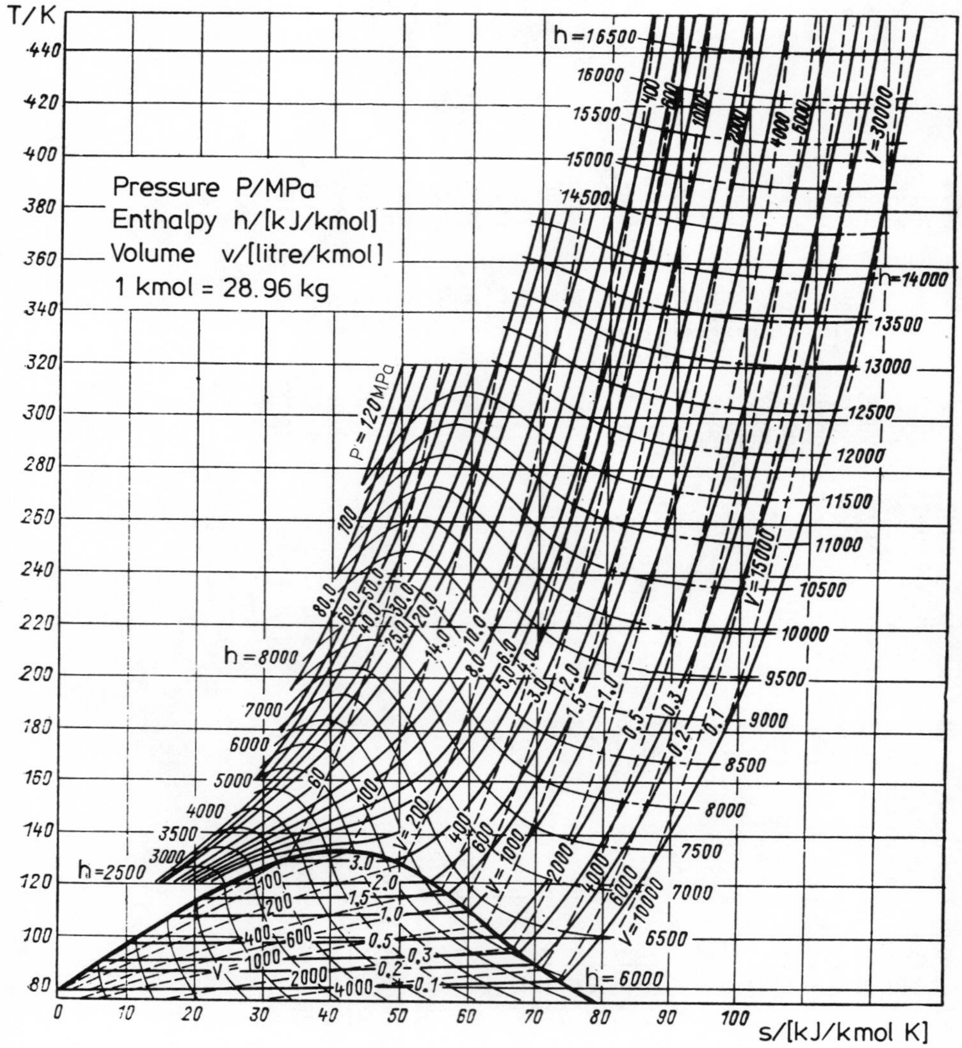

Fig E.1 Temperature–entropy chart for air. (Reproduced from Ref [4.3], by permission.)

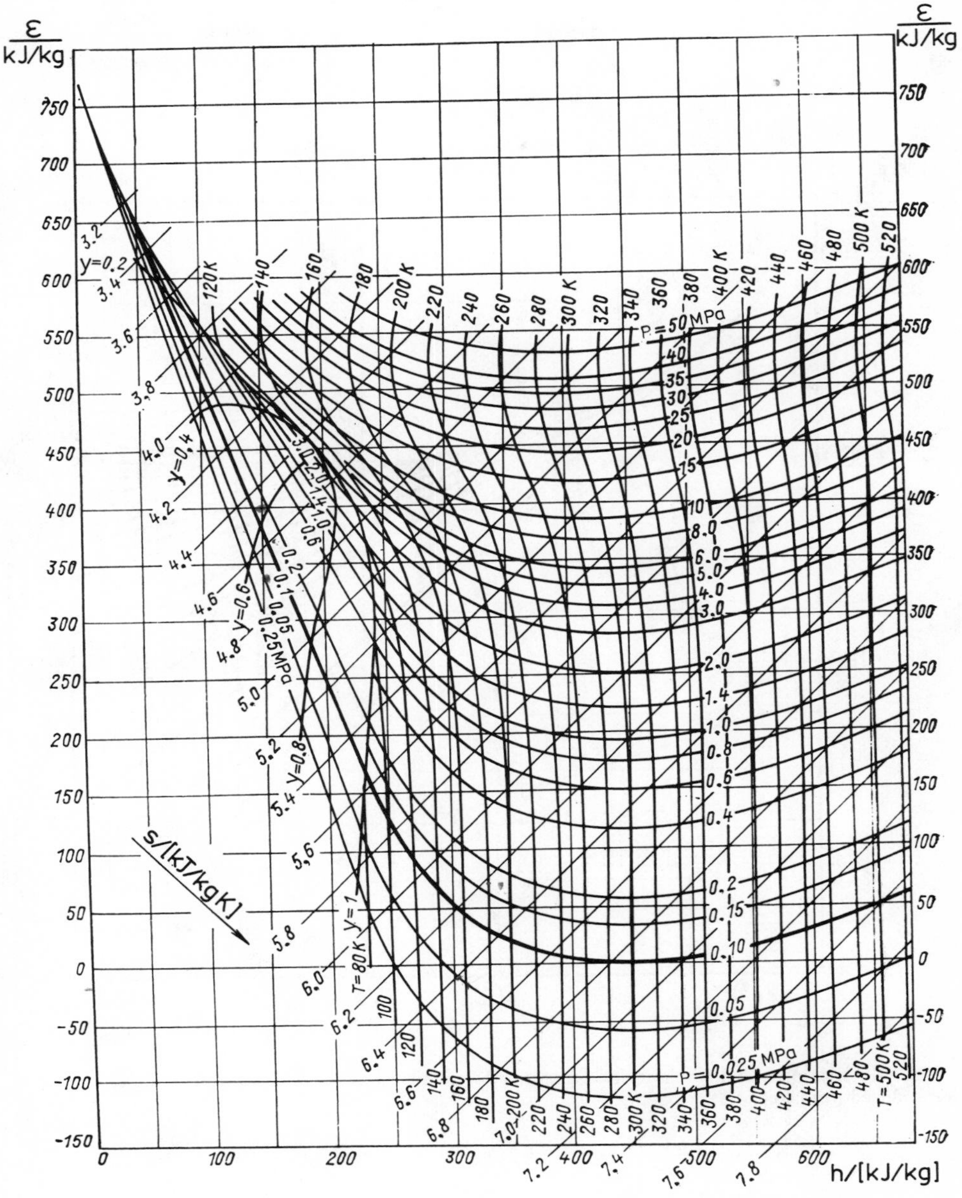

Fig E.2 Exergy–enthalpy chart for air. (Reproduced from Ref [4.3], by permission.)

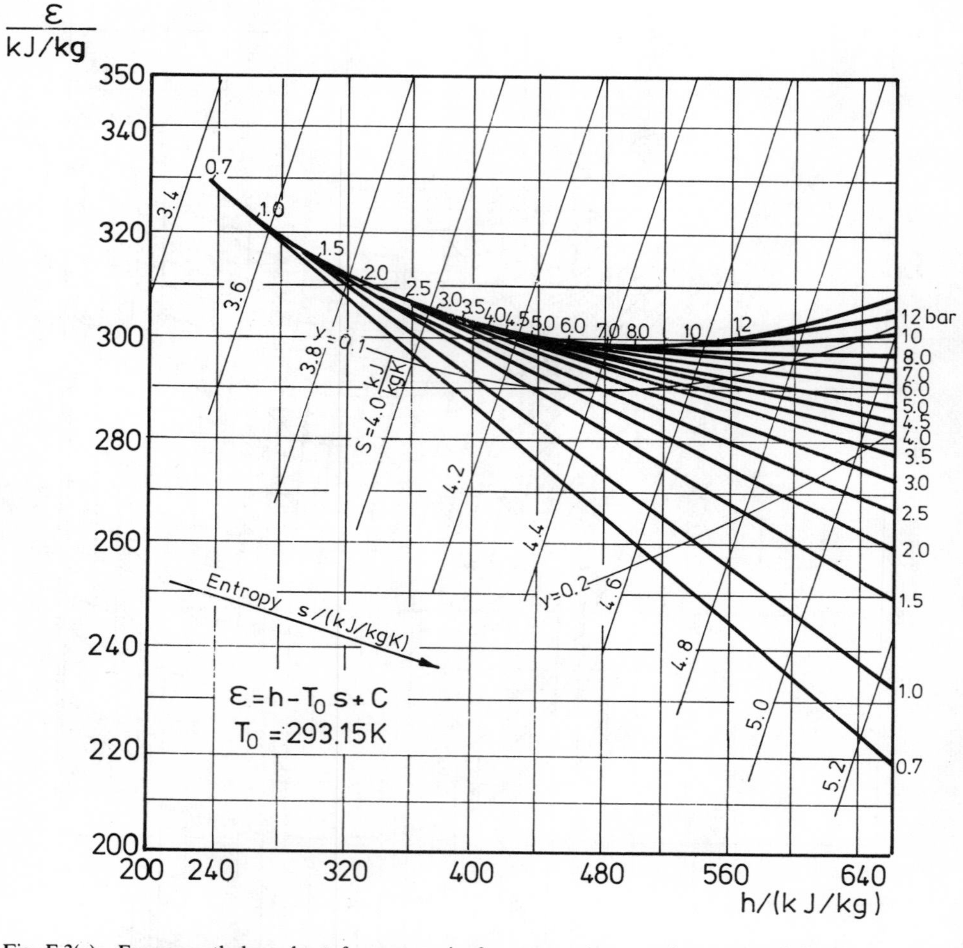

Fig E.3(a) Exergy–enthalpy chart for ammonia for states close to the saturated liquid line $y=0$. (Reproduced from Ref [4.3], by permission.)

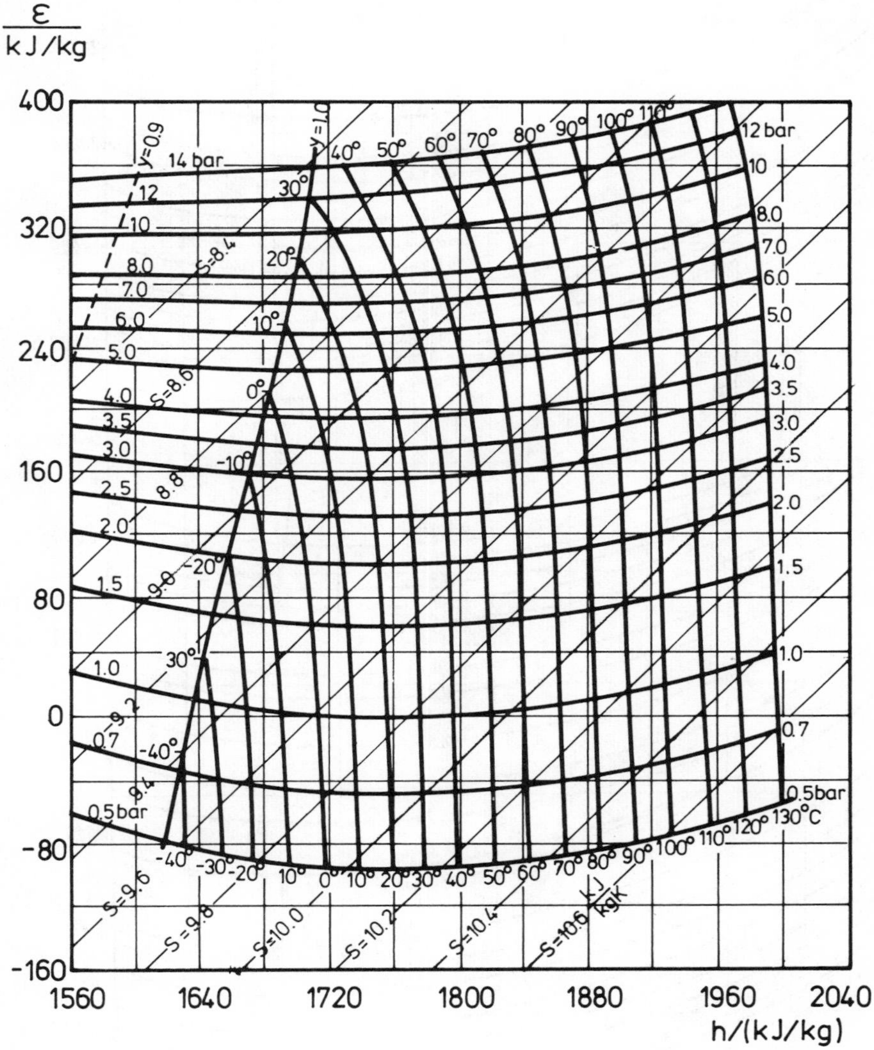

Fig E.3(b) Exergy–enthalpy chart for ammonia for states close to the saturated vapour line, $y = 1$. (Reproduced from Ref [4.3], by permission.)

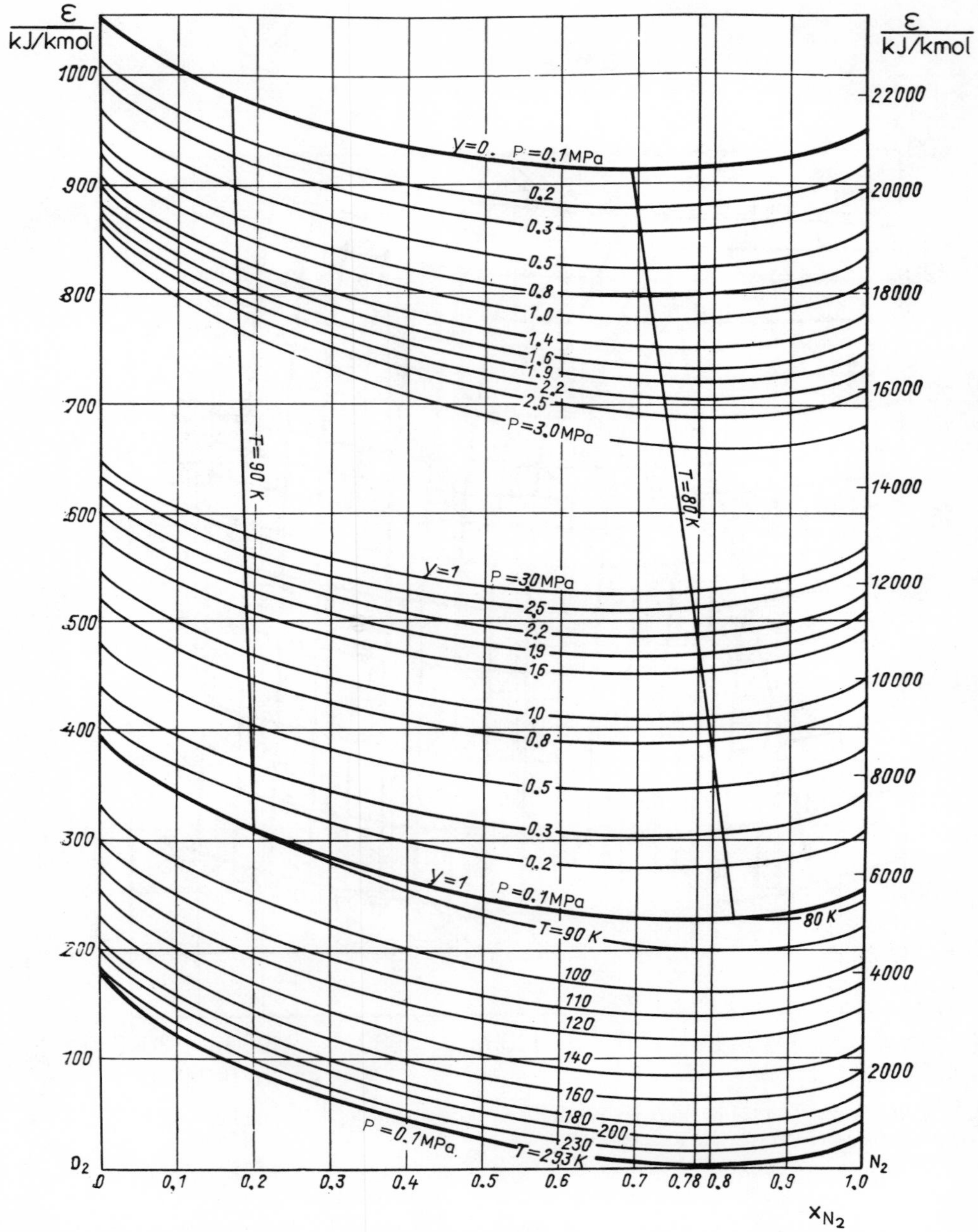

Fig E.4 Exergy–concentration chart for the nitrogen–oxygen mixture, N_2—79 per cent, O_2—21 per cent, by mole. (Reproduced from Ref [4.3], by permission.)

Appendix F Glossary of terms

Avoidable irreversibility rate is the difference between the actual irreversibility rate of a process and its intrinsic irreversibility rate (Section 3.3).

Chemical exergy is equal to the maximum work obtainable when the substance under consideration is brought from the environmental state to the dead state by processes involving heat transfer and exchange of substances only with the environment (Section 2.5).

Coefficient of external bonds (CEB) is the ratio of the change in the exergy flow entering or leaving a sub-system to the corresponding change in the exergy input rate to the plant as a whole, resulting from a change in a selected plant parameter (Eq (6.9)).

Coefficient of structural bonds (CSB) is the ratio of the change in the irreversibility rate of a plant component, to the corresponding change in the irreversibility rate of the plant as a whole, resulting from a change in a selected plant parameter (Eqs (6.1a) and (6.1b)).

Component efficiency defect is the ratio of the component irreversibility rate to the plant exergy input rate (Eq ((3.41)).

Cryoexpander is any expander which has cooling as its primary function (Section 4.1).

Dead state. When a substance is in a state of thermal, mechanical and chemical equilibrium with the conceptual environment it is said to be in the dead state (Section 2.3).

Degradation of energy is the phenomenon of the loss of work potential of a system which takes place during an irreversible process (Section 2.2).

Disordered energy is any form of energy which is stored in a random manner, eg kinetic energy associated with random molecular motion. The transfer of disordered energy is accompanied by changes in the entropies of the interacting systems (Section 2.1).

Efficiency defect is the amount by which the rational efficiency of a plant differs from unity (Eq (3.40)).

Enthalpy of devaluation is the value of enthalpy evaluated for the standard state with reference to selected substances in their dead states (Appendix A).

Entropy production associated with a process is equal to the entropy increase of an isolated system consisting of all systems involved in the process. (Sections 1.9 and 1.10).

Environment as a concept peculiar to the Exergy Method is an idealisation of the real environment which is characterised by a perfect state of equilibrium with no gradients or differences involving pressure, temperature, chemical potential, and kinetic or potential energy. The environment constitutes a natural reference medium with respect to which the maximum work potential (exergy) of different systems is assessed (Section 2.2).

Environmental state is the state of a simple substance defined by the pressure P_0 and temperature T_0 of the environment (Section 2.3).

Exergy, as a general concept, is the maximum work potential of a given form of energy with the environment taken as the reference medium. Exergy is expressed by different thermodynamic functions according to the form of energy under consideration. The unqualified term 'exergy' or 'exergy flow' refers to the exergy of a steady stream of matter (Section 2.3).

Exergy function is a thermodynamic function defined by $(H - T_0S)$ (Section 2.4).

Exergetic temperature is a dimensionless function of the thermodynamic temperature T in the form of the Carnot efficiency in which the other temperature is the environmental temperature T_0 (Eq (2.2)).

Intrinsic irreversibility rate is the minimum rate of irreversibility of a particular process, the limit being imposed by physical, technological, economic and other constraints (Section 3.3).

Irreversibility (in addition to the use of the word as an abstract noun) is a *quantitative* concept peculiar to the Exergy Method. Irreversibility is equal to the amount of exergy lost in a process within the physical space defined by the boundary of the system or the control surface. Irreversibility rate denotes the rate of loss of exergy in an open system (Sections 3.1 and 3.2).

Non-flow exergy is the exergy of a closed system at rest (Section 2.7).

Non-flow exergy function is a thermodynamic function defined by $(U + P_0V - T_0S)$ (Section 2.7).

Ordered energy is any form of energy which is stored in an organised manner, eg the kinetic energy of a spinning flywheel. Ordered energy is fully convertible to other forms of energy and its transfer takes place through work interaction (Section 2.1).

Physical exergy of a stream of substance is equal to the maximum amount of work obtainable when it is brought from its initial state to the environmental state $(P_0 T_0)$, by physical processes involving thermal interaction only with the environment (Section 2.4).

Pressure component of exergy is equal to the amount of work obtainable in a reversible isothermal process by virtue of the *pressure difference* between the stream under consideration and the environment (Section 2.5).

Rational efficiency is a criterion of performance which can be formulated for a plant or a plant component. It is given by the ratio of the exergy transfer rate associated with the plant (or component) output to the exergy transfer rate associated with the corresponding exergy input (Section 3.4).

Reference substance is a selected common environmental substance of low chemical potential with reference to which the chemical exergy and the enthalpy of devaluation of a chemical element are calculated (Appendix A).

Restoring work is the minimum amount of work which has to be done on a system by its surroundings to restore the former to its initial state after it has undergone an irreversible process (Section 2.5).

Standard state is that state for which $P^0 = 0.101\ 325$ MPa and $T^0 = 298.15$ K. When the standard value of chemical exergy or enthalpy of devaluation is used, P^0 and T^0 may be regarded as standard values of the environmental pressure and temperature (Appendix A).

Structural coefficients is the joint name for CSB and CEB (Section 6.1).

Thermal component of exergy is equal to the amount of work obtainable in a reversible isobaric process by virtue of the temperature difference between the stream under consideration and the environment (Section 2.4).

Thermal exergy is the exergy associated with a heat interaction. This term is applicable to closed system analysis (Section 2.6).

Thermal exergy flow is the exergy flow associated with a steady heat transfer rate. This term is applicable to control region analysis (Section 2.3).

Appendix G References

Introduction

0.1 GOUY, G., 'Sur l'énergie utilisable' (On usable energy), *Journal de physique*, 1889, **8** (2nd Series), 501–518.

0.1 STODOLA, A., 'Die Kreisprozesse der Gasmachinen' (Gas engine cycles), *2. Ver dt. Ing.*, 1898, **42**, 1088.

0.2 BOŠNJAKOVIC, F., 'Kampf den Nichtumkehrbarkeiten' (Fight irreversibilities), *Arch. Wärmewirtsch*, 1938, **19**, No. 1, 1–2.

0.4 BOŠNJAKOVIČ, F., 'Güte vom Wärmeanlagen und die Leistungsregeln' (The degree of perfection of thermal plants and load control), *Tech. Mitt. Essen*, 1939, **32**, No. 15, 439–445.

0.5 RANT, Z., 'Exergie, ein neues Wort für "technische Arbeitsfähigkeit"' (Exergy, a new word for 'technical work capacity'), *Forsch. Gebiete Ingenieurwes.*, 1956.

0.6 HEYWOOD, R. W., 'A critical review of theorems of thermodynamic availability with concise formulations'. Part 1: Availability; Part 2: Irreversibility, *J. Mec. Engng Sci.*, 1974, **16**, Nos. 3 and 4, 160–173, 258–267.

0.7 FRATZSCHER, W. and BEYER, J., 'Stand und Tendenzen bei der Anwendung und Weiterentwicklung des Exergiebegriffs' (The present state and trends in the application and further development of the exergy concept), *Chemische Technik*, 1981, **33**, 1–10.

Chapter 1

1.1 ZEMANSKY, M. W., *Heat and Thermodynamics*, McGraw-Hill, 1957.

1.2 REYNOLDS, W. C. and PERKINS, M. C., *Engineering Thermodynamics*, McGraw-Hill, 1970.

1.3 HATSOPOULOS, G. N. and KEENAN, J. H., *Principles of General Thermodynamics*, John Wiley, 1972.

1.4 ROGERS, G. F. C. and MAYHEW, Y. R., *Engineering Thermodynamics Work and Heat Transfer*, Longmans, 1967.

1.5 PLANCK, M., *Treatise on Thermodynamics*, transl. from 7th German ed., Longmans, 1927.

1.6 CALLAN, H. B., *Thermodynamics*, John Wiley, 1960.

1.7 DENBIGH, K. G., *The Principles of Chemical Equilibrium*, Cambridge University Press, 1981.

1.8 KESTIN, J., *A Course in Thermodynamics*, Blaisdell, 1966.

1.9 GUGGENHEIM, E. A., *Thermodynamics*, North Holland Publishing Company, 1977.

Chapter 2

2.1 PETELA, R., 'Exergy of heat radiation', *Journal of Heat Transfer*, *Trans. ASME*, Series C, 1964, **86**, 187–192.

2.2 BRODYANSKII, V. M., 'Exergy method of thermodynamic analysis' (in Russian), 1973, Energiya, Moscow.
2.3 SZARGUT, J., *Teoria procesow cieplnych* (in Polish), Państwowe Wydawnictwa Naukowe, Warsaw, 1973.
2.4 ROGERS, G. F. C. and MAYHEW, Y. R., *Thermodynamic and Transport Properties of Fluids*, 3rd Edition, 1981, Blackwell, Oxford.
2.5 MARCHAL, R., *La thermodynamique et le theoreme de l'energie utilisable*, 1956, Paris.
2.6 ROSSINI, F. D., *et al.*, 'Selected values of chemical thermodynamic properties', 1952, Circular No. 500 (National Bureau of Standards).

Chapter 3

3.1 BOŠNJAKOVIČ, F., *Technical Thermodynamics*, Holt, Rinehart and Winston, 1965.
3.2 FRATZSCHER, W., 'Exergetical efficiency', *B.W.K.*, 1961, **13**, No. 11, 483.
3.3 SZARGUT, J. and PETELA, R., *Egzergia* (in Polish), Wydawnictwa Naukowo-Techniczne, Warsaw, 1965.
3.4 HORLOCK, J. H., 'The rational efficiency of power plants with external combustion', *Proc. Instn Mech. Engrs*, 1963/64, 178 (Part 31), 43.
3.5 KOTAS, T. J., 'Exergy criteria of performance for thermal plant', *Int. J. Heat and Fluid Flow*, 1980, **2**, 147–163.
3.6 SILVER, R. S., 'Reflexions sur la puissance chaleurique du feu', *Heat Recovery Systems*, 1981, **1**, No. 3, 205 (in English).
3.7 BES, T., 'Egzergia w procesach ogrzewania, klimatyzacji i suszenia' (Exergy in heating, air-conditioning and drying processes), *Energetyka Przemyslowa* (in Polish), 1962, **10**, No. 11, 388–392.
3.8 THRING, M. W., 'The virtue of energy, its meaning and practical significance', *J. Inst. Fuel*, 1944, **17**, 116–123.
3.9 TRIBUS, M., *Thermostatics and Thermodynamics*, D. Van Nostrand Co. Inc., Princeton, New Jersey, 1961.

Chapter 4

4.1 KEENAN, J. H., 'Availability and irreversibility in thermodynamics', *Brit. J. appl. Physics*, 1951, **2**, July, 183–192.
4.2 BRUGES, E. A., 'Performance of heat exchangers', *The Engineer*, 1957, **204**, 225.
4.3 SOKOLOV, E. Ya. and BRODYANSKII, *Energy Fundamentals in Heat Transformation and Refrigeration Processes* (in Russian), 1981, Energoizdat, Moscow.
4.4 LARIAN, M. G., *Fundamentals of Chemical Engineering Operations*, Constable & Co. Ltd., London, 1959.
4.5 KING, C. J., *Separation Processes*, McGraw-Hill Book Co., New York, 1971.
4.6 MAH, R. S. H., 'On binary distillations and their idealizations', *Chem. Eng. Commun.*, 1979, **3**, 59–64.
4.7 FITZMORRIS, R. E. and MAH, R. S. H., 'Improving distillation column design using thermodynamic availability analysis', *AICE Journal*, 1980, **26**, No. 2, 265–274.
4.8 FLOWER, J. R. and JACKSON, R., 'Energy requirements in the separation of mixtures by distillation', *Trans. Instn Chem. Engrs*, 1964, T249.
4.9 SZARGUT, J. and ZIEBIK, A., 'The effect of oxygen enrichment of air on exergy losses in combustion processes', *Zeszyty Naukowe Politechniki Slaskiej* (in Polish), 1964, No. 104 Energetyka z.14, 53–63.
4.10 SUSSMAN, M. V., 'The standard chemical availability', *86th National Meeting of A.I.Ch.E.*, Houston, Texas, April 1979, pp. F10–G14.

Chapter 5

5.1 HAYWOOD, R. W., *Analysis of Engineering Cycles*, Pergamon Press, Oxford, 1975.

5.2 BRODYANSKII, V. M. and ISHKIN, I. P., 'Thermodynamic analysis of gas liquefaction process', *Inzhenerno-fizicheskii zhurnal*, October 1963, **6**, 19–26 (English translation: *J. of Engineering Physics*).
5.3 HIMMELBLAU, D. M., *Basic Principles and Calculations in Chemical Engineering*, Prentice-Hall, Inc., Englewood Cliffs, New Jersey, 1962.
5.4 ANON., *Sulphuric Acid Plants*, Simon-Carves Ltd., Simon House, London, 1958.
5.5 COHEN, H. and ROGERS, G. F. C., *Gas Turbine Theory*, Longmans, 1951.
5.6 BRODYANSKII, V. M. and ISHKIN, I. P., 'Application of the exergy-enthalpy diagram to thermodynamic calculations', *Kholodilnaia Tekhnika*, 1962, **1**, 19–24.

Chapter 6

6.1 LINNHOFF, B. and FLOWER, J. R., 'Synthesis of heat exchanger networks': I, 'Systematic generation of energy optimal networks'; II, 'Evolutionary generation of networks with various criteria of optimality', *AIChE Journal*, **24**, No. 4, 633–654.
6.2 LINNHOFF, B. and TURNER, J. A., 'Simple concepts in process synthesis give energy savings and elegant designs', *The Chemical Engineer*, Dec. 1980, 742–747.
6.3 LINNHOFF, B., *et al.*, *A User Guide on Process Integration for the Efficient Use of Energy*, published by the I.Chem.E., Rugby, England.
6.4 UMEDA, T., NIIDA, K. and SHIROKO, K., 'A thermodynamic approach to heat integration in distillation systems', *AIChE Journal*, May 1979, **25**, No. 3, 423–429.
6.5 NAKA, Y., TERASHITA, M. and TAKAMATSU, T., 'A thermodynamic approach to multi-component distillation system synthesis', *AIChE Journal*, Sept. 1982, **28**, No. 5, 812–820.
6.6 BEYER, J., 'Strukturuntersuchungen-notwendiger Bestandteil der Effekivitatsanalyse von Warmeverbrauchersysteme' (Structural investigations—an essential part of the analysis of the efficiency of thermal systems), *Energieanwendung*, Dec. 1970, **19**, No. 12, 358–361.
6.7 BEYER, J., 'Strukturuntersuchung des Warmeverbrauchs in Zuckerfabriken' (Structural investigations of heat consumption in sugar plants), *Energieanwendung*, March 1972, **21**, No. 3, 79–82.
6.8 BEYER, J., 'Zur Aufteilung der Primarenergie kosten in Koppelprozessen auf Grundlage der Strukturanalyse' (Distribution of primary energy costs in multi-purpose processes on the basis of structural analysis), *Energieanwendung*, June 1972, **21**, No. 6, 179–183.
6.9 BEYER, J., 'Strukturwärmetechnischer Systeme und ökonomische Optimirung der Systemparameter' (Structure of thermal systems and economic optimization of system parameters), *Energieanwendung*, Sept. 1974, **23**, No. 9, 274–279.
6.10 BEYER, J., 'Einige Probleme der praktischen Anwendung der exergetischen Methode in Warmewirtshaftlichen Untersuchungen industrieller Produktionsprozesse' (Some problems of the practical application of the exergy method in thermoeconomic investigation of industrial production processes), *Energieanwendung:* Teil I (Part I)—Nov./Dec. 1978, **27**, No. 6, 204–208; Teil II (Part II)—March/April 1979, **28**, No. 2, 66–70; Teil III (Part III)—May/June 1979, **28**, No. 2, 86–90; Teil IV (Part IV)—July/Aug. 1979, **28**, No. 4, 137–139.
6.11 KALININA, E. I. and BRODYANSKII, V. M., 'Basic rules of the method of thermodynamic analysis of complex processes', *Izvestia Vuzov, Energyetika* (in Russian), 1973, No. 12, 57–64.
6.12 KALININA, E. I. and BRODYANSKII, V. M., 'Thermoeconomic method of distribution of expenditure in a multi-purpose technical system', *Izvestia Vuzov, Energyetika* (in Russian), 1974, No. 3, 58–63.
6.13 WONG, F. C., *System-Theoretic Models for the Analysis of Thermodynamic Systems*, Ph.D. Thesis, 1979, University of Waterloo, Waterloo, Canada.
6.14 CHANDRASHEKAR, M. and WONG, F. C., 'Thermodynamic system analysis. I. A graph theoretic approach', *Energy*, June 1982, **7**, No. 6, 539–566.
6.15 KOTAS, T. J. and KIBIIKYO, D. S., *An Example of Heat Exchanger Optimisation by the Exergy Method*, QMC Faculty Report EP 5019.

6.16 KAYS, W. M. and LONDON, A. L., *Compact Heat Exchangers*, McGraw-Hill Book Co., New York, 1964.

6.17 GRATCHEV, A. B. and VOROSHILOV, B. S., 'Analytical calculations of losses due to valve resistance in a reciprocating expander', *Izvestia Vuzov, Energyetika*, 1969, No. 8, 123–125.

6.18 TRIBUS, M., EVANS, R. B. and CRELLIN, G. L., 'Thermoeconomic demineralization', *Principles of Desalination*,Spiegler, K. W., ed., Academic Press, New York, 1966, Chapter 2.

6.19 EL-SAYED, Y. M. and EVANS, R. B., 'Thermoeconomics and the design of heat systems', *Trans. of the ASME, Journal of Engineering for Power*, Jan. 1970, 27–35.

6.20 EL-SAYED, Y. M. and APLENC, A. J., 'Application of the thermoeconomic approach to the analysis and optimization of a vapor-compression desalting system', *Trans. of the ASME, Journal of Engineering for Power*, Jan. 1970, 17–26.

6.21 STOECKER, W. F., *Design of Thermal Systems*, McGraw-Hill Book Co., New York, 1971.

6.22 ASCHNER, F. S., 'Dual purpose plants', *Principles of Desalination*, Spiegler, K. W., ed., Academic Press, New York, 1966, Chapter 5.

6.23 REISTAD, G. M. and GAGGIOLI, R. A., 'Available-energy costing', *Thermodynamics: Second Law Analysis*, Gaggioli, R. A., ed., American Chemical Society, Washington, D.C., 1980, 143.

6.24 SZARGUT, J., 'Anwendung der Exergie zur angenährten wirtschaftlichen Optimirung' (Application of exergy for approximate economic optimization), *BWK*, Dec. 1971, **23**, No. 12, 516.

6.25 SZARGUT, J. and MACZEK, K., 'Thermoeconomic estimation of the optimum exergetic efficiency of refrigeration systems', *XVI-th International Congress of Refrigeration*, Commission B2, Paris, 1983.

6.26 SZARGUT, J., 'Minimization of the consumption of natural resources', *Bulletin de l'Academie Polonaise des Sciences*, Serie des sciences techniques, Vol. XXVI, No. 6, 1978.

6.27 BOUSTEAD, I. and HANCOCK, G. F., *Handbook of Industrial Energy Analysis*, Ellis Harwood, Chichester, 1979.

6.28 FEHRING, T. H. and GAGGIOLI, R. A., 'Economics of feedwater heater replacement', *Trans. of the ASME, Journal of Engineering for Power*, July 1977, 482–489.

6.29 WEPFER, J. W., 'Applications of available-energy accounting', *Thermodynamics: Second Law Analysis*, Gaggioli, R. A., ed., American Chemical Society, Washington, D.C., 1980.

6.30 SZARGUT, J., 'Application of exergy to the solution of thermoeconomic problems in power generation industry', *Zeszyty Naukowe Politechniki Slaskiej* (in Polish), 1965, Energetyka, z.18, No. 129, 101–116.

6.31 SZARGUT, J., *Thermodynamic and Economic Analysis in Power Generation Industry*, (in Polish), Wydawnictwa Naukowo—Techniczne, Warsaw, 1983.

Appendix A

A.1 SZARGUT, J., 'Bilans eksergetyczny procesow hutniczych' (Exergy balance for metal smelting processes) (in Polish), *Archiwum Hutnictwa*, 1961, **6**, No. 1, 23–60.

A.2 SZARGUT, J. and DZIEDZINIEWICZ, K., 'Energie utilisable des substances chimiques inorganiques', *Entropie*, July–August 1971, No. 41, 1; English translation: 'Exergy of chemical substances', Queen Mary College, Faculty Report No. E.P. 5018.

A.3 MORRIS, D. R., *et al.*, 'Energy efficiency of a lead smelter', *Energy*, **8**, 337–349 (Pergamon Press Ltd.), 1983.

A.4 ROSSINI, F., *et al.*, *Selected Values of Chemical Thermodynamic Properties*, 1952, Circular No. 500 (National Bureau of Standards).

A.5 KOTAS, T. J., 'Exergy concepts for thermal plant', *Int. J. Heat & Fluid Flow*, 1980, **2**, No. 3, 105–114.

A.6 SUSSMAN, M. V., 'Steady-flow availability and the standard chemical availability', *Energy*, **5**, 793–802 (Pergamon Press Ltd.), 1980.

A.7 BEYER, J., 'Einige Probleme der praktischen Anwendung der exergetischen Methode in wärmewirtschaftlichen Untersuchungen industrieller Produktionsprozesse'—Teil 1, *Energieanwendung*, Nov./Dec. 1978, **27**, No. 3, 204–208.

A.8 AHRENDTS, J., 'Reference states', *Energy*, **5**, 667–677 (Pergamon Press Ltd.), 1980.

A.9 COOPER, J. R. and LE FEVRE, E. J., *Thermophysical Properties of Water Substance*, Edward Arnold Ltd., 1969.

A.10 SZARGUT, J. and ZIEBIK, A., 'Exergy of chemical compounds in smelting processes' (in Polish), *Problemy Projektowe Hutnictwa*, 1965, **13**, No. 2, 40–49.

A.11 PERRY, R. H., CHILTON, C. H. and KIRKPATRICK, C. H., eds., *Chemical Engineers' Handbook*, 4th Edition, McGraw-Hill Book Co., New York, 1963.

Appendix C

C.1 SZARGUT, J. and STYRYLSKA, T., Angenäherte Bestimmung der Exergie von Brennstoffen, *Brennstoff-Wärme-Kraft*, 1964, **16**, No. 12, 589–596.

Index